REGENERATIVE SOIL
MICROSCOPY

written by

Matt Powers

Send all Inquiries to:
Matt Powers
Matt@ThePermacultureStudent.com

Published and Distributed by PowersPermaculture123
ISBN: 978-1-953005-05-2
website: http://www.RegenerativeSoilScience.com

For Adriana, James, & Oliver.

With Special Thanks to
Tanya Vin
the founding Patron of the R-Soil Database

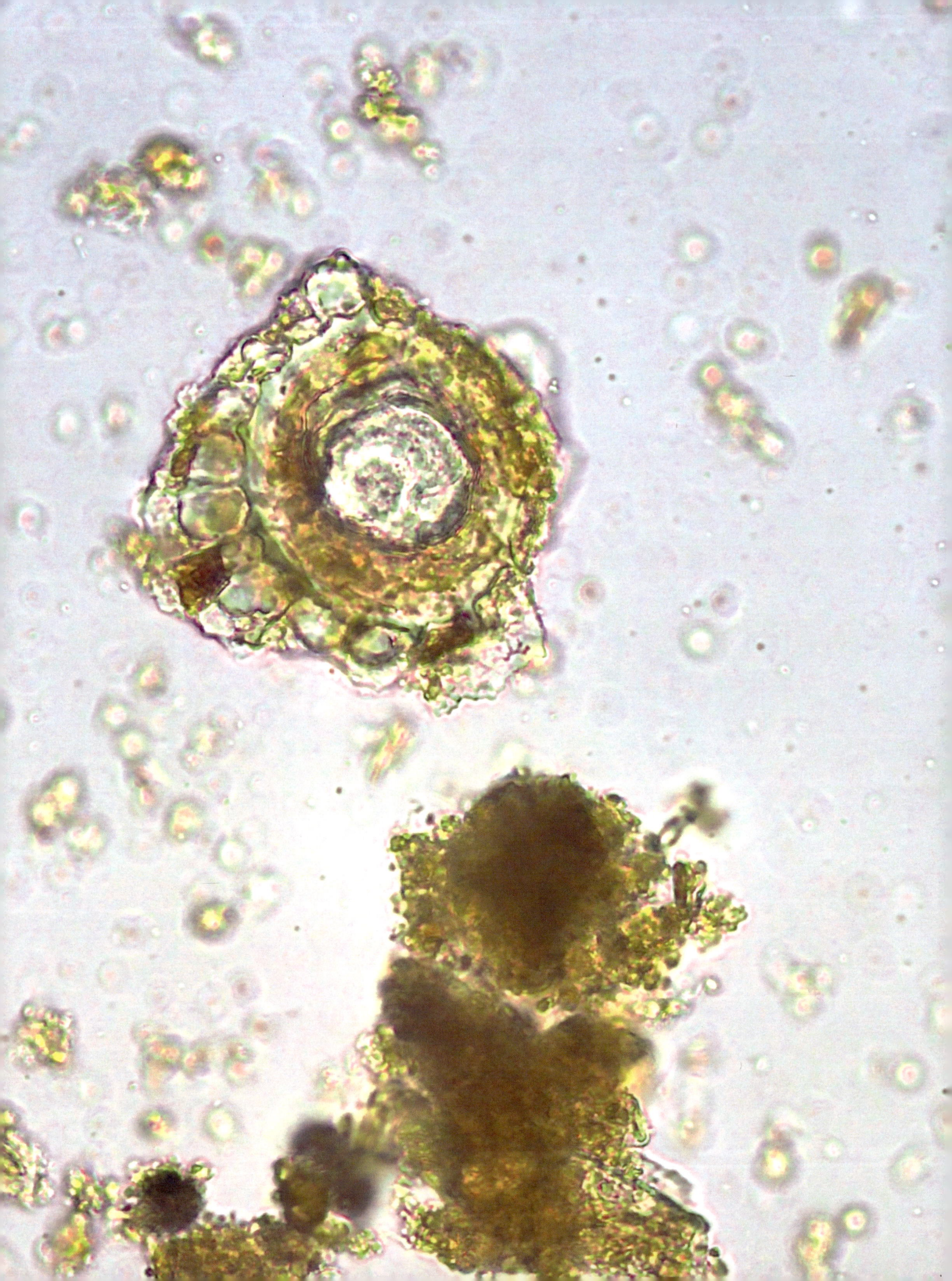

Praise for Regenerative Soil Microscopy

"With my books, you can read about it.

With this book, you can see the microbes.

Not only can you see what should be there, you can learn to use your own microscope to ID microbes, test, and then evaluate soil, compost, and roots.

If you Team With Microbes, you are going to love Matt Powers' ***Regenerative Soil Microscopy***!

- Jeff Lowenfels, the best-selling author
of the Teaming with Microbes series

"I am impressed!
I find ***Regenerative Soil Microscopy*** to be especially timely. This book will fuel the renaissance in microscopy that is currently underway globally.

Regenerative Soil Microscopy is of exceptional quality with instructions and many micrographs that will help explorers of soil biology to use microscopy and interpret what they see in their microscopes. This book must have taken several years to develop."

- Dr. James F. White, the mycologist, research scientist,
rhizophagy expert, and Rutgers University professor

"This book is going to live right next to my microscope! Such an amazing resource that really filled a void in soil microscopy!"

- Matt Wakeman, soil microscopist for @shootingthesoil

"This followup to Matt's invaluable book ***Regenerative Soil*** is perfect for those of us who are inspired to take the next step as citizen scientists.

Matt's ability to combine and connect the variety of perspectives on soil health, microbiology, microscopy, and the holistic thinking that brings it into relevance for normal people is what makes his writing stand out.

Beyond all of the practical information for using a microscope and exploring the living world under our feet, there is an inspiration and wisdom in Matt's way of communicating that makes what could easily be a dry topic into one that pulls you to the edge of your seat as you read.

I'm already excited for the generations of readers who will lead new discoveries into the vast ecosystem of soil informed by this book.

I truly hope many of you will be among them!:"

- Oliver Goshey, designer, educator,
and host of the Regenerative Skills podcast

"I can't express enough how amazing your technique is in presenting such advanced topics without threatening the reader. Kudos to you for being the bridge between advanced scientific rigor and the human experience.

I can honestly say you are an inspiration for generations to come in the way you present the material. The book is going to be a mainstay for microscope enthusiasts in the permaculture realms as well as any soil scientist looking to break away from the mainstream literature. Brilliant indeed, I cannot applaud you any further!"

- David "Diamond" Mauriello, geologist, microscopist & founder of the Oppenheimer Ranch Project

"Matt Powers' new book ***Regenerative Soil Microscopy,*** *the Science and Methods* is a must-own for anyone serious about understanding what is really going on in your soil. Matt builds upon the foundation that many of our mentors have laid before us and builds a whole new level of understanding on top of that amazing foundation others have provided for us. He dispels many common myths about soil science with hard facts and DNA testing that simply has never been done to date to properly confirm observations and assumptions that have become so well established in soil science today and laid bare the raw truth of those assumptions. It is so refreshing to see someone doing the hard work and hard science to confirm the observations that so many of us have made in our work. Matt sets a new standard for what a regenerative soil science book should be and you will not regret digging into this book. A must-own for anyone serious about soil science."

- Stephen Raisner, author, farmer, and owner of PotentPonics

"...**Regenerative Soil Microscopy** is the necessary compendium to the epic **Regenerative Soil** and I'm stoked to hear that Matt will bookend this trilogy with a volume on soil DNA. Through these volumes, Matt has proven his ability as an educator to distill complex information into a form that anyone can understand and apply. That is a rare gift and established by the grunt work Matt has put into researching, networking a topic and THEN doing all of the graphics and layout. I am sure that **Regenerative Soil Microscopy** will open up this largely invisible but massive lithosphere to a whole new generation of folks of all ages who, up to now, have been lacking access to the knowledge of how to make it visible and to know what they are looking at. We highly recommend the **Regenerative Soil series** to anyone who has any interest in that world within our world…"

- Darren Doherty, of REX and Regrarians.org

"This book is essential reading for all people that are serious about creating a truly sustainable world. Matt Powers' **Regenerative Soil Microscopy** is destined to be a classic. Buy one for yourself and one for your grandchildren."

- Geoff Lawton, permaculture expert at GeoffLawton.com

"Only someone who has put in hours of research, teaching, and correspondence with the field's leading scientists and practicing experts could put together such a holistic and nuanced guide to soil microscopy. Matt has opened the door for us all to explore the complexity of the foundation of life with enhanced tools of vision and understanding."

- Dusty Eddy, Erdbaum.com

RSM

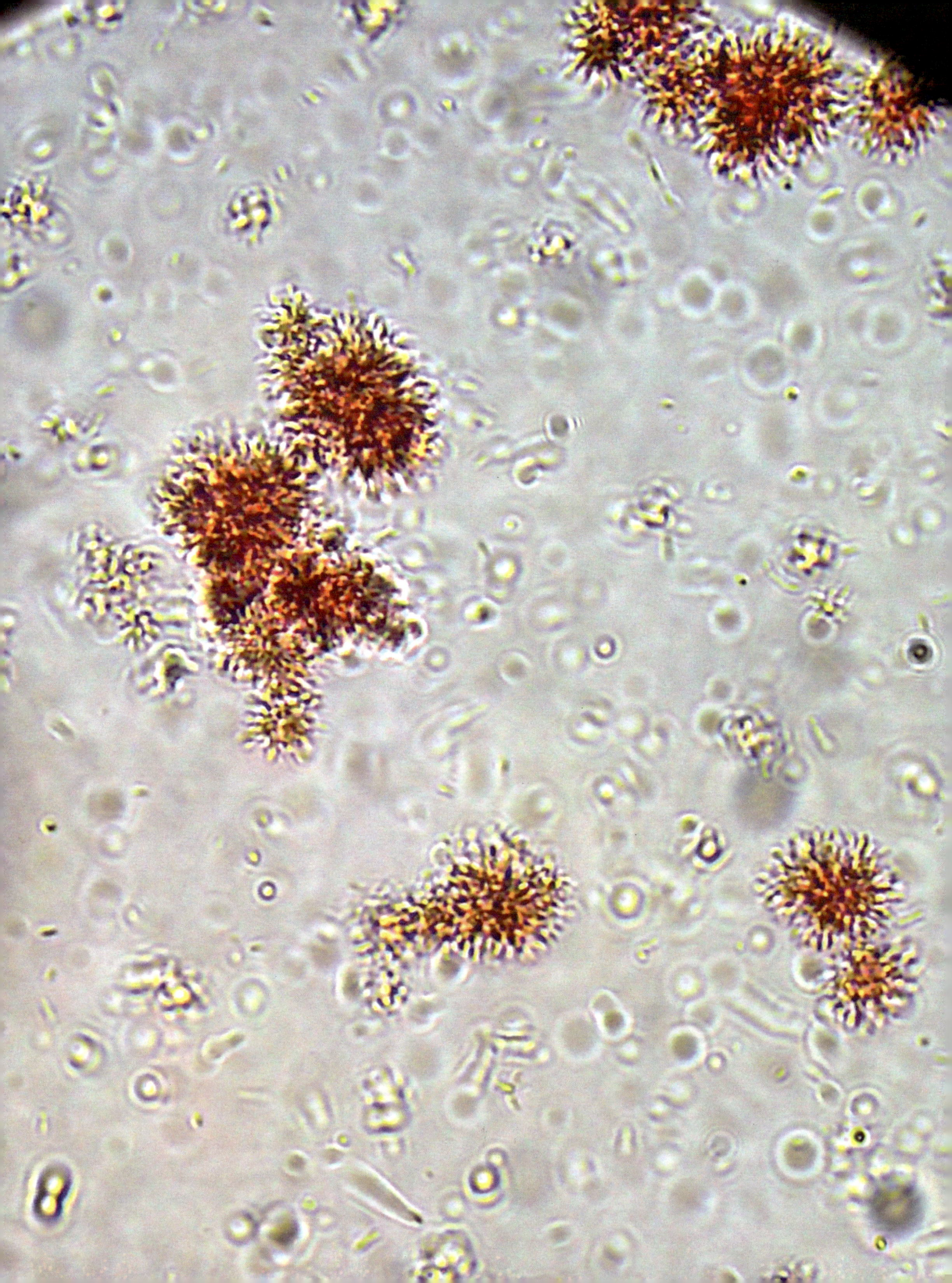

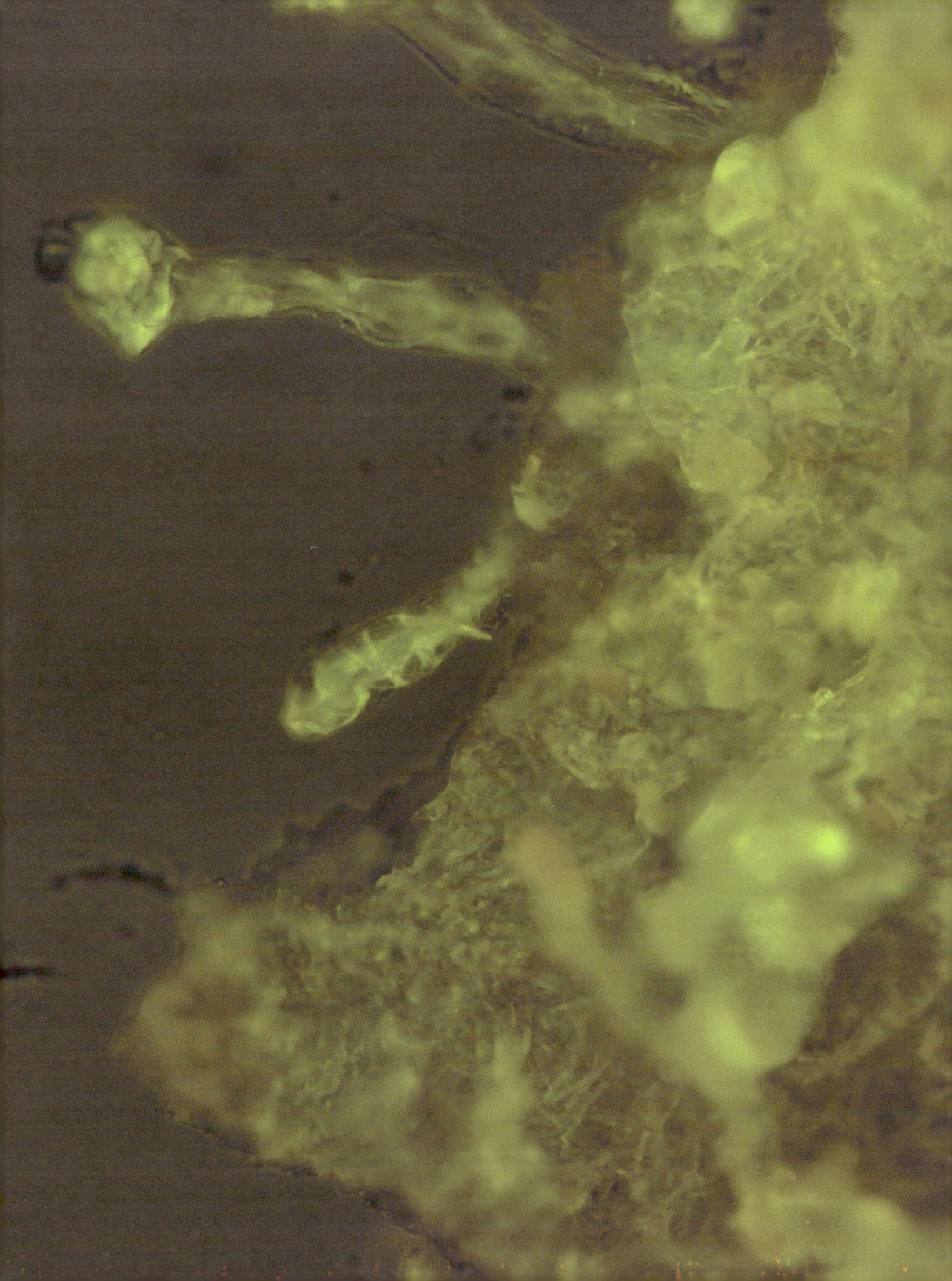

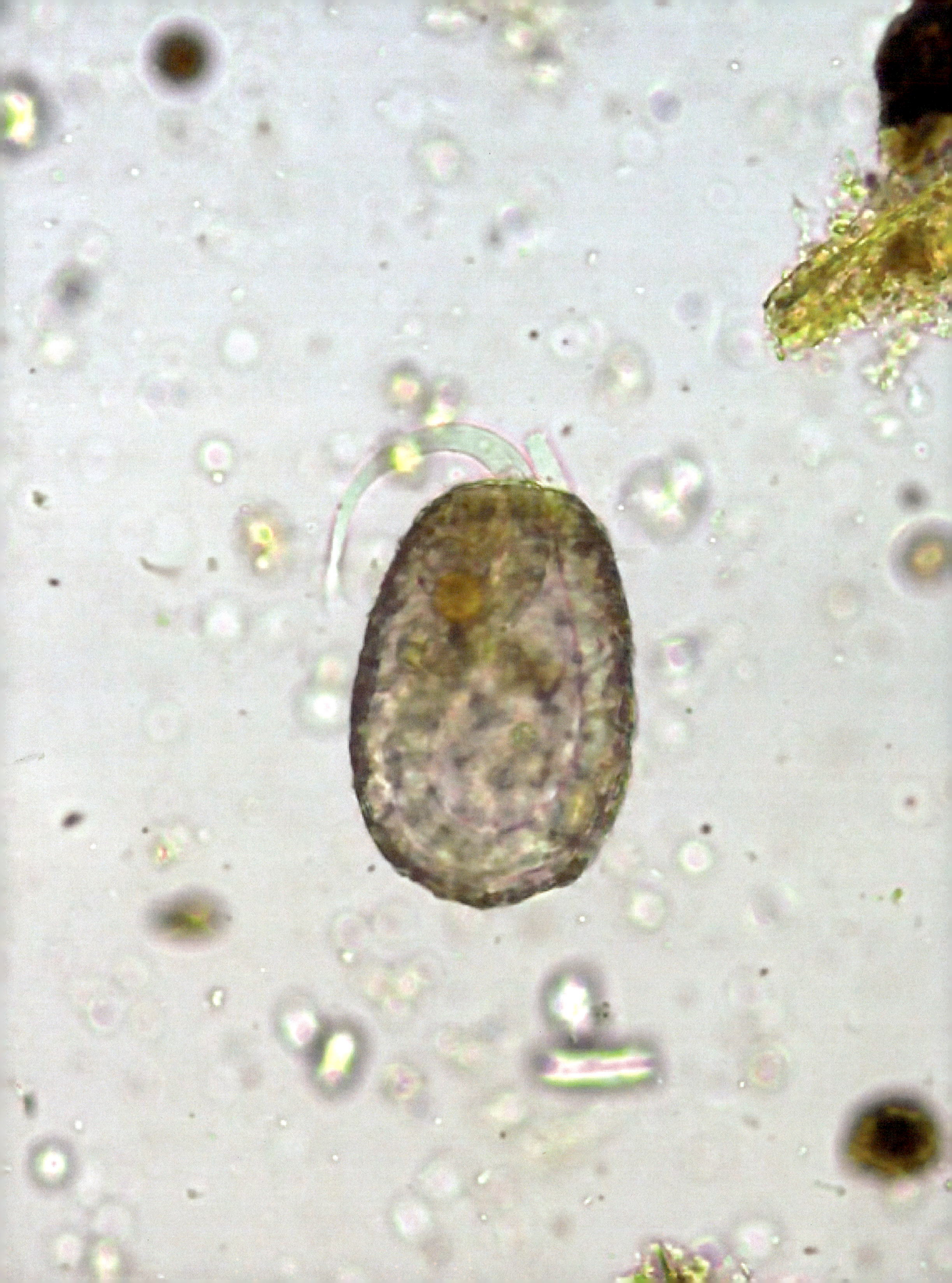

WHY MICROSCOPY?

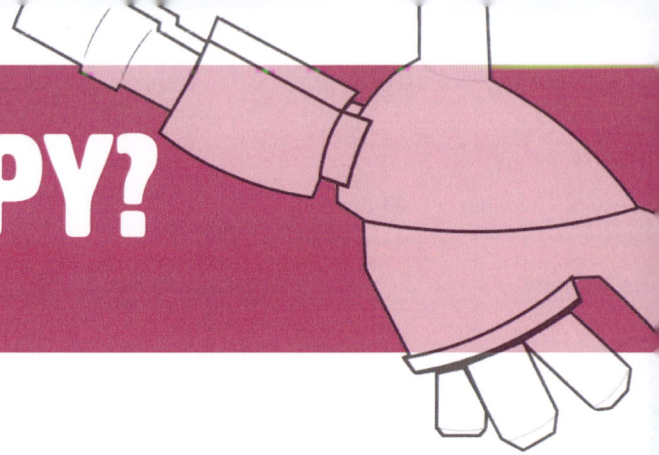

Why Microscopy?

The humble microscope is the lynchpin testing and evaluating tool that ties together all the sciences that focus on soil: the biology, organic matter, plants, water, and minerals. While each domain of focus has their own collection of test methods, they all rely upon the microscope for holistic assessment and specific tasks. If we combine these perspectives and practices, we get a much clearer picture of what is actually happening in our soils, and if we own our own microscope, we can assess everything from the comfort of our home laboratories, and we can do it as much as we need to. The consistency of observation is directly correlated to true comprehension since each slide is just a drop of soil solution from 1 ml of soil per sample, SO the more testing we do and the more kinds of tests we do, the better resolution picture of the situation we'll get.

Soil is dynamic, alive, and constantly evolving and adapting in real-time. It is a world unto itself with vast stretches of area yet to be explored. The microscope is the vehicle that allows us to enter into that world – they are our spaceships into the microcosmos!! And while DNA sequencing has dropped in cost dramatically in recent years with advances like the Oxford Nanopore devices, the power of DNA sequencing must be paired with microscopy to understand context, to map behavior, and to fill in the gaps that DNA sequencing cannot reach. DNA sequencing works exceptionally well on bacteria and archaea, but struggles with fungi: currently most fungi must be amplified and targeted to select sequences using PCR techniques which increases mutations and false reads. William Padilla-Brown, the mycologist, even had pure fungal cultures *read as BACTERIA* when using PCR methods. Part 3 of the Regenerative Soil Trilogy is focused on solving this issue, so we all can sequence the soil fungi in a more reliable way. If we are forced to rely upon PCR, it puts an even greater and more lasting emphasis on the importance of microscopy – though the fact that we cannot visually differentiate or ID the species of bacteria we are looking at with a light microscope means we also need to rely upon DNA testing as a check on the microscopy testing moving forward as well, but first…

Who Am I?

I didn't start out doing science – that's partly why I know anyone can learn to do all this. I was a high school English teacher in Madera County California. I was teaching my students English+Permaculture using seeds, a garden, music, and more. I began to correspond in 2014 with Dr. Elaine Ingham about a book I was writing – up until that point, I'd only learned about soil from the internet or books. I'd struggled with science teachers in high school and had assumed that I was challenged in science – Elaine was incredibly encouraging and worked with me for hours over Skype and helped me 1:1 learn the soil food web science, how it relates to pH, the nitrogen cycle, and so much more. This changed that first book dramatically and folks noticed – it also changed me. I began down a road that led me to the cutting edge of soil science, and because I figured it out, so can you. That is the other part of why I know you can do this: I've been teaching and writing instructional books for children and adults, from K-12 to universities and beyond, for almost 20 years, and I know everyone can participate in the methods and practices in this book to help evaluate and improve their soils, plants, and compost.

These days, I am a full-time citizen scientist, regenerative farmer, and educator – I work with academic researchers, farmers, fellow citizen scientists, gardeners, students, and university professors from all around the world – folks like Dr. Olivier Husson, Chris Trump, Dr. Elaine Ingham, William Padilla-Brown, Stephen Raisner, and Dr. James Francis White, Rutgers University professor and author of over 625 published studies which have been cited almost 20,000 times.

Why This Book?

I could not find any book to teach me how to use a microscope with soil, compost, or plant roots. With the exception of a few morphological reference guides, almost all the books I've found either talk about the anatomy of the microscope itself OR were focused on tests reliant upon harsh chemicals often in conjunction with machines I didn't have and didn't want, and they were all incredibly expensive on top of being references about the tools rather than guides to deciphering the seemingly chaotic world quietly teeming beneath our feet. Even the identification keys I gathered admitted to being "highly artificial" in light of new DNA testing.

I don't want to overload folks with too much information – 95% of what I've read the past 3 years has been irrelevant to the final processes and protocols. It's been like sifting beach sand for wedding rings, and while I've got quite a collection now, it's mind-boggling how spread out the information is and how disconnected it is. I had to read things from all sorts of perspectives to prove things out from multiple angles, to verify the fundamentals of **Regenerative Soil Microscopy** (RSM), and to catch subtle misinformation that IS ever-present in the scientific world especially online, in the commercial marketplace, and even in publication – all are competing for attention, BUT despite the cacophony of white noise in this space, I've come away with a cohesive system that pairs with all the insights found within **Regenerative Soil** as well as a path forward through our community to see even further.

When I was trying to figure out which stains to use I called up some companies and was shocked about their lack of basic knowledge around their products: *"If it works, it works"* was one response I got as to whether that particular stain would work for my UV nanometer (nm) wavelength and excitation ranges. It has dawned on me the further I waded into this space that it is in some ways the Wild West in science – folks will say anything. With only 1–10% of the soil microbes identified and described fully, there's a lot of room for confusion, constant revelation, and, unfortunately, deception.

And it's not just the stain companies, commercial compost in general is also plagued by a lack of understanding and proper testing. Most commercial compost is BAD for our soil and plants. It's not just anaerobic and filled with anaerobes, but the nutrients have gassed off or been locked up. Often these sealed bags sit in the sun outdoors in a pile, cooking in the heat. If they had good microbes to begin with, they don't now, and the plastic is also breaking down and releasing into the compost as well. Most of these commercial operations focus on temperature, C:N ratios, and moisture only – very few look at their microbiology. Catalyst Bioamendments'

Keisha Wheeler is a Soil Food Web School graduate, someone who refers to **Regenerative Soil** regularly, and she monitors their company's compost and mother culture pile, and they have exemplary compost that can serve as a reference for anyone wanting to know what "good" is though one should note the hallmarks of hot composting with a mother compost are less motile protozoa as testate amoebae begin to dominate over time as the heat and turning serves to select for those amoebae that can protect themselves (using a test or shell as shelter to the pH and Eh changes), so hot composting will always beget less mobile amoebae (less naked ones) – and if it's paired with a mother compost, especially so. I suspect judging by the amount of compost companies in **Regenerative Soil**, the online course, this fall (2022) that things will begin to change in commercial composting in the next few seasons quite dramatically.

I can remember my 2nd year gardening, over 10 years ago, buying a truckload of locally "composted" bull manure… it was certainly NOT composted and thrashed my garden that season with an excess of nitrates. It was then that I realized that I needed to understand how things actually worked in the soil. That began my journey into composting at home which led to teas and extracts and then led to EM and mycorrhizae and eventually individual microbes – along the way I learned 1:1 from scientists, researchers, natural farmers, and independent experts more, and I realized that everyone was examining and evaluating soil from a single perspective: mineral, biological, paramagnetic, etc. Folks weren't connecting everything, so for me, it felt like folks were missing the actual reality. **Regenerative Soil** was my long-format, holistic response to that initial question raised by adding hot bull manure to the garden and seeing everything die: *What is going on here?* I had to go down to the fundamental principles, processes, and patterns at work in the soil – it's the only way to teach fluency, and I'm not interested in anything less. I want my students to always understand from a place where they can figure things out, solve problems, and create new solutions and ways of thinking. It's only from a community of fluency that we get the greatest expressions of human creativity and intellect: Bach was born into a family of musicians with music teachers for parents, Shakespeare wrote in a time period where the English language was alive and his audience was aware of the subtleties of his humor, stacked meanings, and extensive vocabulary, and outliers throughout time have been born out of environments of fluency (Malcom Gladwell's **Outliers** book is a great exploration of this.)

My mission is to shed light on this space and the thinking required to navigate it, to give everyone the tools to explore it without bias, and to provide a space for everyone to freely discuss, evaluate, organize, analyze, and document the exploration of the microscopic worlds found in soil, compost, plants, fungi, all microbes, and more. I believe in spreading this knowledge and hosting the R-Soil Database, a public soil testing database, it will foster a fluency at a community level that will generate breakthroughs, deeper insights, and better practice overall over time.

Regenerative Soil, the 1st book in this trilogy, covers what these microbes are, what they mean, how they interact, how to bring them into your system, and how to maintain them. This book is about: how to use a microscope to morphologically ID microbes, how to test your soil, compost, and roots holistically, and how to evaluate those test results + what actions to take given those results.

If you loved **Regenerative Soil** and have seen the real benefits it has to offer, then you're likely here because you want to understand MORE and do more of your testing DIY whether you're a curious gardener, a full-time farmer, or a cannabis cultivator, this book will help you understand what your soil needs to be more fertile, resilient, and regenerative.

How to Read This Book

You can go right to the tests and protocols & begin testing – This book is designed to build your fluency, understanding, and overall practice of working with soil regeneratively. Use it as a primer, a reference, a book of recipes, and an evaluation tool – it's all these things and more.

You can skip to the morphological identification sections - start analyzing your soil and compost samples. It won't take long for you to begin to differentiate the microbes!

You can always start at the back of the book - you can use the Index!

BUT if you want a thorough understanding of the why and how behind the protocols and microscope methods, so you can generate your own protocols and methods, I highly recommend starting at the beginning and reading it through chapter by chapter.

Do I need a microscope? YES!! One that can go 40x – 600x (or 1000x), can hold your slide, and let you move the slide around while filming and taking pictures with ease. More details further in!

Do I need Regenerative Soil the first book? Absolutely – it's the foundational text that is the touchstone for the entire trilogy. It's how to put everything together and decipher it.

With this book, you can learn to use a microscope at home.

You can learn to analyze, characterize, assess, morphological ID, and count your bacteria, fungi, ciliates, protozoa, and nematodes. Most of the important tests are easy especially since there are now online morphological keys open to everyone, and most of the complex, labor-intensive, and chemical-laden tests are unnecessary these days. The new breakthroughs have allowed us to leapfrog over an entire legacy of chemically-dependent laboratory methods that are outdated and often counterproductive to getting a good read on what's actually happening.

Most tests done in a "standard" soil laboratory begin with drying the soil out, usually by baking it. This action shrinks the microbes by dehydrating them (making them harder to examine) and kills many of them at the same time. Most lab tests are dependent on culturing (in a petri dish), so that's not a natural state for most soil microbes. Thus, many do not grow, are not observed, and therefore don't get represented in the final testing results). Many tests require an open flame to heat-fix smears of bacteria-rich agar or broth to slides, only then to bath them and soak them in different dyes and harsh chemical washes, and after all that, we decide who's living or dead… *Do you see the problem here?* It's shocking to read what universities are teaching these days despite the obvious flaws wrapped up in the formalities of lengthy almost ritualistic preparations. My work at times is the exact opposite – we'll learn how to put a naked root, no cover slip, no water drowning it, under the microscope using the once dubbed Light-field technique (quite different from the Light Field Microscope objective as you'll see), and we'll see the root in a more natural state with NO WORK which means no disturbance, no alteration, and a more accurate representation of what is actually going on. I'm shocked at how little pragmatism there is to be found in the university textbooks focused on soil testing – it's all voodoo essentially (don't *voodoo is real* me – you know what I'm saying!) It's WILD what they're getting away with:

If the idea is to understand the way soil life actually works, shouldn't we be examining them under conditions that are as natural as possible?

That's the focus of this book: *to equip you with the most clear window into this territory.* We have identified and described an estimated 1–10% of the soil life thus far – we know more about space than we do about our world's soils. The methods in this book are chosen to hit the sweet spot between cost and efficacy – there's always nicer microscopes and always more expensive testing methods making wild promises (remember it's the Wild West), and I often share what alternative methods cost and their limitations and/or advantages.

Are You Ready for Regenerative Soil Microscopy??

WHAT IS RSM?

What is Regenerative Soil Microscopy?

Regenerative Soil Microscopy (RSM) is a holistic approach to microscope-based assessment and evaluation of soil, compost, microbes, mycorrhizae, biofertilizers, and plant roots – it can also be easily applied and adapted to broader mycology and other fields.

RSM is a combination of accepted and adapted microscope methodologies and techniques, brought together to streamline the soil testing process and, at the same time, fit RSM into a larger constellation of testing that will give us far more insight than any one testing methodology ever could. The future is not a one-button-click solution on mobile device with thumbs up or down – it's going to be more testing to create a more realistic and fleshed out understanding of what is happening in terms of cycles, flow, and interaction in our soils and plants.

REGENERATIVE SOIL MICROSCOPY *IN A NUTSHELL*

Soil, Compost, IMO, Tea, Extract, & Biofertilizer Analysis & Evaluation

#1 Initial Analysis & Characterization of Bacteria, Fungi, Protozoa, Nematodes, Microarthropods, Soil Minerals, and Humic compounds

#2 Hemocytometer Bacteria Counting & Viability Testing with New Stain + Epifluorescence

#3 Analysis & Characterization of Nutrient Cycling

#4 Any Additional Test Methods: NPK, Salinity, pH, DNA, Clay/Sand/Silt, etc.

#5 Soil & Compost Rubric & Evaluation (i.e. what is the state of your soil or compost?)

#6 Develop a Holistic Soil Management Plan using Actions & Recipes in **Regenerative Soil**

Plant Root & Leaf Analysis & Evaluation

#1 Initial Analysis & Characterization of Roots & Leaves with Stain or Epifluorescence

#2 Grid Overlay & Inoculation Calculation

#3 Plant Root & Leaf Rubrics & Evaluation

#4 Comparison of RSM Plant Root & Leaf Evaluation Results to Soil Evaluation Results

#5 Develop a Holistic Soil Management Plan using Actions & Recipes in Regenerative Soil

We also need to separate measurement activities from behavioral assessments and this means more types of testing rather than just more tests of the same variety. *What do I mean by that?* Sometimes I crush organic matter under the slide from compost and view it – the slide may be uneven and lopsided but that's not the point. This does throw off our ability to calculate things properly – especially if we add in extra drops to get it to float decently, but it's more about capturing the scene in various states to triangulate meaning. I may not be measuring and counting to the drop at those times, but I can SEE how things behave in different ways and that is much more useful than another number. The same is true with a LIVE/DEAD stain – you can see things easily for counting in a hemocytometer but it doesn't give you a good idea of how they are distributed in the substrate or their behavior because it's so thinly compressed and we shook things up until they were diffusely distributed. It's best to also do a plain slide perspective (with a cover slip) and in some cases even a well slide (mites and other microarthropods tend to get crushed under a regular flat slide and cover slide setup). We need to think dynamically to properly capture different states of the soil and the life found therein.

The numbers game can also be deceiving because we are taking small samples from the field and then taking even smaller samples from those samples, diluting them 1:10 or 1:100, only to take drops from that diluted sample of a sample to analyze. This can skew our perspective as most life tends to develop in pockets, especially fungi, so we are always going to see more or less fungi than is present on average because we are only collecting pin points of a galaxy of information and differentiation. Even if we don't see a nematode, ciliate, or fungi in our samples, that does not mean it's not there in that field of soil in low numbers or even in high numbers but in isolated pockets.

What's New & Different?

Traditional chemistry methods relied upon many steps, dangerous chemicals, and, often, changing the very nature of the substance being studied. The standard methods almost all rely upon a myriad of chemicals, extensive laboratory equipment, and automated estimation after a series of selective events occur (usually killing the microbes or changing the sample environment radically). MPN, most probable number, involves evidence of gassing off, growth, or color change in a petri dish or test tube. They rely upon color and number charts to find their answers quickly. It is hardly efficient or accurate.

Soil food web certified laboratories are even finding it difficult to come to the same conclusions on F:B ratios with the same exact sample analyzed in tandem LIVE over Skype with each other and their microscopes, matching each other's arm motions and sampling depth, and STILL they disagreed. When there is inaccuracy in both the traditional and independent fields, it stirs my curiosity, and one of those laboratories asked me personally to spread the word for them, so that everyone can know: we must be more careful and transparent as a community, and by adding in other soil tests, we generate a more comprehensive and realistic albeit more complex view of our soil or compost. The days of doing one test type, like biological assessment using bright field microscopy, are over. The complexity must be dealt with – it is what the plants actually face and so must we.

These numerical methods have led us astray quite often in many branches of science – it's a form of abstraction. In this book, we want to close the gap, see things as they are naturally as much as possible, pair the numbers with the context, simplify the entire process, and get useful data that is readily understandable and actionable. *That's what this book and method offer* – it is also an open door for you to follow the same path I've walked down: to experiment, modify, adapt, and improve upon what I introduce here. I've expanded, contracted, compared, and contrasted methods in my efforts to better understand what works best, what can we streamline, and how can we deepen our understanding through combining methods.

What Does RSM Reveal?

The microscope is the laboratory lynchpin to so many scientific branches because it's so fast and easy. The simplicity of it also begets experimentation and variation: this is why most labs develop their own protocols. In order to get new information, we have to formulate new experiments, new formulas, and new methods. RSM combines new and old methods with new research-based insights and technology.

With RSM, we will be doing some counting but primarily characterization of microbes, organic matter, and minerals, further testing the soil with additional tests, and comparing our results using rubrics to describe soil condition, fertility, & the cycling pathways (how are nutrients reaching your plant) using primarily bright field and epifluorescence microscopy + dark field at times and an adapted bright field technique that I've previously dubbed "light field" on social media but is not to be confused with the light field microscope objective (they are different) – in this book I will be calling it the Manual Lighting technique because it is a combination of light sources from different angles and can be accomplished with any wavelength of light but is done manually (by hand usually with a flashlight).

Microbes are the plant-preferred pathway to acquire nutrition, and microscopes are the fastest way to monitor and assess this process, but as we've learned in recent years: plants feed in many ways, so we have to observe all those sites in a variety of ways to generate a holistic picture. You will learn in this book how to examine the leaves, roots, and even stems of plants to add in their inoculation rates and states into your overall observations and final analysis.

By combining all this information within the soil pH, soil REDOX, CEC, SOM%, soil mineral test results, soil compaction, soil DNA, OM%, plant BRIX readings, plant sap analyses, and Bionutrient meter readings, we can understand our findings under the microscope at new depths, previously never considered. You may not have all these test capabilities now, but with the **R-Soil Database**, communities will begin to value this information, and community laboratories will begin to carry these devices and offer these services more and more – especially as they are becoming more and more affordable.

What are the Advantages?

- **Knowing If Your Farming or Composting Methods and Practices are Truly Beneficial**

- **Knowing If Your Soil or Compost is Truly Complete & Beneficial for Your Plants**

- **Knowing What's Missing in your Soil or Compost**

  Do you have enough of x, y, and z?

- **Knowing Where You are in Succession**

  Is this soil ready for a garden or a food forest to be planted here? Is it in need of cover crops first?

- **Knowing If Something is Anaerobic or Heading that Direction**

  Is my soil sick? Compacted? Water-logged? etc.

- **Knowing If you have Pathogenic or Disease-causing Conditions**

- **Knowing If you have Potentially Pathogenic or Disease-causing Microbes**

  Root Feeding Nematodes, Oomycetes, & Imbalances

- **Knowing How Nutrients are Cycling *or NOT Cycling***

  Identifying the members of soil food web in the soil or compost and mapping the nutrient cycles

- **Knowing Your Organic Matter Is Harboring Beneficial Life & Fertility**

  Visualizing the behavior and distribution of your microbes and organic matter

- **Knowing If Your Fungal Inoculation Actually Worked**

  Visualizing the roots of our plants to see if the mycorrhizae we bought or made is actually making those symbiotic relationships.

- **Knowing WHAT TO DO & what NOT TO DO to Help Your Soil**

- **Knowing How to Integrate the RSM test findings with All Other Testing Modalities**

  for the most complete understanding and deepest insights

This is your Guide

This book is a reference guide on how to use a microscope in general, what supplies you'll need, where to get them (including the microscope), how to morphologically identify microbes, how to use a microscope to analyze soil, compost, microbes, biofertilizers, roots, and leaves and stems, and how to evaluate and integrate your findings into a spectrum of test results and perspectives.

Regenerative Soil Microscopy (RSM) gives us modalities that open up specific insights that cannot be found in other areas. Even as powerful as DNA sequencing is, it cannot tell you how things will behave or how they are laid out: it only tells you a chunk of who's there. Just as the microscope connects the different branches of science, the readings and numbers from non-microscope-based testing find their connections to the functioning reality in the findings we generate with the microscope – the microscope creates a deeper, more cohesive understanding because we SEE the cycles and processes of soil, soil life, water, and minerals at work in real-time. The numbers are just an abstraction that helps us better understand the reality which we can see with our microscopes. I can still remember the first time I set out to see rhizophagy myself: I grew out inoculated pumpkin seeds, examined their large roots, and was AMAZED: their root hairs were expelling bacteria right on cue!! I was astounded and overjoyed that my first attempt yielded incredible images of rhizophagy at work.

If you're already wondering about some of the vocabulary I'm using, I highly recommend picking up a copy of **Regenerative Soil**, the first book in this trilogy, and please know that **you can do this**. The words become 2nd nature in time, and the actions are much more important in all reality, and everyone can do the actions and be involved – plus there's a glossary too! *You can use a microscope at home to understand your soil and compost, plants, microbes, and more!!*

Now let's get you the supplies and microscope you'll need for your home laboratory.

WHAT YOU'LL NEED

While these lists seek to give you exactly what you need, there's always more: more expensive microscopes, more special and costly stains, more gadgets, etc., so keep in mind, I've vetted these tools and requirements with you in mind. I want them to be the most effective, cost-efficient, safe, reliable, and insightful, and so that's a sweet-spot that I've sought to hit squarely, but even here I want to share a range as you can get started with used equipment often for very little cost, but you can also spend thousands on a professional system that can serve as the cornerstone for a professional lab or consulting service. Please note that I'm not describing or defining things here, just listing things; the definitions, descriptions, directions, and rubrics come later in the book. This area will serve as a checklist & overview, so those details and definitions have context as you encounter them later on.

The Minimum Requirements

This is to get you started and familiar with the microscope, the basic tests, and the processes involved.

- Bright field Trinocular Microscope with 4x, 10x, 40x, & 60x (or 100x) Objective Lenses with an adjustable stage with lever, coarse & fine focus, dimmable LED light source, adjustable condenser, & iris diaphragm
- A Microscope Dust Cover (this can be a sheet, but it's a MUST!)
- HD or 4k Microscope Camera
- HD or 4k Monitor that connects to your Camera (HD:HD, 4K:4K)
- Computer with Keynote, GIMP, or PhotoShop style graphic editing and layering App & the ability to play back and view all the videos and images you'll acquire
- Methylene Blue Stain
- Standard Slides & Slide Cover Slips (18mm x 18mm – 22mm x 22mm), 15 ml Test Tubes, Lens cleaner, Lens cleaner tissues, Soft Lens Polishing Cloth, Test Tube Rack, Immersion Oil, & 3 ml Pipettes

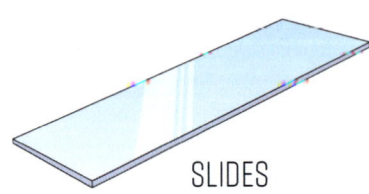
SLIDES

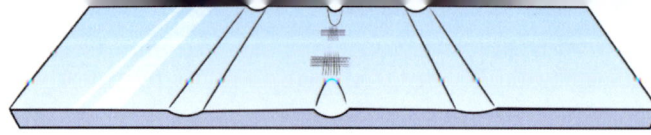
A HEMOCYTOMETER SLIDE

SLIDE COVERS

STAINS

TEST TUBES

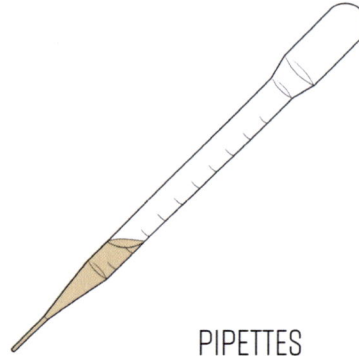
PIPETTES

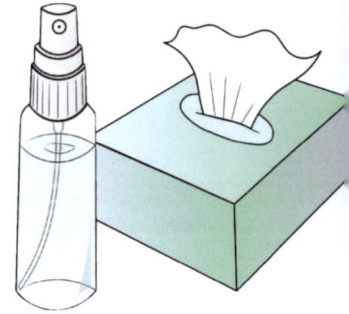
LENS CLEANER + TISSUES

- pH test strips

- NPK + pH test kit like La Motte's

- Handheld flashlights of plain white light of variable brightnesses

- Sterile, Distilled, or Purified Water

- Ziplock baggies for samples

- Permanent Marker for marking tubes and samples

- Space to have your laboratory always setup at home

The Professional RSM Requirements

In addition to the above list, this list goes further and showcases everything you'll need for a professional microscope-based home, community, or commercial consultation laboratory.

- Hemocytometer slide (high quality only *or use a overlay grid*)

- Well slides and covers (for mites, macroarthropods, roots in nutrient solution, etc.)

- Dark Field Condenser (preferably the oil-immersion condenser – sometimes better for visualizing)

- Epifluorescence Lamp, Objective, or Kit (required for many tests - it is the gold standard for fungal assessment)

- Acridine Orange Stain (for viability test & counting)

- Specialized Microscope Camera like the 4K BioVid with light compensation, measurements, and more

Recommended Additional Test Modalities

This list showcases a wide range of additional equipment and tests you can add to your repertoire. Some you'll find to be more affordable than others like the BRIX meter.

- The Microbiometer – time-saving way to calculate F:B ratios + verification of/check on our own counts
- ORP (REDOX) meter
- Salinity (EC) Meter
- Phillip Callahan Paramagnetic Meter
- Solvita Respiration, Ammonia, & Amino Nitrogen Test Kits
- Oxford Nanopore Technologies MinIon DNA Sequencing Device (or you can send it in to a lab)
- BRIX meter with temperature correction (for plant and root saps)
- Bionutrient Meter (for plants)

The R-Soil Database

Soil science has the power to correct all science and liberate people from suffering, sickness, malnutrition, and more. The profundity of soil life is that it ties everything else into its patterns and cycles: all animal and plant health… Yet, we don't treat it that way. We treat soil like dirt as a culture, even globally speaking, and it may be the reductionist manner in which we teach and communicate about it that leads to this disconnection. We are the soil; the soil is our future and our past – so completely unescapable as to be indecipherable from ourselves in any given moment in all reality, BUT if we began to collectively test, share, and showcase our results and interpretations, we can begin a process of community learning that will give birth to fluency at levels we can't approach as individuals or even in the pyramid structures of universities and gurus. It has to become something we discuss, we accumulate perspective on, and as we turn it over and pass it between, new insights, language, and understanding will always emerge. That's what's been missing: the social element of learning and the holistic feedback loop. This book has the holistic tests and protocols to give you that feedback connection and the R-Soil community is where we can share our results and grow in our understanding further than in this book. The future is bright especially if we share information that is as important as this because it's not the norm.

Currently one company (Wiley.com) is privatizing everything it can touch in the microbiological, bacteriological, and mycological worlds. This is making it harder for everyone to understand what has happened and what is currently happening in these branches of science. For the most part, students, teachers, and labs worldwide rely upon online keys and standard guides like Bergey's massive volumes on Bacteriology (which has shifted to wiley.com and is no longer in print) or *Illustrated Genera of Imperfect Fungi (4th ed)*. Increasingly as DNA sequencing is spreading, these keys are becoming less relevant, the paradigms upon which they were based are shifting, and the new results and insights are mostly behind massive paywalls and access is through select universities and by permission only. This is because we are seeing morphological identicals with divergent genes and the inverse as well: we are finding divergent species are actually one species. The entire mycological family tree is still in chaos due to DNA-based revelations that began years ago but have only continued to compound to this very day. Does this mean that microscopic identification is worthless? Not at all – it instead means we need to map out a new paradigm for how genetic and morphological expression occur. We will always need the microscopic perspective to give context to our testing – it's not a perfect picture of the space, but everything else is far abstraction in comparison.

Why Online Keys? It is the current standard process – it is how laboratories are quickly morphologically identifying microbes all over the world. There are too many attributes, too many types, and too many variations for anyone to memorize, so we use our pattern recognition (which constantly improves) with morphological keys systems which are very intuitive and easy to use. Researchers also know that all their identifications from mixed or wild samples are all tentative, not definitive, by nature of the method and medium of evaluation: we can usually ID to the genus level tentatively, so don't feel bad if something is hard to identify.

Where can these books and studies be found?

The books listed here and in the References are anywhere from $50-$500 each but some can be online for free through places like on Sci Hub: https://sci-hub.ru, some are on Amazon, and some are only available through Springer, Nature, and other online scientific journals. They are primarily university and graduate school textbooks or collections of published studies.

Which Microscope is Right for You?

Starter Microscopes can cost anywhere between $250-500, and you can find new workarounds for cameras that use mobile devices and old NIKON cameras even, and the promise and hope of even cheaper microscopes can be seen in the *Foldscope*, a paper microscope manufactured for only $1.75 each, that can go down to 400x or even 1000x. *Though it's not compatible with iPhones and many other devices*, it is a window into what is possible.

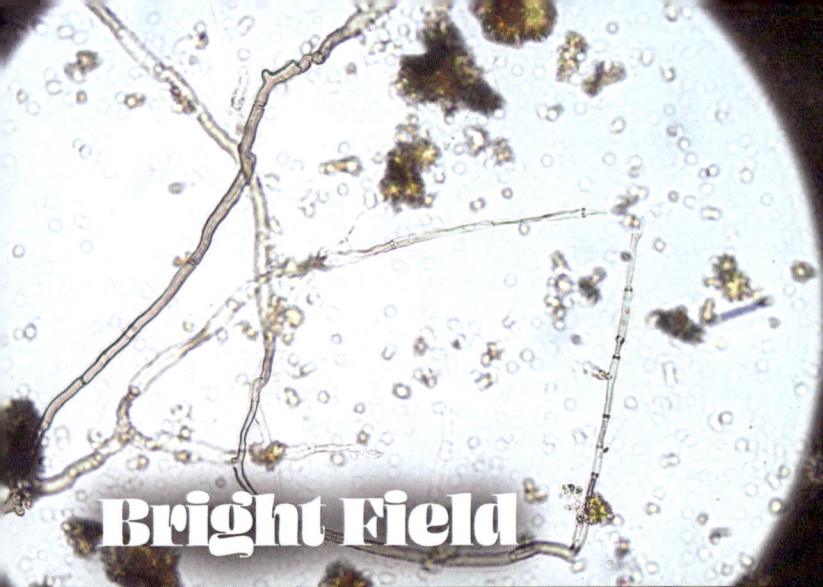
Bright Field

Manual

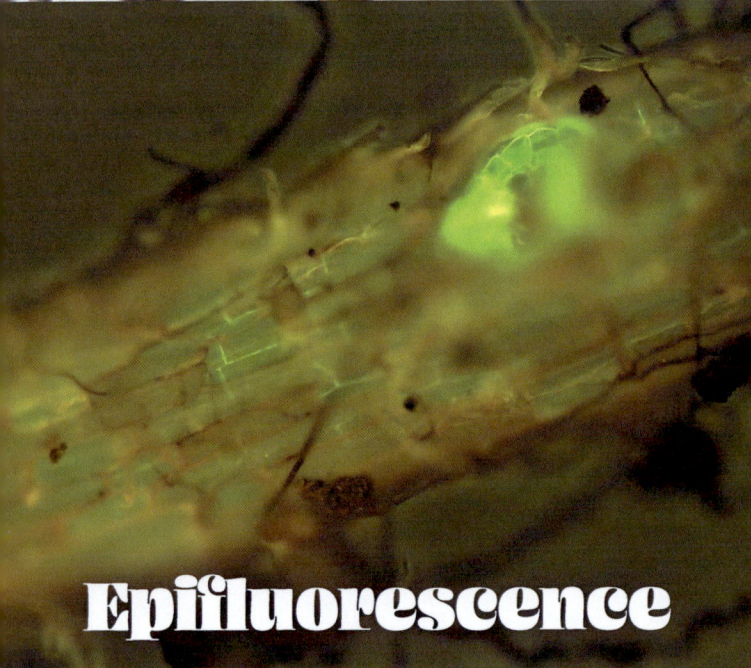
Epifluorescence

Dark Field

Matt Powers ©

THE BRIGHT FIELD

Within the world of light microscopy, the method that has the most overlap between disciplines would be bright field. This is system where the light shines from below, you adjust how much light comes through with the condenser, and you shine the light *through* your specimens. This is what Dr. Elaine Ingham teaches and what we all used in high school in biology, environmental science, and chemistry classes. It is in many ways very simple – the older models would even use natural light and a small mirror, but you can imagine how that could get your eyes in trouble if the sun hits that lens. Now with the ubiquity of the LED light, we have immense new opportunities opening up with LED epifluorescence, LED bright field, and even LED handheld flashlights. We can visualize everything more easily.

With a bright field microscope, we can morphologically correlate all the microbes in the soil, solutions, and compost to their role in the soil food web and ID them tentatively to their genus, we can use Blue Methylene as a stain to observe fungi, bacteria, and roots, we can map a great deal of the soil food web relationships, we can rudimentarily count our bacteria and measure our fungi, and we can view thin cross sections of plant leaves, roots, and stems. It is the foundational space that is only enhanced by the addition of other microscopy methods as they are simply different ways of visualizing the same exact space.

THE DARK FIELD

Dark field is simply bright field with a condenser over the light source with its center blacked out – this prevents the direct light from reaching the slide and only diffuse light that reaches the microbes at an angle causing everything to look a lot like outer space. It's actually quite stunning.

In the dark field, we can see things in a gentle, more natural lighting setting. The organic matter colors look much more true to the macro world. The microbes are often translucent like deep-sea creatures, and so they are in many ways. Organic matter can be easier to characterize and examine in the dark field – I've never seen with more accuracy how microbes feast on the surface of organic matter than in the dark field: you SEE them individually moving and feasting on the surface. Fine details of interaction differ visually in the dark field, and so little work has been done in this space – it remains largely unexplored, though extraordinarily beautiful.

We can swap our condensers without moving the slide (though if we have an immersion-oil condenser, you can't swap back and return to working with that slide: the underside will be covered in oil at that point). You can also 3D print plastic dark field condenser covers – this is actual something Ray Milidoni is doing in AU right now and may have them for sale by the time this book is complete. In this way we can view the same slide or even the same field of view under bright field and then dark field. If you have an epifluorescence system, you can view it a 3rd way without moving it or changing anything at all as well.

EPIFLUORESCENCE

Did you know that fungi glow when we shine 485nm light at them (it's cyan blue light) and when we view the light that reflects back through a 510nm barrier filter? Our eyes only perceive certain ranges of light, and if we apply the same principle to the microscope, we can create some very interesting effects. Epifluorescence reveals the world of phosphorus bearing compounds and fungi primarily because fungi are the primary phosphorus mobilizers and solubilizers: they release it from the soil minerals, forming calcium-rich phosphorus-bearing crystals (oxalic), and transport P through themselves like glowing straws. With the flip of a switch, you can see how inoculated your plant roots are, how fungal your decomposition pathways are in your compost, the viability of a sample with acridine orange as a stain, and so much more: it allows us to SEE fungi like nothing else. If you are a mycologist interested in soil, this is a must-have.

Epifluorescence microscopy used to be restricted to university and commercial laboratory settings – they were tens of thousands of dollars and relied upon hazardous mercury vapor light bulbs (metal halide) that had the tendency to explode near the end of their lifespans of only a few thousand hours. The new LED-based epifluorescence microscopes are a massive breakthrough in terms of cost and safety – for less than ten thousand dollars (and, with the 37% off discount my students receive, less than five thousand dollars), anyone can now own a epifluorescence microscope in their home or community lab. What's even more amazing is these LWscientific epifluorescence lamps open the door to non-destructive epifluorescence viewings of roots and fungi for the first time. Before in order to get those beautiful branching images of arbuscular mycorrhizae, they'd have to do an extensive and destructive process with harsh chemicals. In fact, I am of the opinion that the classic branching image is in part a product of the process and doesn't look quite like that in natural states. I should know: I look at unadulterated roots primarily. The arbuscules may be the remaining and thickest part of the inoculation that remains after the many chemical preps. The way it looks when I put it fresh under the microscope is much more like pools of digestion and inoculation spreading out evenly, finer and finer. This is another reason why I am constantly seeking to observe roots, soil, compost, and microbes in as natural a state as possible. Bright field is blasting microbes with light at levels they never naturally encounter, so how can we expect them to behave naturally? Samples are drowned in water, so how can we expect them to behave naturally when they are in the middle of a 100 year flood simulation or some other natural disaster? The ability for us to take a root and, with no alterations to it, place it on a slide gently under the epifluorescence microscope is incredibly powerful. Because the light comes from above, we don't have to shine through the root: we can see the surface and into the root deeply because roots look like they are all made of glass in this wavelength of light (and when we apply light from above and at different angle using a flashlight).

I was amazed when talking with a former university professor that works with PNW forests and native fungi about how he longed to use an epifluorescence microscope again – he asked me how I'd afforded the one I had. When I shared the retail price ($6,000) and how I had gotten a 40% discount on that, he'd been astounded and told me the ones he worked with were $30,000 each!! He then went on to ask me how I was doing it all – he was doubly shocked to learn that I was doing everything non-destructively without any harsh chemicals or even stains. That's when I realized that what I was doing at that time was revolutionary: very few people knew about this new technology and even fewer were embracing it, even among trained professionals.

This feeling was only deepened when Mike Thomas of LW Scientific reached out to me with this email – little did he know that Elaine was my original soil mentor:

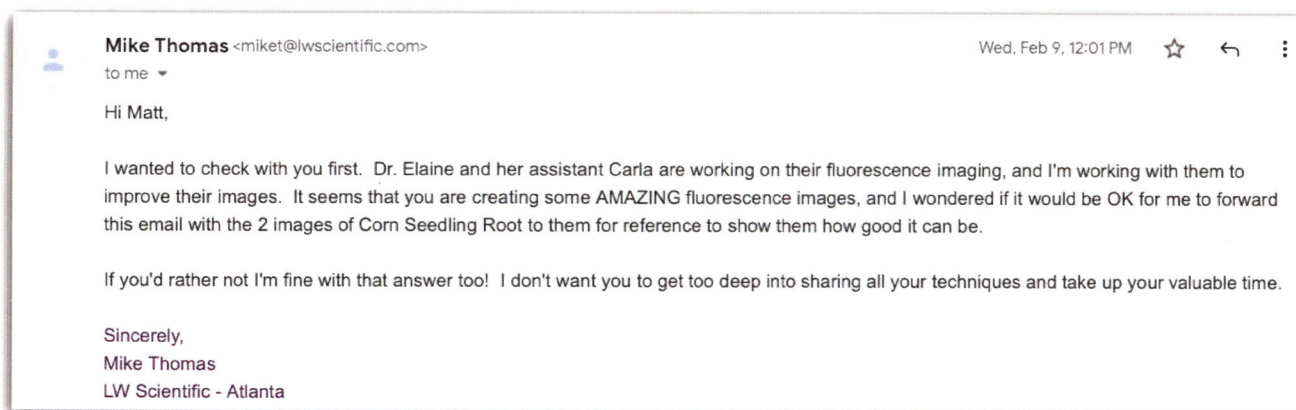

Mike Thomas <miket@lwscientific.com>   Wed, Feb 9, 12:01 PM
to me

Hi Matt,

I wanted to check with you first. Dr. Elaine and her assistant Carla are working on their fluorescence imaging, and I'm working with them to improve their images. It seems that you are creating some AMAZING fluorescence images, and I wondered if it would be OK for me to forward this email with the 2 images of Corn Seedling Root to them for reference to show them how good it can be.

If you'd rather not I'm fine with that answer too! I don't want you to get too deep into sharing all your techniques and take up your valuable time.

Sincerely,
Mike Thomas
LW Scientific - Atlanta

I realized that I needed to map out the full potentials and create protocols for working with these new LED epifluorescence microscopes. This email was 6 months before the Kickstarter campaign launched, but it was a huge motivation and a main reason why the book was mostly written and fully outlined by the start of the

Kickstarter in August 2022. It spurred me to read more, to buy the incredibly expensive books, and to experiment more and more.

Other Professional Microscopes

There are so many types of microscopes out there and some are so expensive that they aren't feasible for community or individuals to acquire – only businesses, organizations, or universities with large budgets can afford things like an electron microscope which takes up an entire room of space. That being said, there is another level to what we are doing, and that is the DIC style microscope which Alan Rockefeller, the mushroom hunter, cobbled together with used equipment off eBay for only $15,000 – he even figured out how to make it work with LED lights, not the standard metal halide bulbs! Youtube sensation, *Journey into the Microcosmos*, relies upon a DIC microscope as well. It is absolutely the next level in clarity and color.

This type of microscopy has many names: DIC (Differential Interference Contrast) microscope is also known as a NIC (Nomarski Interference Contrast) or Nomarski Microscopy. Based on the same principles underpinning phase contrast and polarized microscopy, DIC uses two different polarized lenses to simultaneously view a sample and then recombines the two light paths into one image for the observer - this generates sharper details, revealing invisible features, and reveals remarkable colors in unstained samples. If you are a fan of *Journey into the Microcosmos*, you know how amazing this microscope method is. It's quite expensive yet the images are unparalleled.

There's also stereo microscopes which shine light from above in a generalized blast, not angled like manual lighting offers. They are low resolution always (typically 90x but it can reach 180x) - it is a method often used to photograph plant roots and leaves. It is similar to a macro lens on a camera in terms of its resolution. Using the manual technique using a flashlight with our regular microscopes, we can get a much more sophisticated result than stereo microscopes can offer. I don't plan on ever getting one – I may eventually get one but I don't feel any real need to do so (especially after figuring out the power of angled light).

Polarized microscopes, microscopes with polarized lenses, usually are standalone machines with rotating stages with degrees on their lenses and accessory plates. Everything is mapped out to make it easy for you to identify minerals using their color and brightness – I may get a polarized mineral microscope because it is the only way to get a truly accurate read on the minerals using a microscope because I do not like opening my microscope to install and then uninstall the polarized lenses required to get my i4 microscope to have this functionality. The system of polarized lens and the 1st accessory plate I have through LWscientific can be used on their bright field i4 model, but it does not have the degrees on it. Using known references we can map things out roughly though it's not as easy. In addition, the reference slides are expensive, and things are color-based, so if you have an accurate Michel-Levy color chart and have a good eye for color differentiation, you can differentiate and identify the minerals in any given sample, but if not, it could be a lot of frustrating attempts. *Be Warned:* I got a used set of prepared mineral slides on eBay for under $200, and while a few slides are impressive references, many were damaged and scratched, so be wary of cheaper prepared slides. I have invested in the sets offered by Fisher Scientific (each set is over $400 each), and I will showcase them in the R-Soil database as frames of reference. As my lab expands, I'm going to invest in a polarized mineral microscope – the ease, speed, and accuracy with which they work is just too valuable to pass up.

Infrared microscopes are also becoming more and more widely used though like the DIC systems, they are very expensive. They can visualize the heat from microbes which is very interesting especially because in one paper I read they saw the heat signature before they could see the bacteria!! That's enough to intrigue me to want one, but for the price, it's not worth it yet. On top of that, an IR microscope does not have the same breadth of application as a DIC microscope does. I've been researching IR stains and dyes that correspond with specific light wavelengths, similar to epifluorescence, to try and develop a kit for home IR microscopy experiments with regular bright field microscopes. The most amazing thing I've discovered about current science in general is: there's way more possible than what is being done currently. It all just requires someone determined enough to push things through for them to become reality.

THE LIMITATIONS OF RESOLUTION

There is a limit to resolution: 1000x. We can't sharpen the image any further after that. So, why do companies like Omax advertise 2000x magnification? Because they are banking on you not knowing this: after 1000x, we are just zooming in, which is why it's so important that we seek out the highest clarity optics we can afford and then rely upon HD, 4K, or better cameras to take our pictures and capture our videos. Cheaper microscope objectives especially at 40x – 100x can't compare with the more expensive ones because you can zoom in further and the image holds definition longer, and the same idea holds for the lower quality cameras vs the higher quality ones. Just one aspect being lower quality, lowers all others in this scenario, so we have to take care to match things in terms of quality as much as we can – including our cables! Putting an expensive camera on an AMscope doesn't make sense; just as putting a cheap camera on an expensive microscope doesn't either.

Achromat, Plan, Semi-Plan

Lenses naturally generate visual artifacts that are not actual representations of what is in the sample. This can be a colorful greenish purplish halo effect around microbes as with achromat lenses (as seen in my images at times), or this could be the center of your field of view (FOV) being in focus and the edges being out of focus. There's a whole host of corrections that can be made, so it's important to note which objectives your images or videos are made with, so the color interpretations are bound to that context.

- **Achromat** - 65% of the FOV is in focus
- **Semi-Plan** - 80% of the FOV is in focus *(my epifluorescence microscope is semi-plan)*
- **Plan** - 95% of the FOV is in focus
- **Plan Fluorite** - corrects 2–3 colors and their associated spherical aberrations – these are considered ideal for fluorescence microscopy.
- **Plan Apochromat** - the most color correct (corrects 4–5 colors) and least prone to the haloing effect.

What's best for you depends on what work you are doing, what you can afford, and what your microscope can support.

Where to Buy Everything?

It's important to remember that availability for everything depends on where you live and as of late external factors like supply chain disruptions are making things harder – some things are expensive to ship, some exchange rates are astronomically high right now, and some things just aren't available currently. That being said: most everything is available still online and can be shipped most places in the world although it can get pricey at times.

For microscopes, all my online **Regenerative Soil** and **Regenerative Soil Microscopy** course students have a 37% discount on LWscientific microscopes and 10% off microscopes.au microscopes and supplies as of the year 2023 (discounts are subject to change year to year). LWscientific has superb hemocytometer slides, dark field condensers, epifluorescence lamps, microscope cases, and more – they are the best! Amscope and Omax on Amazon have affordable entry level microscopes in the $300 – $500 range that work well but lack the sharpness, precision, and clarity of LWscientific's i4 microscope which is typically $1400 but with a 37% discount is over $500 off! My LWscientific i4 microscope has their Lumin epifluorescence lamp and 4K Biovid camera – it is truly an amazing system that is unparalleled for that price point and for the work it can do. Only a DIC microscope can compare which is usually over 2–4x the cost, and often they have a metal halide lamp for the epifluorescence lighting, and while they come with a series of filter cubes, it is a mercury vapor lamp that only has 2,000 hrs of safe usage, and it can explode the nearer it comes to the end of its lifespan. There are many companies offering these toxic options: Zeiss, Leica, Olympus, Reichart, etc. I'm grateful for LWscientific's safer and longer lasting LED solution.

For supplies, it can be harder to buy them from commercial lab suppliers because they often refuse to deliver to residential addresses, PO Boxes, or UPS rented mailboxes for businesses. They want an established lab only or they won't let you buy anything from them. I've been dealing with this for years now. Often we have to go to Amazon, Fisher Scientific, Lab Alley, Ebay, etc. to find the things we need – that's just the reality we are currently in unfortunately. That being said, isn't it great we still can get access to these things, really for the first time ever, as citizen scientists? Even the 1800s "gentleman" scientists were very limited in what they could get access to – today we can order almost anything. A quick google search and you can find all the supplies listed, but take your time and compare the prices as different sites have wildly different qualities and price points on everything.

THE ANATOMY OF THE MICROSCOPE

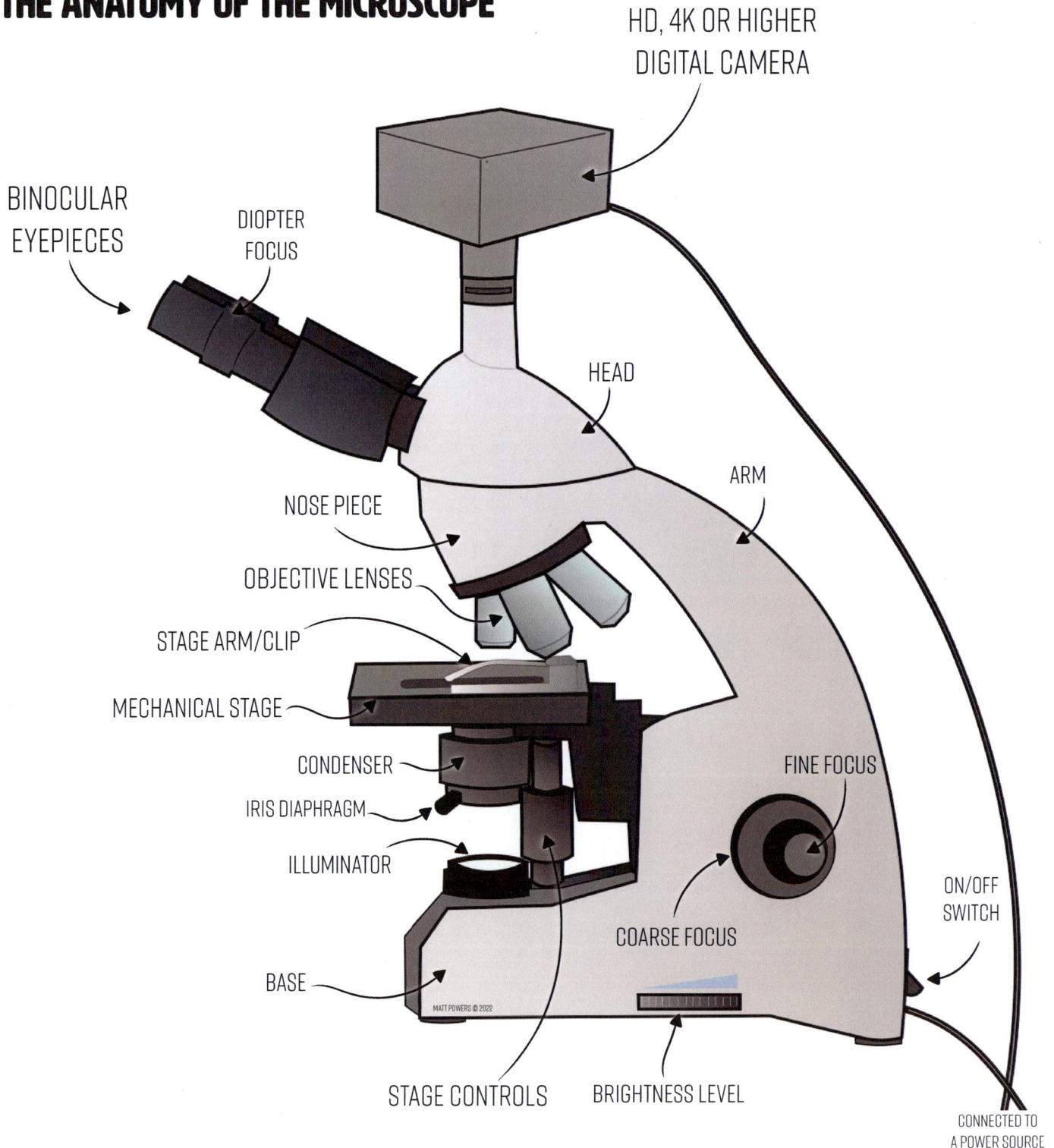

HD, 4K OR HIGHER
DIGITAL CAMERA

BINOCULAR
EYEPIECES

DIOPTER
FOCUS

HEAD

ARM

NOSE PIECE

OBJECTIVE LENSES

STAGE ARM/CLIP

MECHANICAL STAGE

CONDENSER

IRIS DIAPHRAGM

ILLUMINATOR

FINE FOCUS

ON/OFF
SWITCH

COARSE FOCUS

BASE

MATT POWERS © 2022

STAGE CONTROLS

BRIGHTNESS LEVEL

CONNECTED TO
A POWER SOURCE

How to Operate the Microscope

First we want to make sure everything is clean, dust-free, and ready to go. Is everything plugged in? Do we have all our supplies? Can we lay them out so we can use them logically and smoothly to make the process easier, faster, and more guaranteed to be uniform in our operation? Have your samples diluted and prepped, your slides cleaned, and your pipettes at the ready with a well for rinsing. Have your screen and camera setup – skip the eyepieces altogether!

Next, turn your microscope ON, and adjust the brightness to just bright enough. Turn your camera ON. Let your screen boot up, plug it in, or turn it on. Now you have 2 schools of thought – some folks have been teaching to use a card on the stage and to adjust the condenser until it's a hard edge, BUT I just adjust the condenser to a sweet spot just below the slide, just before the image reveals the condenser in the field of view (FOV). From there while you are viewing a slide, you can adjust the condenser to a more precise height. I close the diaphragm entirely on the condenser and then let it out until the light flashes and then I pick the sweet spot just after the flash. You can always close the diaphragm to thicken edges or to view only the glowing fungal activity, phosphorus bearing minerals, and fungi exhibiting those features (not all fungi glow which might imply a variety of things). Depending on the sample you can adjust your Brightness, resolution, and sensitivity on the Biovid camera as well to affect the lighting – it's important to give yourself the freedom to gently and subtly adjust the image lighting in a variety of ways.

After the condenser height is properly set, slides featuring samples are placed on the mechanical stage and held in place by the stage arm (or clip depending on your system). Light is shown on the slide, typically from below as is the case in bright field and dark field (though manual lighting can come from any angle and my epifluorescence system shines from directly above. Viewing through the 4x objective, pull the slide into focus using the course focus knob and then the fine focus. Set whatever light you are using to a level that is comfortable and clear for viewing – flare it up to see if it reveals a different picture, and then take the lighting low again and view it: the camera's brightness will compensate, but different images will appear, showing us different views on the same scene. Remember, you aren't putting your eyes on the eyepieces, so you can play with the light much more.

And it should be repeated more clearly: we always start at our lowest objective magnification and get the sample slide into focus with the coarse focus knob, then we work our way deeper by switching to the next higher objective lens. All deeper

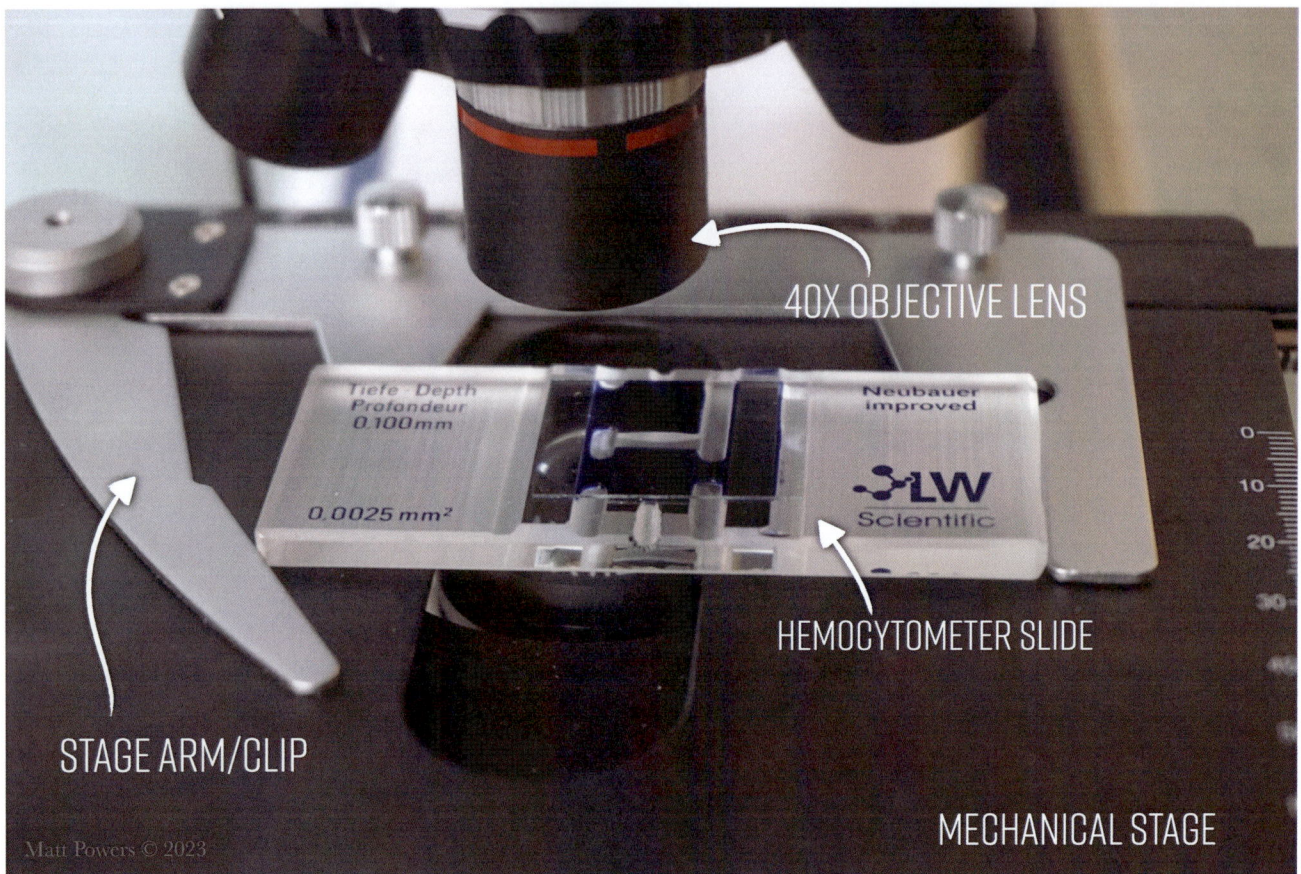

40X OBJECTIVE LENS

HEMOCYTOMETER SLIDE

STAGE ARM/CLIP

MECHANICAL STAGE

Tiefe · Depth
Profondeur
0.100mm

0.0025 mm²

Neubauer
improved

LW
Scientific

Matt Powers © 2023

layers are brought into focus using the fine focus knob. Most identification and evaluation occurs at 400x–600x magnification (with 1000x being the deepest possible viewing with light microscopy). The reasoning behind this method is we avoid breaking slides, damaging our objectives, and hurting our eyes – it's also easier.

While we'll cover these steps in more depth and detail later in this book, it really is this simple. It's light, magnification, and basic actions that anyone can learn. Dr. Elaine Ingham was working with her father in graduate school laboratories at age 6 doing graduate school level work counting bacteria, and my 12 year old is currently my lab assistant in this work – gratefully so, everyone can learn to use a microscope at a professional level!!

THE MORPHOLOGICAL GUIDE

THE MORPHOLOGICAL GUIDE TO SOIL MICROBES & MINERALS

Use this guide as a reference, primer, and touchstone in your work and exploration of soil, compost, roots, mycorrhizae, and more.

Why "Morphological"?

Before there was DNA testing, we simple went by how organisms appeared, reproduced, and behaved. This is why in the past 20 years, we've seen so many changes in the names of plants, phylum, groups, and families – this is why actinomycetes became actinobacteria: we figured out, despite it looking like filamentous fungi, that it is bacteria and actually inhibits fungal growth. Morphological identification, i.e. visual assessment, is still the bedrock of science: *seeing is believing* after all. We trust our eyes; we describe what we see – others can verify quickly and corroborate our testimonies by simply looking as well.

Many of our best tools are just enhancements of basic observations – like the way a ruler enhances our understanding of size, a microscope expands the magnification of our visual range, or how we can capture pictures and video to share what we are seeing with others across space and time – it's an incredibly useful schema, but it does have its limitations. There are things we cannot see easily or at all.

WHAT ARE THE DRAWBACKS?

Obviously there are pretenders and lookalikes in nature – *we can be fooled by our eyes*. In fact, much of nature is reliant upon that fact. DNA testing has opened new doors, but also raised even more questions. In the shuffle of reorganizing the microbiological family tree according to DNA, we are beginning to see complexities we've never considered prior. Arbuscular mycorrhizal fungi collects nuclei from other fungi across species and phylum.

In fact, all bacteria and fungi are absorbing genetic material and incorporating it into their genes all the time through horizontal gene transfer (HGT). They're consuming fragments of DNA from dead organisms from the environment (it appears that up to 40% of the soil and compost profile can be comprised of this primordial DNA soup). They're also conjugating – that's exactly what it sounds like! They're exchanging DNA with each other using microphalluses called sex piluses – they look like round bunny rabbits mating in space under the microscope. Microbes are also infected by viruses as part of HGT – all these pathways lead to faster adaptation to their environment than Darwinian evolution can offer. This is upending our maps of evolution as we see real-time genetic transformation of microbes in response to their environment. Darwin was doing his best with the tools of his day, and the main modality then was visual observation: this is why today we must look beyond what we can see with our eyes to see further.

WHAT'S BETTER?

Morphological identification alone can and likely will lead to mistakes in evaluation and application of soil, compost, and more. I always recall the soil food web certified tomato farm from South Africa that had ideal F:B ratios in their compost but saw lowered yields in areas with more compost and healthier plants with less compost (Malherbe and Marais, 2011). Not just any fungi or bacteria will do – we need specific fungi and bacteria for ideal conditions, and while we can do so much with a microscope, we can never say for sure what specific bacteria we are looking at under the microscope unless it's a purified sample. This is why so many lab based techniques are isolating what grows on petri dishes and then sequencing those dominant and adaptive microbes only: they can cut out a cube of just that area dominated by that microbial growth, and then observe and sequence that. It's a radical simplification and selection process in reality – it makes it easier, but we're fooling ourselves if we think this is accurate observation.

In many ways, we are at just the beginning of soil science and soil microscopy. We may soon have microscopes that allow us to deeply and gently explore the soil around our roots without leaving the garden, farm row, or orchard lane, or we may have new petri dish preps that imitate the soil environment more precisely – what if we combined our agar with sterilized lignin dust, humic acids, biochar dust, and sterilized and purified clays and sands?

In this current environment of rapid change and growth, it's important to ground ourselves in multiple modalities of understanding or we risk losing our bearing. That South African farm initially hoped the F:B ratios were a perfect standalone guide, but nothing is: we must combine perspectives to form a holistic comprehensive understanding, and that is done only through multiple tests across a range of methodologies consistently over time.

WHAT'S BEST PRACTICE?

Combining the RSM protocols with mineral, sap, DNA, and spectrometry testing allows us to understand what we are seeing under the microscope in new ways. We can test our tests, we can cross examine, we can trace cause and effect, and we can map the interactions between microbes, soil, and plants in ways never before possible. This is the very cutting edge of soil science: combining as many testing modalities as possible for a 360° perspective to better understand and explore the space, BUT it all starts with first knowing what to look for at a very basic level, so that's where we're going to begin, because at a certain point in depth, we must always use a key or reference to better explore and tentatively identify what we are looking at. Unless familiarizing yourself with a distinctly unique microbe in purified samples, we must acknowledge that it is impossible for us to differentiate between microbes of the same morphological type – just nematodes represent 2-3 million species and almost every genus has pathogenic members. The idea is not to memorize the details of individual microbial species or strains but familiarize ourselves with the general characteristics and hallmarks signs of who's who morphologically in our samples because if you have a microscope and you can identify the types of nematodes and protozoa your sample has, you can diagnose the soil nutrient pathways. If you can morphologically ID your fungi, you can determine if you have beneficial fungi in your compost or an airborne pathogen like Aspergillus fumigatus (don't lean in and smell that pile deeply…)

There's immense value in being able to morphologically identify and differentiate microbes in soil, compost, and plants. It gives context to all our other testing, and it helps understand the interactions, nutrients, and reactions at work in our soil or compost sample. *If soil is the lynchpin to all life, then microscopy is the lynchpin to all soil science.*

Meet the Microbes

As below, so above: microscopic shapes and organisms share patterns with the universe, planets, galaxies, our bodies, and the natural world around us. Nematodes are simply tiny worms. Spores look like tiny planets or snowflakes. Pollen can look like coffee beans or a spiky burr. Fungi can look like dendritic roots, branches, or even twigs. Testate amoeba look like hermit crabs without exoskeletons and prettier shells. Ciliates move like salmon or fast grazing silvery headless manatees… *okay*, so maybe the last one is a stretch, but they do look like deep sea creatures. Some, like stalked ciliates, look like coral reef organisms. The microscopic environment of the soil, compost, and plant roots looks like a cross between deep space and the deepest parts of the ocean. Bacteria and archaea are likely the least open to metaphorical representation. They are the unicellular beings that form the foundational layer of the soil food web which feeds and powers all the other trophic layers above and adjacent to it.

BACTERIA

Bacteria are single-celled prokaryotes that come in a wide range of sizes – they are difficult if not impossible to definitively identify. We can morphologically describe them to a degree but sometimes they are just so small, nondescript, and/or similar to those around them – it's too weak a method to make any kind of hard conclusions with bacteria. It's also been observed that some bacteria demonstrate pleomorphism and can charge their morphology in response to the environment. There are some hallmarks we can visually see that can alert us to some harmful bacteria, but we are reliant upon DNA testing for detecting exact pathogenic strains of bacteria and viruses. The thing is: bacteria are incredibly small, and some are beyond the reach of light microscopy. Electron microscopy can see what we cannot with light microscopy, BUT we can only see what we've seen and identified definitively. The great majority of bacteria are not classified yet – there are only 9,000 – 10,000 prokaryotes identified currently (that's archaea + bacteria), but it is estimated that there are between $10^7 - 10^9$ (100 million to 1 billion) species of bacteria. Another way folks put it is: we've only described and identified 1–2% of all bacteria. It's pretty humbling when we think about these simple truths, and it shows how important it is for us to be very careful about our assumptions and conclusions.

Building on that further, the other humbling aspect to take in is horizontal gene transfer (HGT), the exchange of genetic material directly between microbes, is a non-reproductive transference of DNA that can happen in split second. For our purposes, HGT means that if we create disease causing conditions (anaerobic, stinky compost for instance), we will find viruses, diseases, and pathogens begin to appear and increase in number. Does this mean that if we have perfect conditions, we can erase pathogens from inputs we add to our soil? It depends on the concentrations and the minerals, foods, and microbes present for that succession or shift to occur. We have plenty of examples of folks remediating soils, but there are also studies showing tropical islands can have pathogenic e.coli spread via the water tables and soil until it becomes ubiquitous. Since e.coli is an endophyte, pathogens can take up residence inside plants and be passed on to infect the consumers of that produce later on (this is how we get e.coli outbreaks). The reality is we need to occupy these spaces with beneficial microbes, dominating whatever space we are working in, and then we must keep the conditions ideal: a goldilocks level of oxygen, moisture, carbon, biology, minerals, and management as described in the first book in this trilogy must be maintained. This is a lot like the emerging conversations around germ vs terrain theory – it does not seem like an either/or paradigm at all, but a complex dance of competing forces.

BACTERIA SIZE

How small do bacteria get? Our light microscopes fail to see the smallest of bacteria – electron microscopes can view the gap, and over time the accepted size range of bacteria has gotten broader and broader with some even being so large as to be visible to the human eye. On average, spherical cocci bacteria range from 0.5 – 2.0 μm in diameter while rod-shaped bacilli, filamentous, and spirilla bacteria range in size between 1 – 10 μm (microns) in length and 0.25 – 1.0 μm in diameter. This means most archaea are the same size as the average bacteria.

That being said, bacteria can be as small as 200 nm. Viruses range in size from 5 – 300 nm but Giruses, giant viruses, can be as large as 400 nm which means we can see them with a light microscope! It is important to note that bacteria and archaea are indistinguishable unless we are specifically testing for the difference in their cell walls – this is why for a long time archaea were thought to be bacteria! Archaea are extremophiles so we won't see them as much in our soils though they can represent ~1% of a soil or compost sample in my experience.

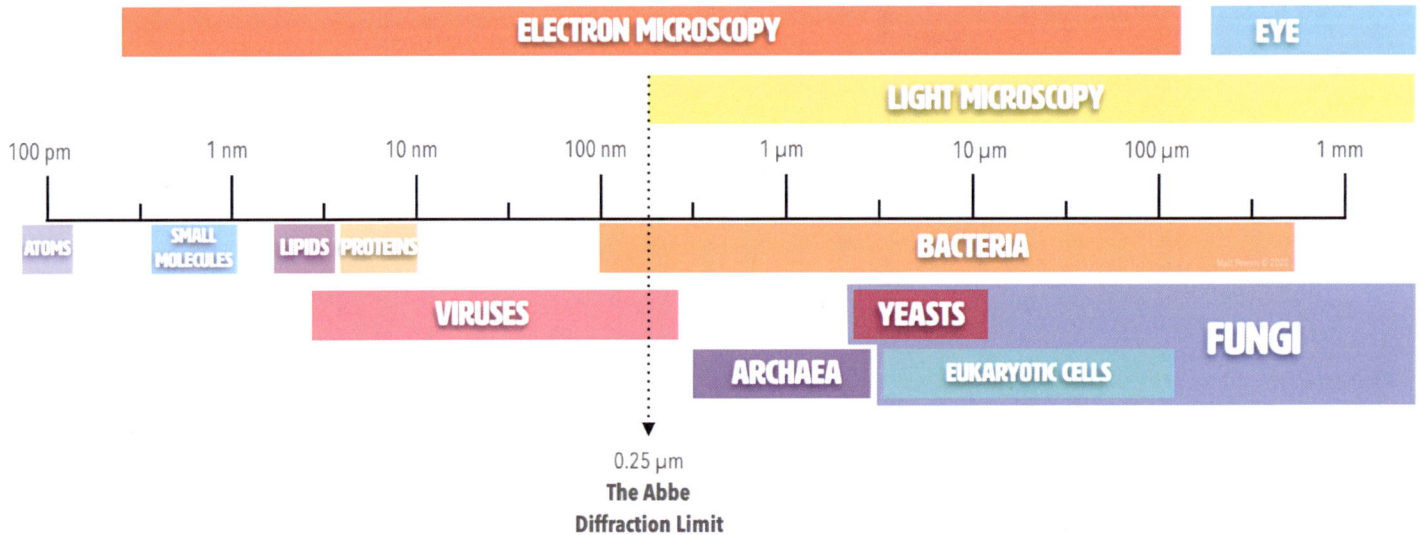

Bacteria are often so small that, even at 1000x, we can see smaller specimens teasing us at the edges of resolution. It has the feel of infinity to it, and in some ways, especially when we consider horizontal gene transfer, it may be so.

BACTERIA SHAPES & MOVEMENT

Are They Round? Rod Shaped? Spirals? Clusters? Still? Moving? Common? Uncommon?

In the morphological identification of bacteria, each shape has a name, and movement, whether they be motile and nonmotile, is also noted. They can also have appendages or be in the process of budding – this can appear like an arm reaching out of the bacteria, or that the bacteria is tethered, or a smaller version of themselves can be attached to their side.

Common Forms:

- **Cocci** – round, spherical, and sometimes deflated basketball shaped bacteria that are ubiquitous in living soils of all types. They are typically 0.5 – 2.0 µm in diameter. Sarcina configurations of bacteria are found in soils, mud, and animal stomachs but they can also be found in compost such as Methanosarcina (these release methane but I've only found them at 0.01% of thermophilic compost). Diplococci are two connected, often encapsulated, cocci like Streptococcus pneumoniae and antibiotic resistant Enterococcus – keep an eye out for these forms; they are not desirable on the whole. Though human microbiomes do benefit from diplococci, plants and soils don't, so if you see

COCCI

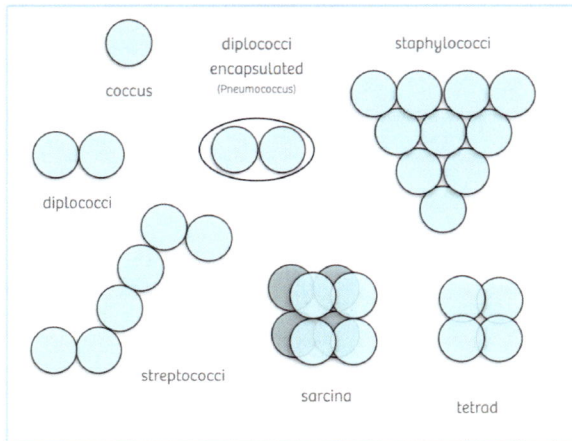

coccus

diplococci
encapsulated
(Pneumococcus)

staphylococci

diplococci

streptococci

sarcina

tetrad

ROUND

BACILLI

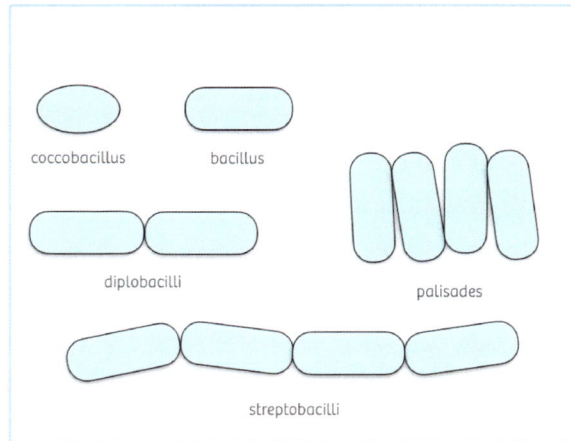

coccobacillus

bacillus

diplobacilli

palisades

streptobacilli

ROD-SHAPED

SPIRILLA

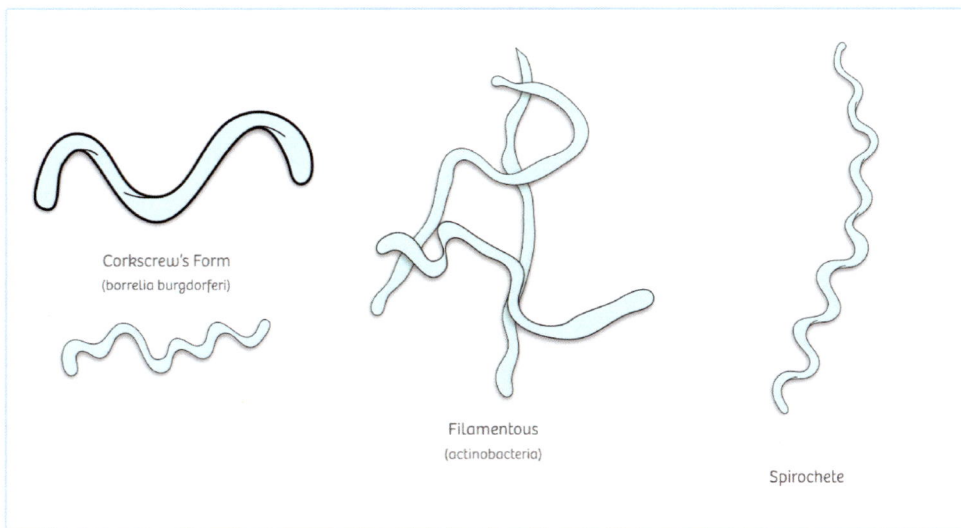

Corkscrew's Form
(borrelia burgdorferi)

Filamentous
(actinobacteria)

Spirochete

SPIRAL-SHAPED

them in large numbers, you need to compost your pile further and perhaps at a higher heat. You may see one or two every thousand bacteria and that's ok, but if you see more than one in a field of vision at 400x or find it in every field of view, that signals a problem usually. Chains of cocci bacteria are called streptococci, and while some are nonpathogenic like the one used to make Emmentaler cheese, most are very dangerous and always found in association with animal feces and infections – they can be motile, so watch out for moving chains of bacteria; they are not good. There's no known streptococci that benefit plants specifically, so not seeing them is always good. Staphylococci are even worse – they are what causes staph infections: these are found on birds and in infections in animals, so signs of these forms are all bad and signs that your compost needs more time if you are composting chicken bedding for instance. Tetrads are found in dairy products, fermented meats, and in wastewater – you may see them but not in any large numbers. If you do, this most likely indicates a problem or a need for further composting/fermenting.

- **Bacilli** – rod or terminator shaped bacteria that can be skinny or thick. Dr. Elaine Ingham has said the large, fat, almost square rods are best as they indicate high oxygen conditions though in low oxygen conditions, the long, slender lactobacillus can serve as a biocontrol in your ferments. Bacilli typically range 1 – 10 µm in length and 0.25 – 1.0 µm in diameter. In large numbers and in favorable conditions, bacilli can form chains, streptobacilli, but these are usually shorter than streptococci. There are also fat, short rods that can appear rounded like cocci called Coccobacillus – they are well known for causing the plague, whooping cough, and sexually transmitted diseases. I have not found any evidence of them in soil or compost in any samples I've sequenced and examined.

- **Spirilla** – spiral, twisted, and corkscrew shaped bacteria as well as dendritic, ribbon, or root-like bacteria. Sometimes spirilla indicate anaerobic conditions as many spirilla prefer a low oxygen environment. While there are several disease-causing spirochetes like *Borrelia burgdorferi* which cause Lyme disease, many are just free-living aerobes found naturally in soil and water – they consume organic matter. They do not always indicate pathogenic conditions, but they often do. Incredibly, spirochetes can range in length anywhere from 3 – 500 µm and their diameters can be 0.09 – 3 µm in size: they cover a vast span. They are also highly motile, appearing like snakes swimming across our field of vision.

Uncommon Forms:

While these are commonly featured on diagrams by microscopy teachers the world over, even experts in soil microscopy, they are only somewhat relevant. They appear in average soils at a 0.01 – 0.2% concentrations, so 2 out of a 1000 bacteria at most but primarily 1–2 out of 10,000 bacteria are in these forms.

- **Vibrio** – while these are anywhere from 0.01% – 0.2% of healthy soil and compost, they are often mentioned by microscopy teachers – the fact is we will rarely see these in healthy soils and compost though you may see Herbaspirillum in plant saps of C4 grasses and high sugar-content plants. They look like curved rods and have flagella they use to motor around. They are found in association with the ocean and food-borne illnesses, often in shellfish, but they are also part of the sulfur reduction pathway in soil environments, especially in waterlogged soils. They can

UNCOMMON FORMS

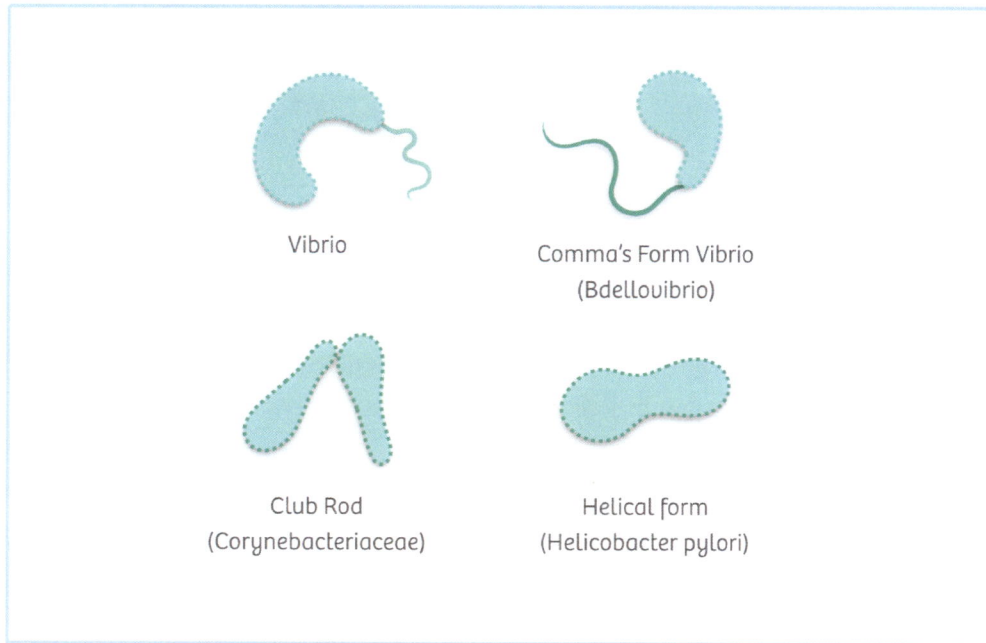

Vibrio

Comma's Form Vibrio
(Bdellovibrio)

Club Rod
(Corynebacteriaceae)

Helical form
(Helicobacter pylori)

be in salt water or fresh. They are commonly 0.5 – 0.8 μm across and 1.4 – 2.6 μm long – we'd have to be using our 100x objective to view them. If you are working with seawater, shellfish, or are growing in halophytic conditions, stay on the look out for vibrio.

- **Club Rod** – the model microbe for club rod is corynebacteriaceae which often forms into palisades – it is a bacilli and technically a member of the Actinomycetes order and Actinobacteria phylum, and it can be found at 0.1 – 1% of healthy soil and compost, and it is the size of average bacteria (0.5 μm across and 2 – 6 μm in length) so it will mix in and hide easily.

- **Helical Form** – comprising on average 0.002% of heathy soil and compost, the model microbe for helical form is helicobacter – it is incredibly rare. The well known helicobacter, *H.pylori*, has not turned up in any of the samples of compost or soil that I've tested, but it is common in the stomachs of mammals, so it could theoretically be found in some situations where there was mismanagement or where further composting is needed.

GRAM STAIN, SMEARS, & PREPARED SLIDES

These methods have been standardized and taught down through the decades, but what are they really?

With gram-positive vs gram-negative, we differentiate between the thickness and makeup of the outer cell walls of bacteria, but it has been discovered that some bacteria don't absorb the dyes and some can change which color they

Each rod-like or spherical speck is a single bacteria feasting on the surface of this wood fragment.

absorb, so it's largely been replaced with genetic testing. It is also worth noting that gram staining does not work with archaea or fungi. Despite all this, it is still common for the pink and purple Gram stain to be how microbes are preserved and presented. This is actually a state of dehydration, and the microbes are smaller than usual which makes them harder to see though this does give us a glimpse into what can happen to microbes in dehydrated states.

Smears are most often heated agar that is combined with the sample microbes, and then it is heat affixed to the slide with an open flame… which is a bizarre way to try and view microbes in my estimation – for many of us, this gives context to Dr. Elaine Ingham's exasperation with classical soil science methodologies. Again, this is a dehydrated and unnatural state for the microbes to be viewed. I don't teach these methods but I have a few pictures depicting microbes in those states so we at least have a lab-verified reference for that microbe. Some of these microbes are quite rare and are virtually impossible to find in a soil sample by eye. *Needle in a haystack* is about right when it's 0.002%!

Prepared slides of biology are dehydrated states of microbes as well – while they are useful for reference, it's best to get laboratory verified living samples as a reference which I have, to the best of my abilities, acquired for this book, my online courses, and the R-Soil Database. Not everything can be found at a reasonable price or at all currently, but I've been able to gather an incredible collection to share with everyone. As we upload these references in long format to the R-Soil Database, it will encourage others to add their images and video of purified cultures to the database as a reference as well, expanding community access to this information. All this being said, the images generated by these methods is subpar when compared to when distilled or purified water is used in a mix with our soil and compost samples.

INDICATORS OF BENEFICIAL BACTERIA

Remember this is just for bacteria – yeasts like to clump!! And remember this is on the whole, it's ok if you see one or two of the concerning microbes per drop or small expressions of actinobacteria every field of view (FOV), but usually any more than that, it signals a problem though in the case of actinobacteria it would have to dominate for it to be a significant problem. This goes for both soil and compost.

- The Simple Bacilli and Cocci Forms Dominate – *Primarily Non-Clumping*
- No Streptococci, No Staphylococci, but there could be bacilli forming some smaller chains (especially in a bacilli brew)
- No Encapsulations & No Pairing (not to be confused with asexual reproduction like budding)
- Tetrads & Sarcina Are Few or Absent
- Filamentous Bacteria Are Not Dominant (it is very normal to have streptomyces at 1–2% of your compost)

INDICATORS OF HARMFUL BACTERIA

It's important to remember that this is all contextual. Lactobacillus is biocontrol in ferments for instance. You may be using an anaerobic ferment, growing rice in facultative muds, or be doing a specialized aquaculture systems, and remember this is the bacteria strictly we are talking about here. It's really about reading the room – is there a balance or is there a particular imbalance to the scene under your microscope? With experience you'll be able to recognize an out-of-ordinary soil or compost sample instantly. All these forms are concerning if they are anything but rarely seen.

- Uncommon Forms Dominate (or are present in any significant #'s)
- Chains
- Clumping
- Encapsulation
- Pairing
- Simple Coccus Form Is Rare

- You Have To Dilute Your Sample past 1:1000
- Tetrads & Sarcina Are Many

While some "experts" even those with a phd after their name may tell you, corkscrew shaped spirillum or helical form bacteria indicate pathogenic e.coli and salmonella, this does not mean that e.coli and salmonella are spirilla – they are both bacilli shaped. Spirilla often prefer anaerobic conditions but can be found in aerobic conditions as well – in other words, if you see a few, no worries, but if you see them everywhere, there's likely a problem or you're doing something specific in the anaerobic space.

If you see two cocci cells encapsulated or in chains, that is likely from manure – it can cause pneumonia, bronchitis, and other unpleasantnesses. E.coli are ubiquitous and primarily commensal, but they can also be a problem if the environment is anaerobic OR if they are being added from an anaerobic environment somehow like leaching from upstream feedlots as it was the case for some organic farmers in California.

BACTERIAL LOOKALIKES

The most challenging part of working with soil bacteria is they are notoriously alike. All the bacilli tend to look the same as we cannot see their flagella with light microscopy even at 1000x, and according to Dr. Elaine Ingham, oxygen levels affect their diameters (lactobacilli are thin and long, while according to her, more aerobic bacilli are short and fat), but it depends on the species as they have specific ranges of diameter and length – for example, Klebsiella is also usually more short and thick than e.coli, so we must be careful. The beautiful thing about most of these microbes is we have long traditions in science of cultivating them successfully and mapping out the temperature and pH ranges for optimal growth. It's also a common graduate school project to map the optimal growth ranges of a microbe across different growth mediums. Even with a hemocytometer or a μm ruler, we cannot differentiate bacteria beyond their forms simply because the different species overlap in size, their features are too small for light microscopy, and many biofertilizing microbes come from families of microbes typified by pathogens – Klebsiella, for instance, is responsible for pneumonia, meningitis, sepsis, and many more terrible illnesses. E.coli is the same – you cannot tell pathogenic e.coli by sight using light microscopy. The only way to know which bacteria are truly in our samples is to sequence the DNA.

ELECTRON MICROSCOPY

EYE

LIGHT MICROSCOPY

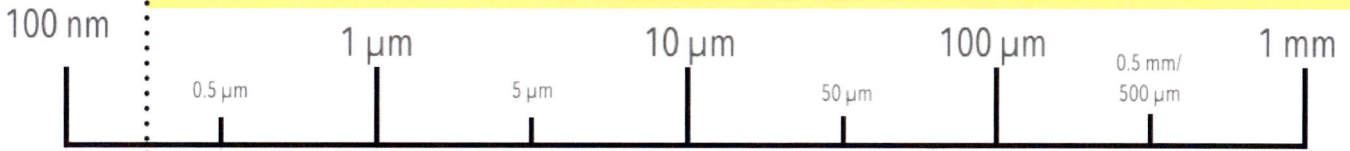

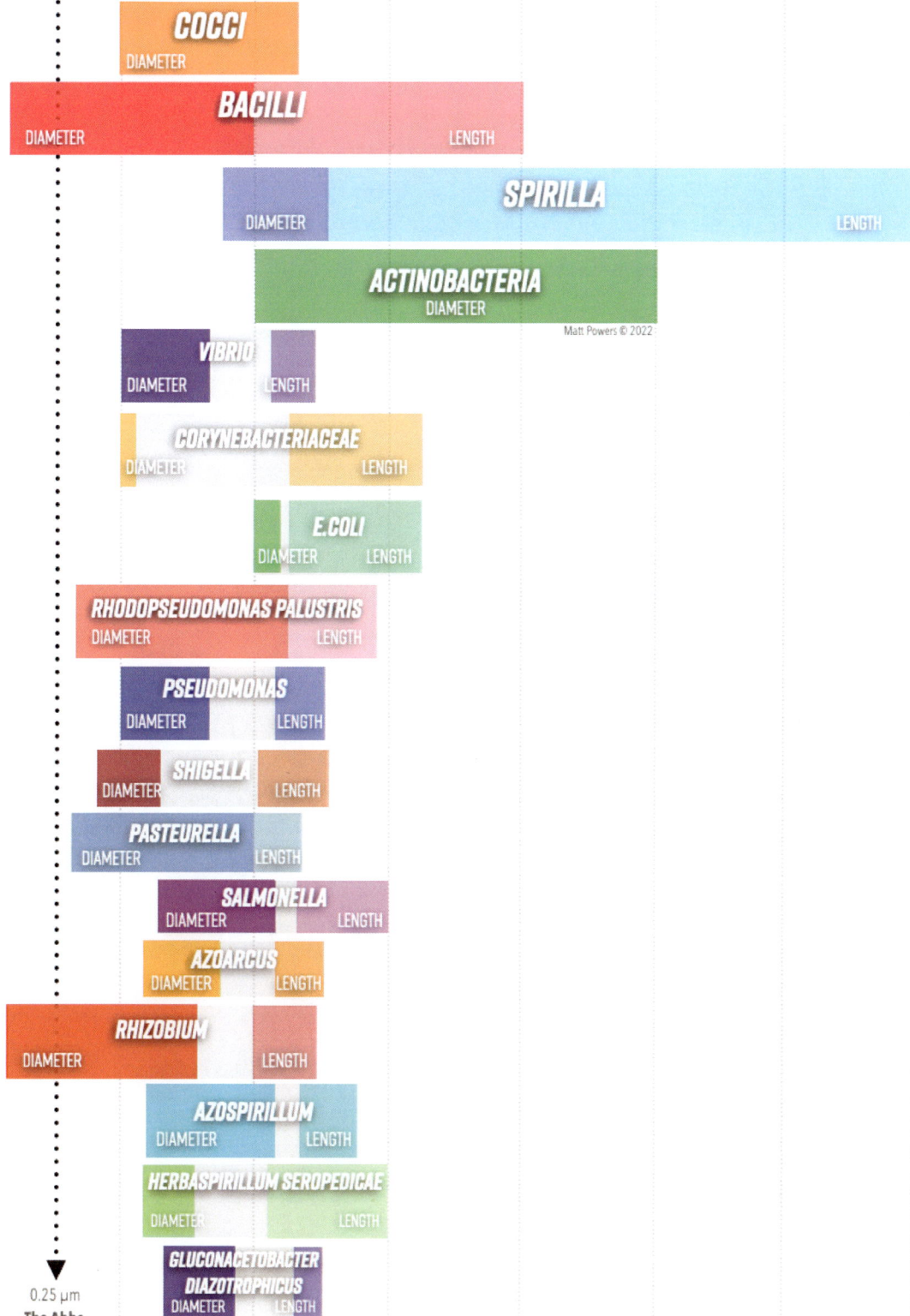

100 nm   1 µm   10 µm   100 µm   1 mm

0.5 µm   5 µm   50 µm   0.5 mm/ 500 µm

COCCI
DIAMETER

BACILLI
DIAMETER   LENGTH

SPIRILLA
DIAMETER   LENGTH

ACTINOBACTERIA
DIAMETER

Matt Powers © 2022

VIBRIO
DIAMETER   LENGTH

CORYNEBACTERIACEAE
DIAMETER   LENGTH

E.COLI
DIAMETER   LENGTH

RHODOPSEUDOMONAS PALUSTRIS
DIAMETER   LENGTH

PSEUDOMONAS
DIAMETER   LENGTH

SHIGELLA
DIAMETER   LENGTH

PASTEURELLA
DIAMETER   LENGTH

SALMONELLA
DIAMETER   LENGTH

AZOARCUS
DIAMETER   LENGTH

RHIZOBIUM
DIAMETER   LENGTH

AZOSPIRILLUM
DIAMETER   LENGTH

HERBASPIRILLUM SEROPEDICAE
DIAMETER   LENGTH

GLUCONACETOBACTER DIAZOTROPHICUS
DIAMETER   LENGTH

0.25 µm
The Abbe Diffraction Limit

42

SPECIES	GRAM-STAIN	FORM	MOTILE/NON-MOTILE	SIZE	BENEFICIAL/DETRIMENTAL	% OF HOT COMPOST CATALYST BIOAMENDMENTS % OF 284,854 READS	% OF EM-1 TERAGANIX % OUT OF 5,109 READS	% OF JOHNSON-SU COMPOST MICHAEL STANGL % OUT OF 142,468 READS	% OF FRONT YARD SOIL SAN MARCOS, TX % OUT OF 36,518 READS	PHYLUM	AEROBIC/ANAEROBIC/ENDOPHYTIC
E.COLI	−	BACILLI	M	1.1–1.5 µm X 2–6 µm	+	18.4%	17%	16%	14%	PSEUDOMONADOTA	FACULTATIVE ANAEROBE
STREPTOMYCES	+	FILAMENTOUS	NM	0.5–2 µm	+	1.2%	0.2%	1.2%	2.4%	ACTINOBACTERIA	AEROBE
BRADYRHIZOBIUM [COMMON DNA SEQUENCING CONTAMINANT]	−	BACILLI	M	0.5–0.9 µm X 1.2–3 µm	+	0.9%	0.05%	0.8%	2%	PSEUDOMONADOTA	AEROBE
PSEUDOMONAS	−	BACILLI	M	0.5 – 1 µm X 1.5–5 µm	+	0.8%	0.5%	0.6%	1%	PSEUDOMONADOTA	AEROBE
BURKHOLDERIA	−	BACILLI	M	0.5–1 µm X 1.5–4 µm	+	0.5%	0.01%	0.5%	0.7%	PSEUDOMONADOTA	AEROBE
MESORHIZOBIUM	−	BACILLI	M	0.4–0.9 µm X 1.2–3 µm	+	0.6%	0%	0.5%	0.8%	PSEUDOMONADOTA	AEROBE
SALMONELLA	−	BACILLI	M	0.7–1.5 µm X 2–5 µm	−	0.3%	0.4%	0.3%	0.1%	PSEUDOMONADOTA	FACULTATIVE ANAEROBE
MUVIRUS		VIRUS		0.027–0.03 µm	+/−	0.3%	0.02%	0.3%	0.3%	UROVIRICOTA	E.COLI & SHIGELLA PHAGES
SPHINGOMONAS	−	BACILLI	M	0.2–1.4 X 0.5–4 µm	+	0.3%	0.04%	0.3%	0.4%	PSEUDOMONADOTA	AEROBE
SORANGIUM	−	BACILLI	M	0.8–15 x 1.2–40 µm	+	0.3%	0%	0.3%	0.2%	MYXOCOCCOTA	AEROBE
RHIZOBIUM	−	BACILLI	M	0.5–2 µm X 0.1–0.8 µm	+	0.2%	0%	0.3%	0.3%	PSEUDOMONADOTA	AEROBE
AZOSPIRILLUM	−	BACILLI	M	0.6–1.7 µm X 2.1–3.8 µm	+	0.2%	0%	0.2%	0.2%	PSEUDOMONADOTA	AEROBE
RHODOPSEUDOMONAS PALUSTRIS	−	BACILLI	M/NM	0.3–2 µm X 2–10 µm	+	0.1%	0.01%	0.1%	0.3%	PSEUDOMONADOTA	FACULTATIVE ANAEROBE
AZOARCUS	−	BACILLI	M	0.4–1.5 µm X 1.1–4 µm	+	0.1%	0%	0.1%	0.2%	PSEUDOMONADOTA	ANAEROBE/ENDOPHYTE
SHIGELLA	−	BACILLI	NM	1–3 µm X 0.7–1 µm	−	0.1%	0%	0.07%	0.06%	PSEUDOMONADOTA	FACULTATIVE ANAEROBE
HERBASPIRILLUM SEROPEDICAE	−	BACILLI	M	0.6–0.7 µm X 1.5–5 µm	+	0.04%	0%	0.03%	0.03%	PSEUDOMONADOTA	ENDOPHYTIC
AZOTOBACTER	−	COCCI	M	1.6–2.7 µm X, 3–10 µm	+	0.02%	0%	0.02%	0.002%	PSEUDOMONADOTA	AEROBE
PATHOGENIC SPHINGOMONAS	−	BACILLI	M	0.2–1.4 X 0.5–4 µm	−	0.01%	0%	0.004%	0%	PSEUDOMONADOTA	AEROBE
ACETOBACTER	−	BACILLI/COCCI	M/NM	0.6–0.9 µm X 1–4 µm	+	0.01%	0.7%	0.009%	0.01%	PSEUDOMONADOTA	AEROBE
PAENIBACILLUS	+	BACILLI	M	~0.3–5 µm × ~0.5–1.5 µm	+	0.1%	0%	0.2%	0.1%	BACILLOTA	AEROBE/FACULTATIVE ANAEROBE
P. FLUORESCENS	−	BACILLI	M	0.5 – 1 µm X 1.5–5 µm	+	0.01%	0%	0.02%	0.03%	PSEUDOMONADOTA	AEROBE
GLUCONACETOBACTER DIAZOTROPHICUS	−	BACILLI/COCCI	NM	0.5–1 µm X 2.6–4.2 µm	+	0.009%	0%	0.01%	0.005%	PSEUDOMONADOTA	AEROBE
LACTOBACILLUS	+	BACILLI	M	0.5–0.9 µm X 1.5–9 µm	+	0.002%	17.6%	0.005%	0%	BACILLOTA	FACULTATIVE ANAEROBE
SYMBIOBACTERIUM	−	PLEOMORPHIC/FILAMENTOUS	NM	0.7–0.8 X 2.7–7.7 µm 5–100 µm in length filamentous	+	0.03%	0.03%	1.2%	0.02%	BACILLOTA	AEROBE/FACULTATIVE ANAEROBE
NOCARDIOIDES	+	PLEOMORPHIC/FILAMENTOUS	M/NM	0.5–0.8 µm in diameter	+/−	0.06%	0.03%	0.2%	0.7%	ACTINOBACTERIA	AEROBIC & ENDOPHYTIC
MYCOLICIBACTERIUM	+	BACILLI	M/NM	3–5 µm long	+/−	0.2%	0%	0.3%	0.6%	ACTINOBACTERIA	AEROBE
THERMOBACILLUS	−	BACILLI	NM	0.5–0.7 µm X 2.0–5.0 µm	+/−	0.03%	0%	0.4%	0.002%	FIRMICUTES	AEROBE
THERMOBISPORA	+	FILAMENTOUS	NM	0.5–0.8 µm in diameter	+/−	0.06%	0.01%	0.3%	0.01%	ACTINOBACTERIA	AEROBE
PASTEURELLA	−	PLEOMORPHIC	NM	0.3–1 µm X 1–2 µm	−	0.0007%	0%	0.001%	0%	PSEUDOMONADOTA	FACULTATIVE ANAEROBE
PATHOGENIC E.COLI	−	BACILLI	M	1.1–1.5 µm X 2–6 µm	−	0%	0%	0%	0%	PSEUDOMONADOTA	FACULTATIVE ANAEROBE

Final Thoughts on Bacteria

Let's look at both the bacterial size and comparison charts. Notice how we have beneficial bacteria of all forms and pathogens represented – we can easily see why DNA testing is so important. Not only is there massive overlap in appearances and size, but did you notice how gram staining shrinks things dramatically? That makes identification even harder. Did you notice how similar they almost *all* look? The pathogens are just as similar. There are beneficial anaerobes and aerobes as well as beneficial microbes that we categorize as gram-negative and gram-positive, spirilla, bacilli, cocci, filamentous, non-spore-forming, endospore forming, endophytic, free living, nodule forming, and more! With our light microscopes, we can only guess at who's really there in terms of bacteria and archaea using the context and DNA testing.

Did you notice that while most of the beneficials are aerobic, they are a minority of the bacterial community as a whole? Facultative anaerobes dominate the space, specifically e.coli., and while most soil and compost bacteria are pseudomonadota, key bacteria are not: streptomyces being the most prominent in this regard. The other surprise for many may be the prominence of rhizobia family bacteria in compost (B*radyrhizobium*, *Mesarhizobium*, and to a much lesser degree *Rhizobium*), but likely the most worrisome surprise is that *Salmonella* is present at about the rates of most aerobic bacteria in all compost and EM samples – though, all things considered, 0.3% is a relatively low number (3 in a 1000 bacteria). What's more: in all my DNA sequencing for compost, e.coli was most prominent followed by streptomyces with pseudomonas, bradyrhizobium, burkholderia, and mesorhizobium all competing for the 3rd most present in compost samples – these 4 tend to be found at about the same % levels with some higher and lower in various arrangements depending on the pile size, ingredients, management, and composting type.

Also something to note: *Muvirus*. This is a prominent reminder of how HGT occurs through viral exchanges of DNA between microbes in our soils and compost constantly all the time. While the majority of these phages were for e.coli, 9/25 were for shigella. This was perhaps the pathway of their senescence (shigella is in low concentration), i.e. a way that compost controls the pathogens – it is not clear, but the reason I say that is: there are zero pathogenic e.coli present & fewer shigella present in ratio to their Muvirus phages which implies they were perhaps more present in the past. Something not mentioned in the chart is that two species of archaea are as numerous as pseudomonas (0.6%) in Johnson-Su compost samples (both are common in thermophilic settings like hot springs and composting) – they are mostly absent from all other samples of soil and compost.

ESCHERICHIA COLI (E.COLI)

Gram-Negative Facultative Anaerobe, Coliform Bacteria

FAMILY: Enterobacteriaceae

ORDER: Enterobacterales

PHLYUM: Pseudomonadota

FORM: Bacilli

SIZE: 1.1 – 1.5 μm diameter, 2 – 6 μm long

APPEARANCE: Straight rods, singly or in pairs, non-sporulating, non-motile & motile with peritrichous flagella (evenly distributed across their surface)

ABUNDANCE: _The #1 most abundant bacteria_ found in all soil and compost samples (comprising 18–25%) but primarily described as found in the lower intestines and feces of warm-blooded animals – millions of e.coli strains are commensal and endophytic, but a well-known few are pathogens (O157:H7, O104:H4, O121, O26, O103, O111, O145, & O104:H21)

FOODS: Carbon and energy – both aerobic and facultative (fermentative) digestion

PH & TEMP PREFERENCES: Prefers pH 5.5 – 7.5, 37°C but it can handle extreme environments as well

REPRODUCTION: Asexual Reproduction (Binary Fission) & HGT (especially Conjugation) – though high Muvirus levels in compost means that HGT is occurring via that pathway as well.

USES: It is a commensal and endophytic microbe that is often supporting animals and plants. If you have pathogenic e.coli, it indicates the potential for other pathogens to be present, for the pathogen to spread/harm people or animals, and that there is an anaerobic or disease causing condition somewhere in the cycle.

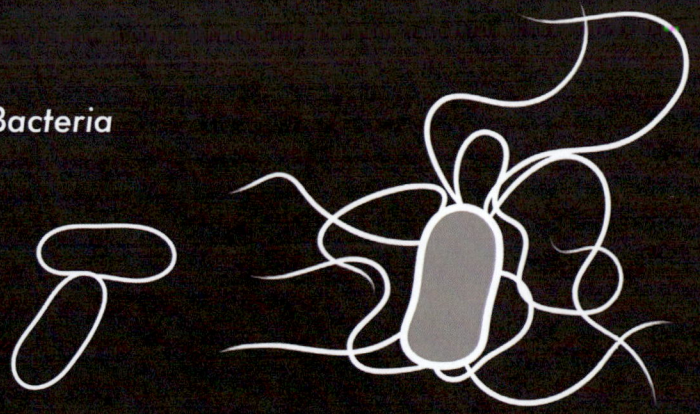

LAB VERIFIED E.COLI PREPARED SLIDE — 1000X

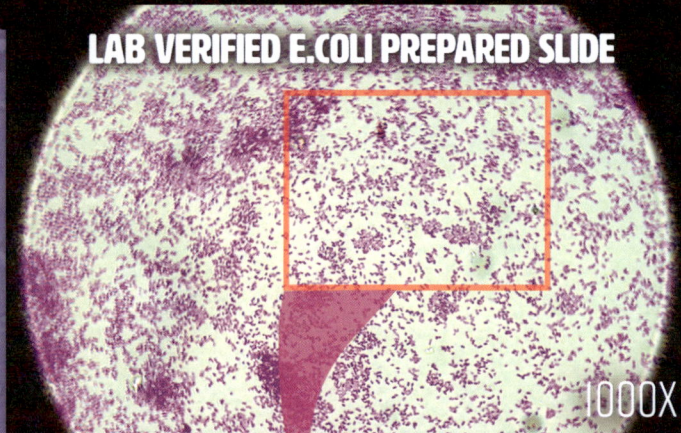

DNA TESTED EM-1 SAMPLE — 400X

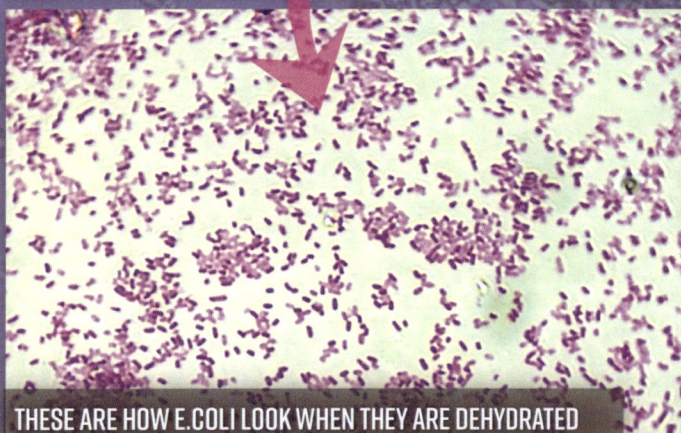

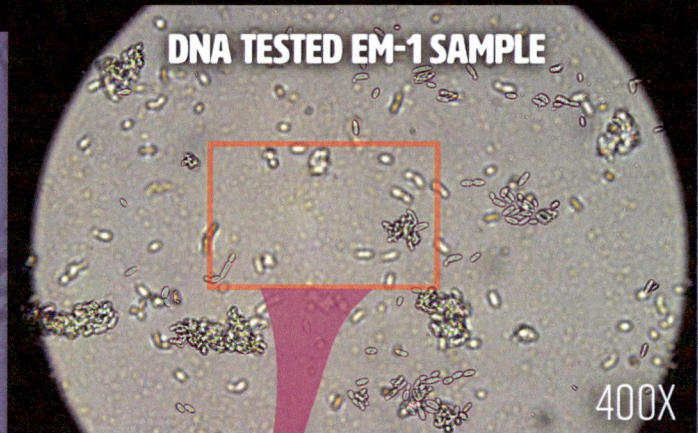

THESE ARE HOW E.COLI LOOK WHEN THEY ARE DEHYDRATED

BACTERIA

YEAST

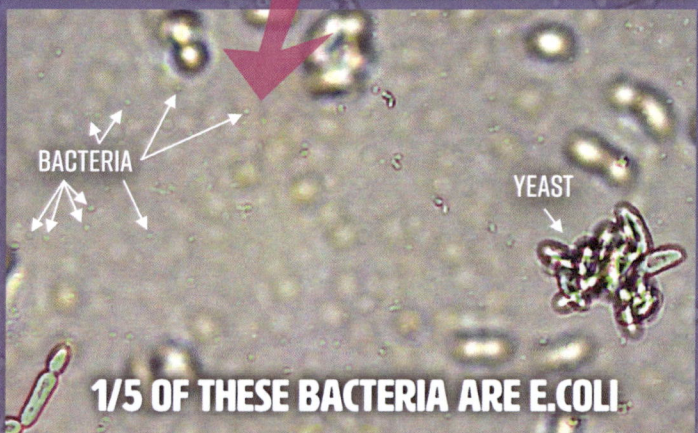

1/5 OF THESE BACTERIA ARE E.COLI

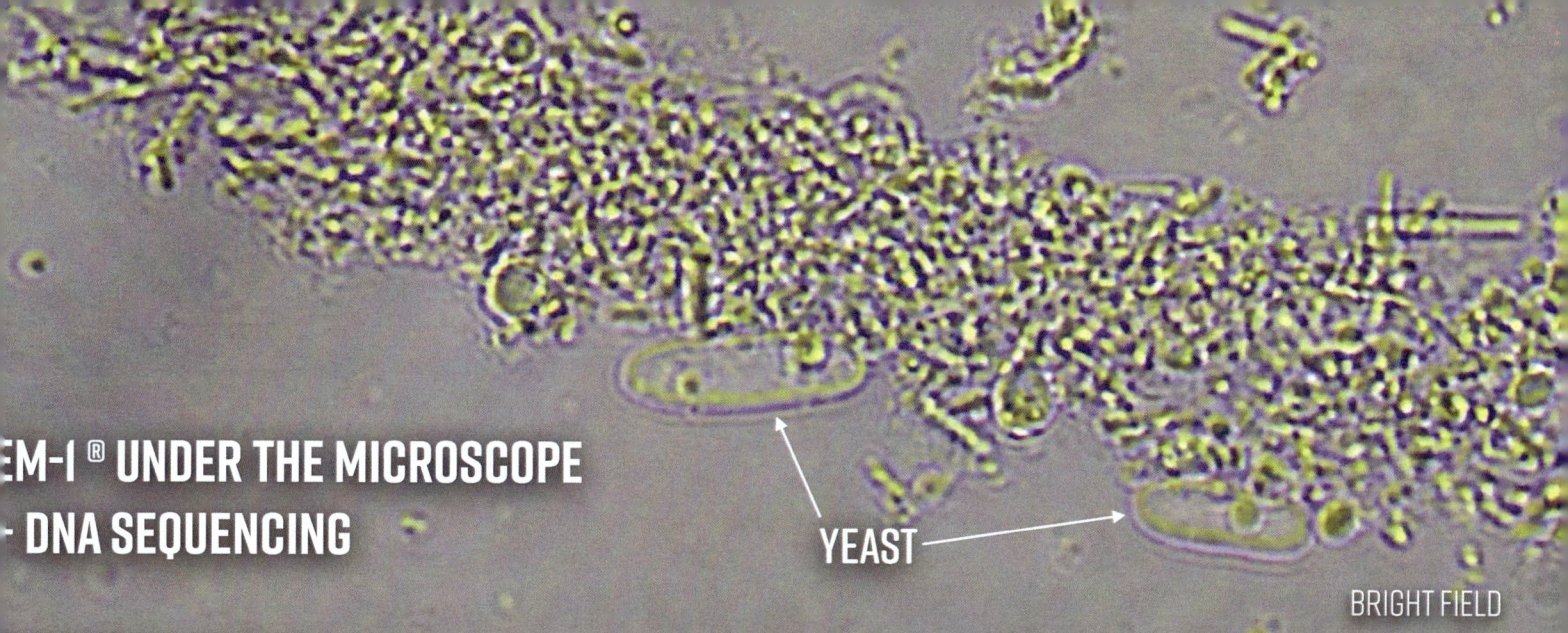

EM-1 ® UNDER THE MICROSCOPE + DNA SEQUENCING

YEAST

BRIGHT FIELD

ZOOMED IN ON 1000X

EPIFLUORESCENCE

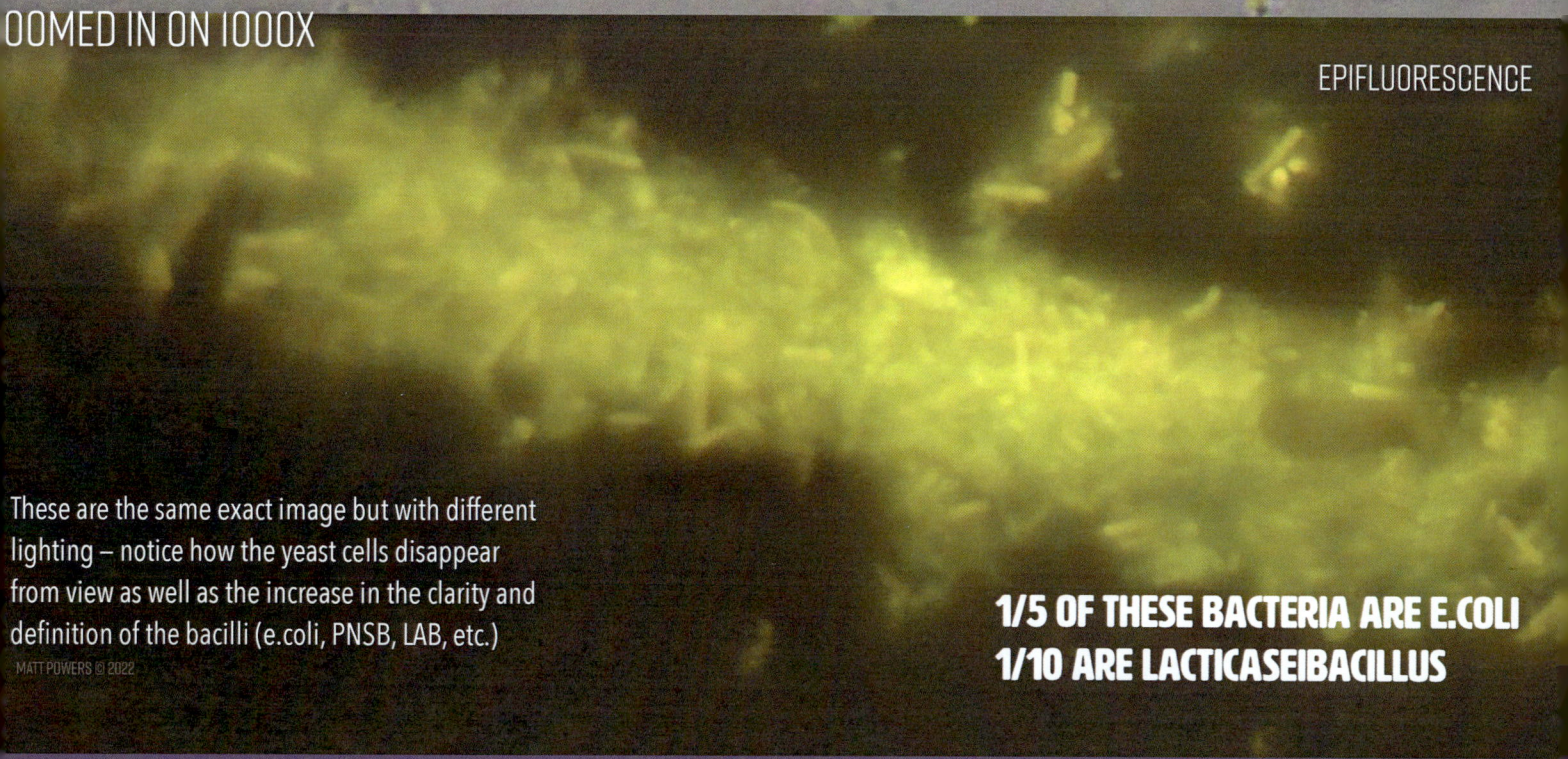

These are the same exact image but with different lighting – notice how the yeast cells disappear from view as well as the increase in the clarity and definition of the bacilli (e.coli, PNSB, LAB, etc.)

MATT POWERS © 2022

1/5 OF THESE BACTERIA ARE E.COLI
1/10 ARE LACTICASEIBACILLUS

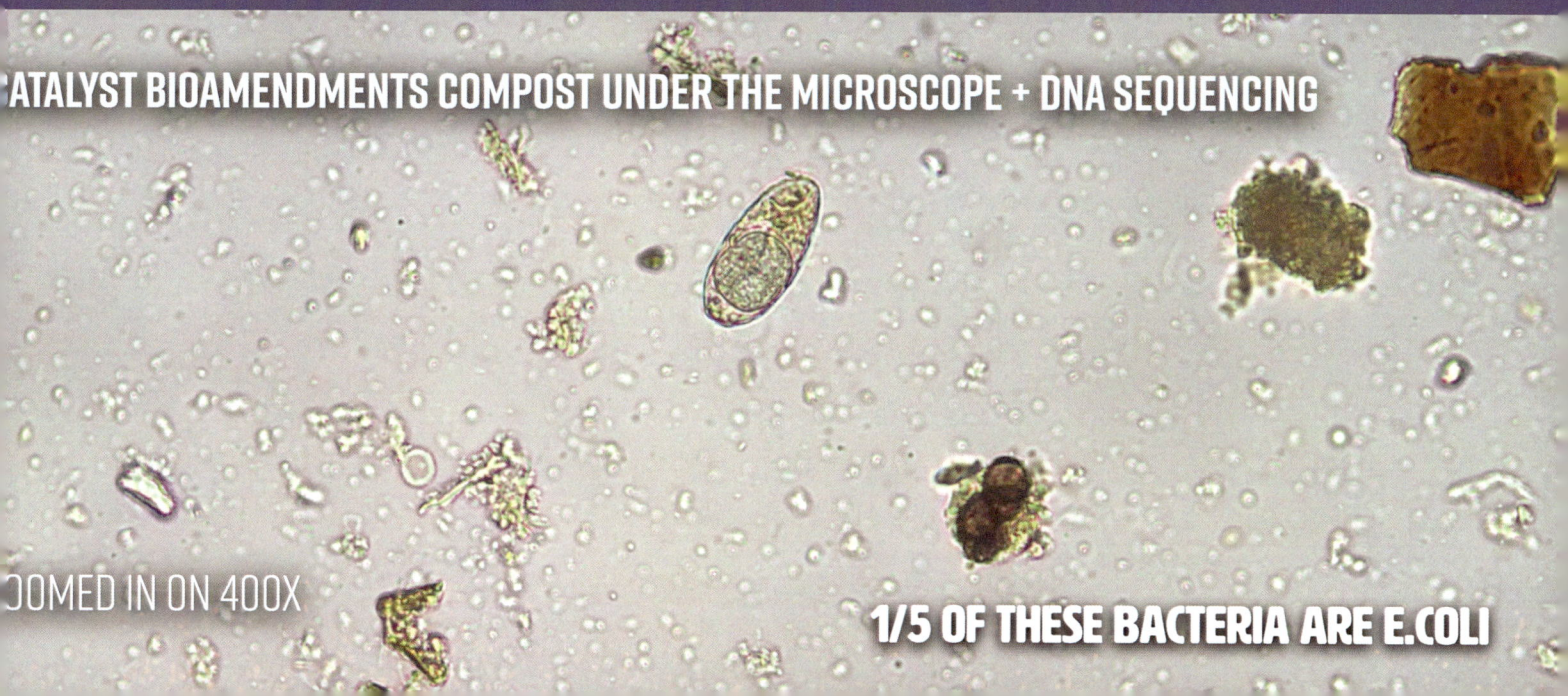

CATALYST BIOAMENDMENTS COMPOST UNDER THE MICROSCOPE + DNA SEQUENCING

ZOOMED IN ON 400X

1/5 OF THESE BACTERIA ARE E.COLI

STREPTOMYCES
Gram-positive Aerobic Mesophile

FAMILY: Streptomycetaceae

ORDER: Actinomycetales

PHYLUM: Actinomycetota

FORM: Filamentous

SIZE: 0.5 – 2 μm diameter vegetative hyphae with 1 – 2 μm length spores

APPEARANCE: Produces a brown melanin, has highly irregular branched mycelia, produces large spore chains (up to 100 long) in various patterns including spirals from its aerial mycelia. Spores can be smooth, hairy, spiny, warty, or even wrinkled.

REPRODUCTION: Conidiospores & HGT

LOCATION: Fresh and salt water, soil, compost, vermicompost, inside the plant **endophytically**, & the rhizosphere

ABUNDANCE: Found at rates of 1 – 5% in compost and soil (making it the 2nd most common bacteria found in exemplary compost), and typically represent 50% of actinobacteria present

FOODS: Organic matter, nitrogen in various forms, & a wide array of sugars (carbon)

PH & TEMPERATURE PREFERENCES: pH 6.5 – 8 and 25 – 35°C, but they are found even at pH 5 – 11, 15 – 40°C) – heat shock (30 – 45°C for 30 – 60 min) will active their spores preferentially over other actinobacteria (*think hot composting*).

USES: Composting, PGPR, & Endophyte

LAB-CERTIFIED LIVE CULTURE

100X

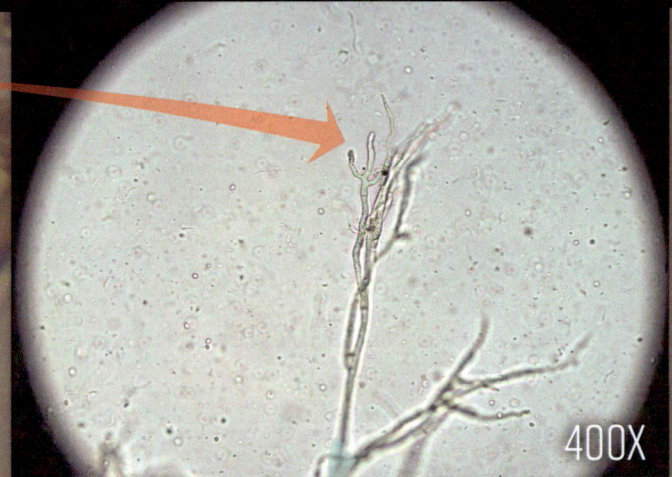

400X

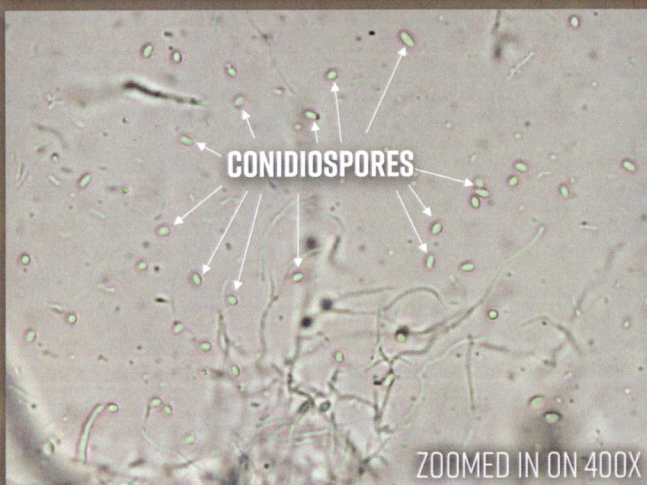

CONIDIOSPORES

ZOOMED IN ON 400X

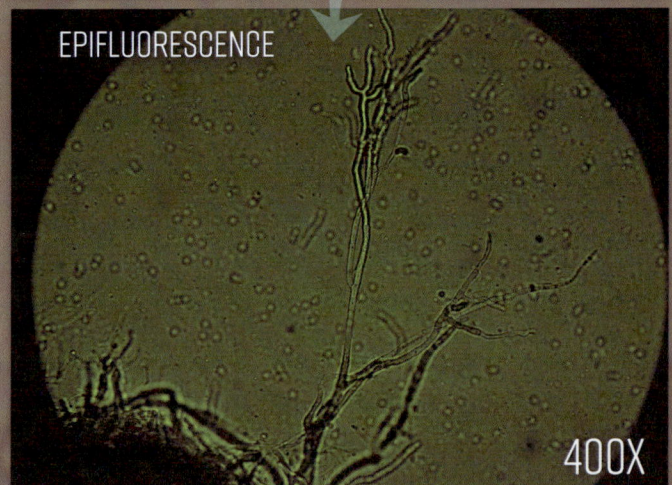

EPIFLUORESCENCE

400X

STREPTOMYCES

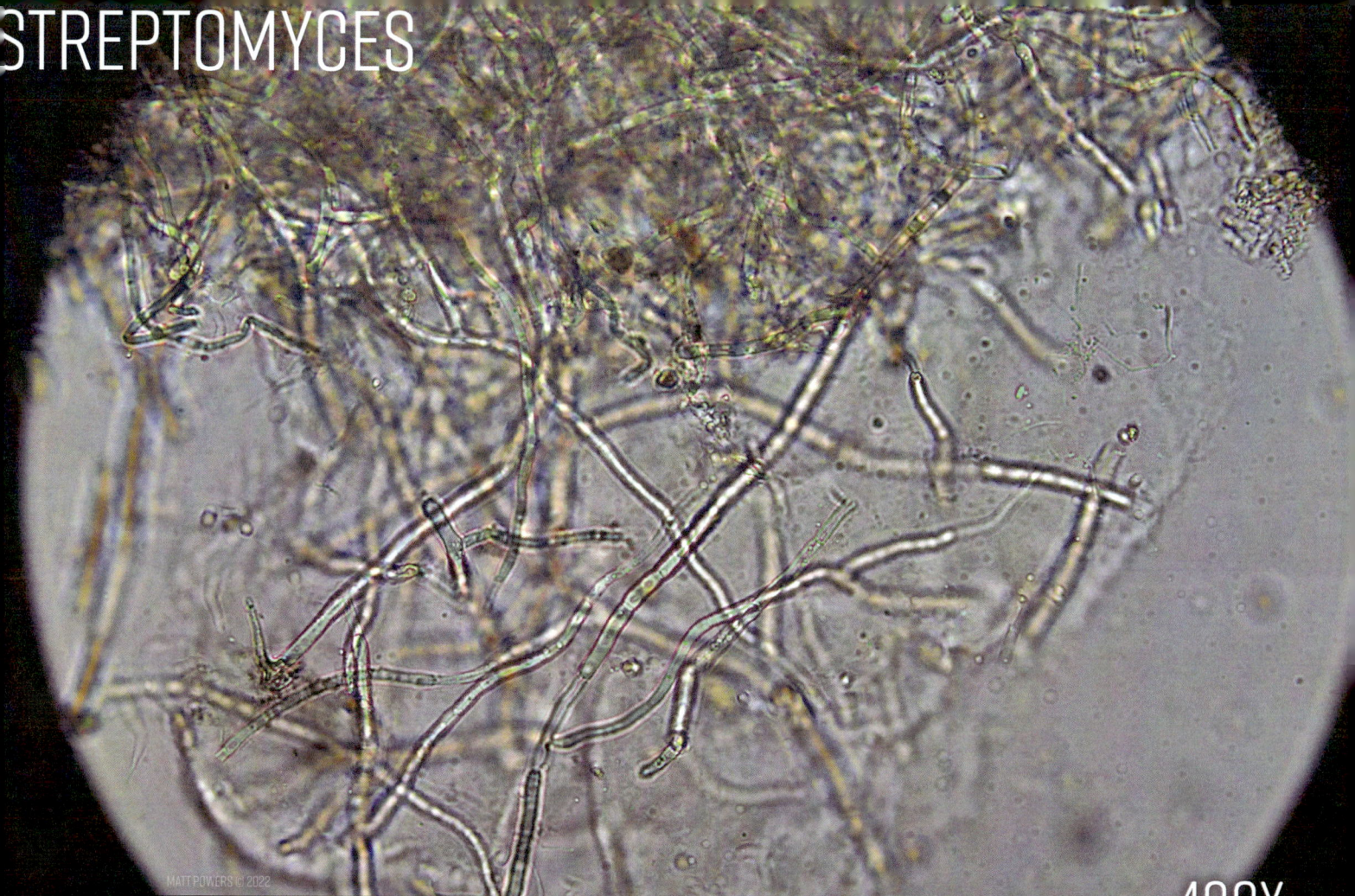

400X

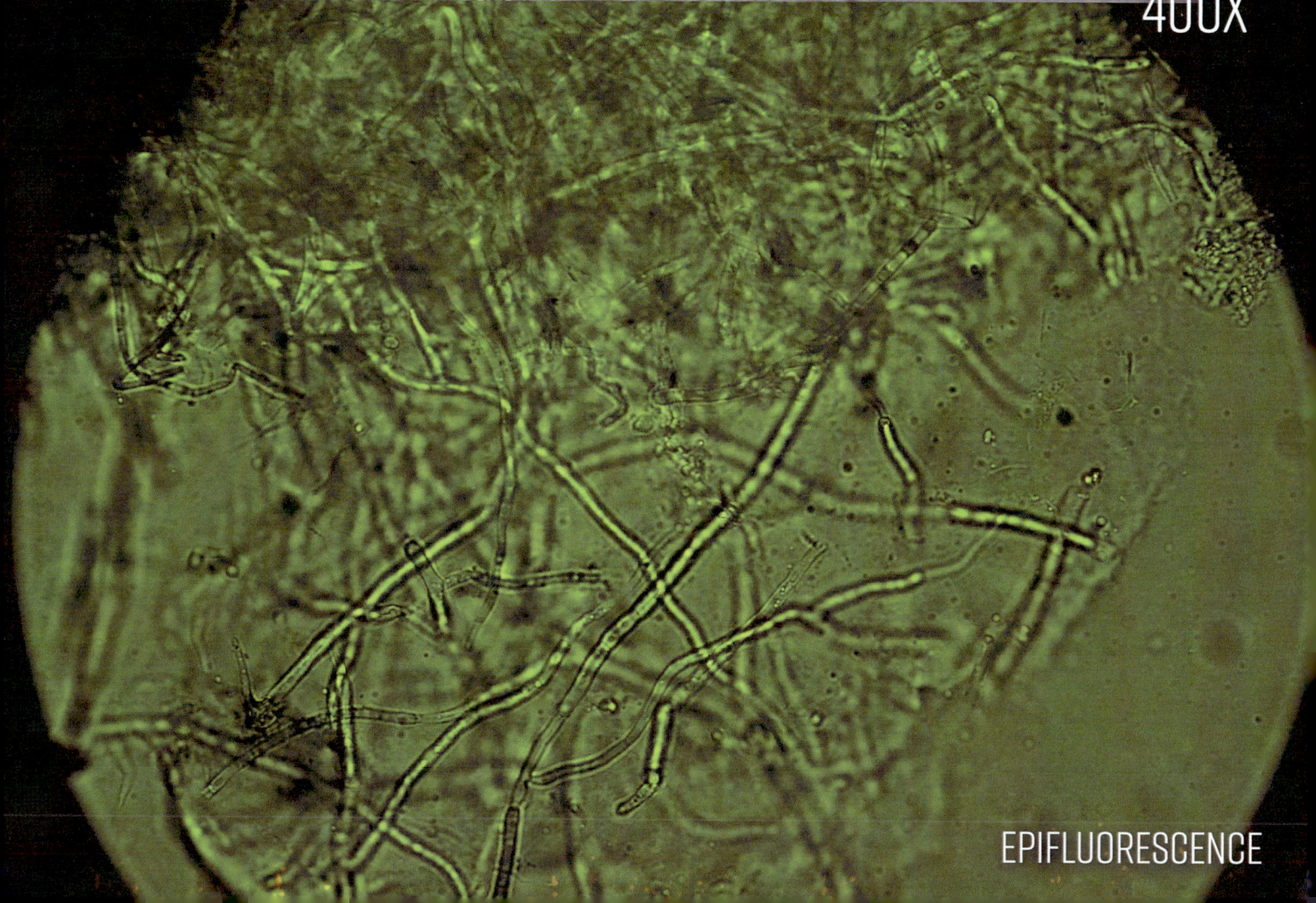

EPIFLUORESCENCE

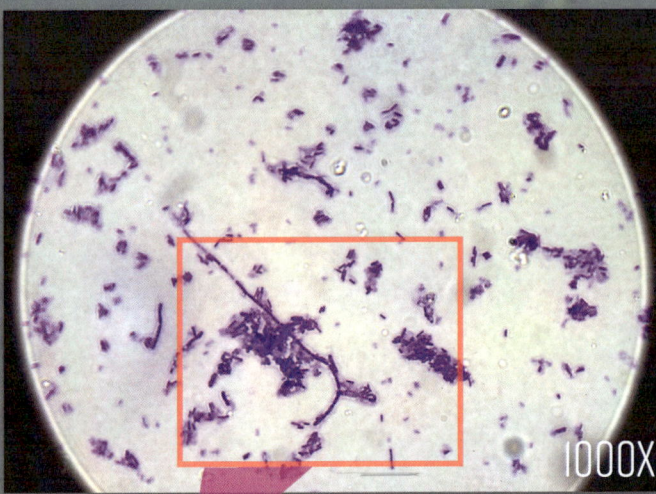
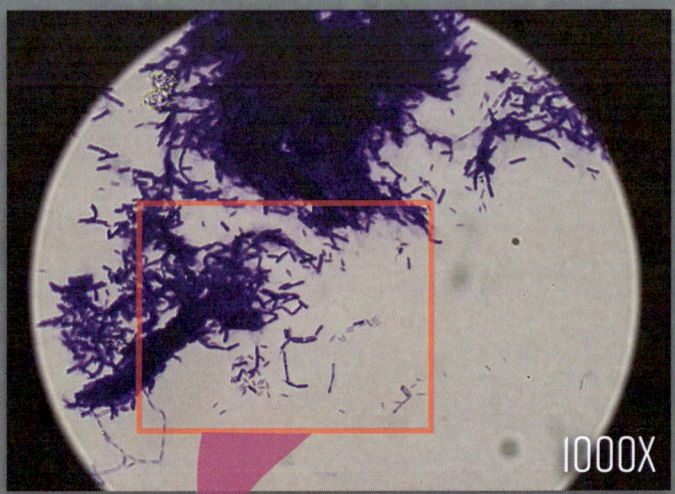
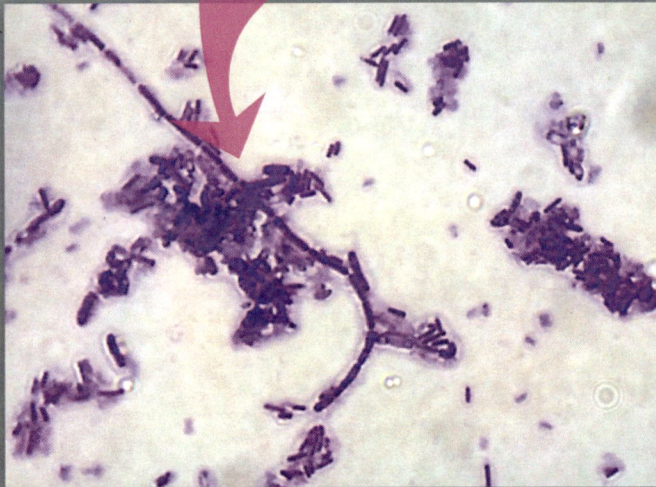
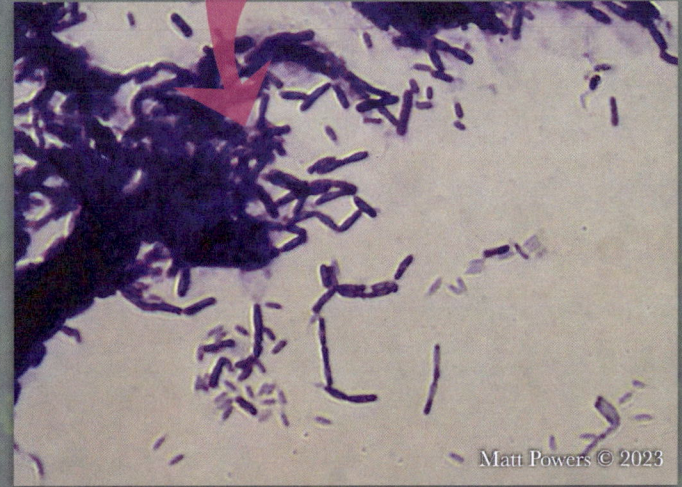

1000X

1000X

Matt Powers © 2023

BACILLUS SUBTILIS

Gram-Positive Obligate Aerobe/Facultative Anaerobe

FAMILY: Bacillaceae

ORDER: Bacillales

PHYLUM: Bacillota

FORM: Bacilli

SIZE: 0.25 – 1 μm wide, 4 – 10 μm long

APPEARANCE: Rod-shaped, flagellated, & highly motile – sometimes with endospores in nutrient poor conditions. Found in bunches, chains, and solitarily. Looks like most bacilli species.

ABUNDANCE: Ubiquitous endophytically and in the rhizosphere but especially in grass and hay

FOODS: Plant exudates, sugars, & organic matter

PH & TEMPERATURE PREFERENCES: pH 5.5, 10 min at 80°C germinates endospores, 37°C thereafter

REPRODUCTION: Symmetric and Asymmetric (Endospores)

USES: PGPR and biocontrol

ENTEROBACTER
Gram-Negative Facultative Anaerobe

FAMILY: Enterobacteriaceae

ORDER: Enterobacterales

PHLYUM: Pseudomonadota

FORM: Bacilli, non-spore-forming

SIZE: 0.6 – 1 μm diameter, 1.2 – 3 μm long (as with most Enterobacteriaceae).

APPEARANCE: Motile straight rods with evenly distributed flagella

ABUNDANCE: Found ubiquitously in soil, water, plants, insects, sewage, manure, & animal intestines

FOODS: Plant exudates, lactose, & glucose

PH & TEMP PREFERENCES: pH 5 – 8, 30°C is the optimal growth temperature (though clinically typically 37°C)

REPRODUCTION: Asexual Reproduction (Binary Fission) & HGT

USES: PGPR, N-Fixation, P-Solubilization, ISR, Siderophore Production, Biocontrol, IAA, Increasing Selenium Uptake, Bioremediation, ACC Deaminase

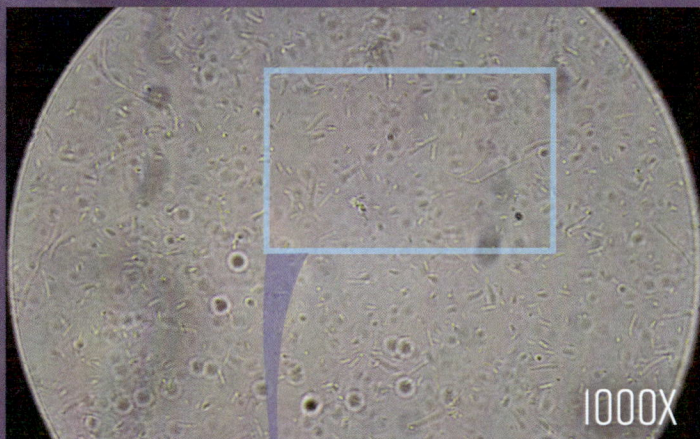

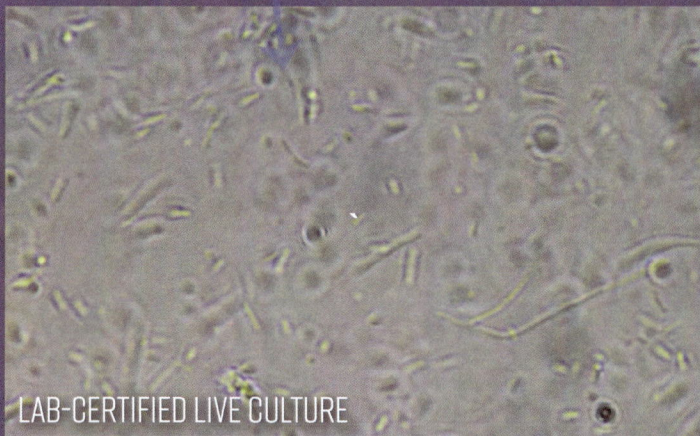

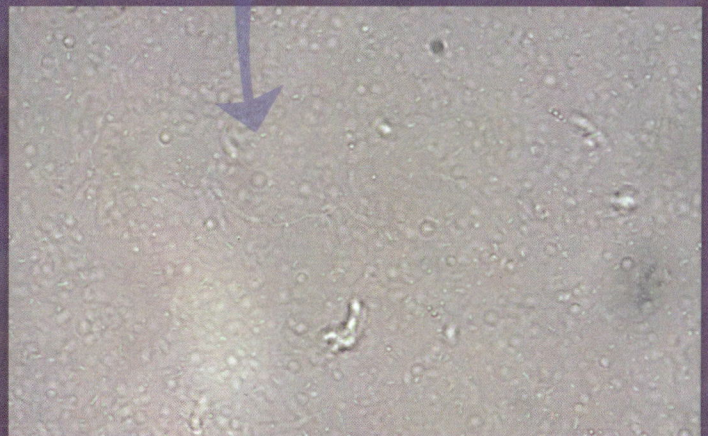

1000X

100X

LAB-CERTIFIED LIVE CULTURE

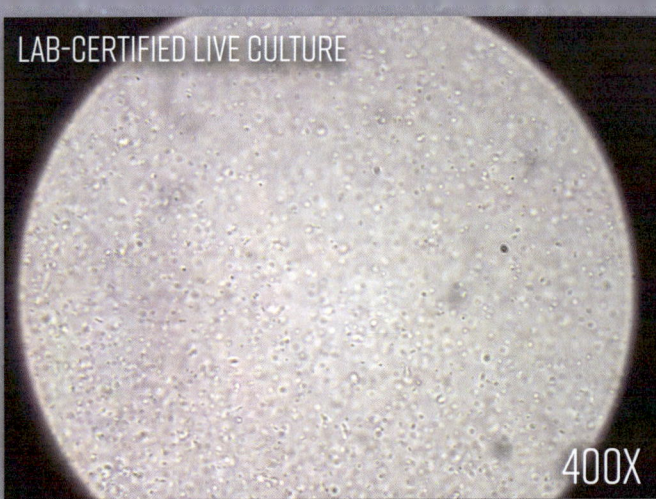
400X

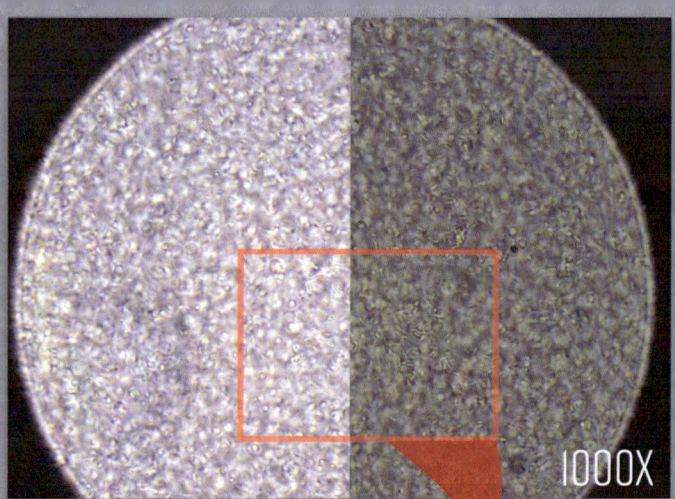
1000X

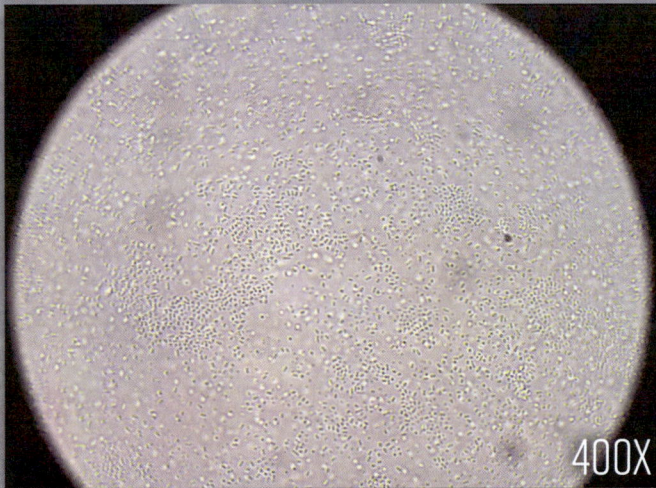
400X

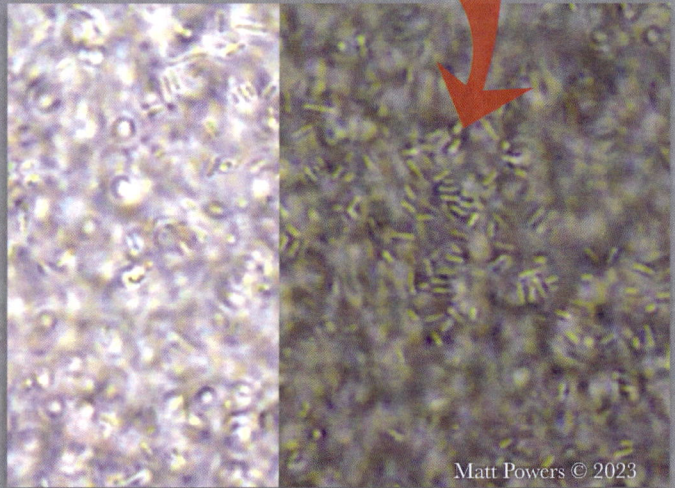
Matt Powers © 2023

PSEUDOMONAS FLUORESCENS

Gram-negative obligate aerobe

FAMILY: Pseudomonadaceae

ORDER: Pseudomonadales

PHYLUM: Pseudomonadota

FORM: Bacilli, non-spore-forming

SIZE: 0.5 – 1 μm wide, 1.5 – 5 μm long

APPEARANCE: Motile straight or slightly curved rods with one or more flagella

ABUNDANCE: Found ubiquitously in soil and water

FOODS: Plant exudates & amino acids

PH & TEMPERATURE PREFERENCES: 25 – 30°C, pH 7 – 8

REPRODUCTION: Asexual Reproduction (Binary Fission) & HGT

USES: PGPR & Siderophores

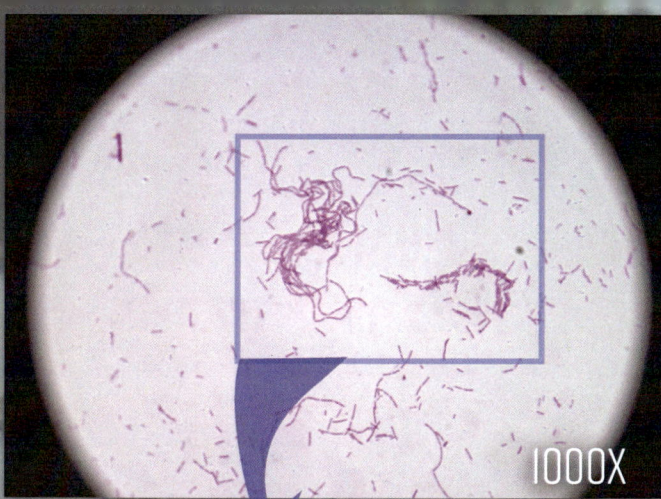
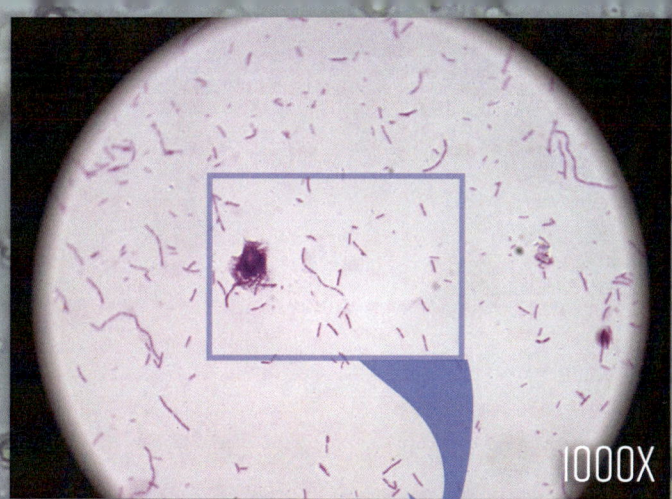

1000X

1000X

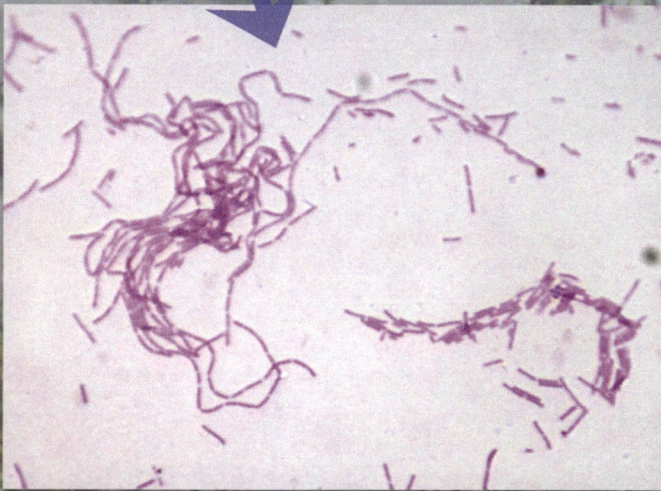
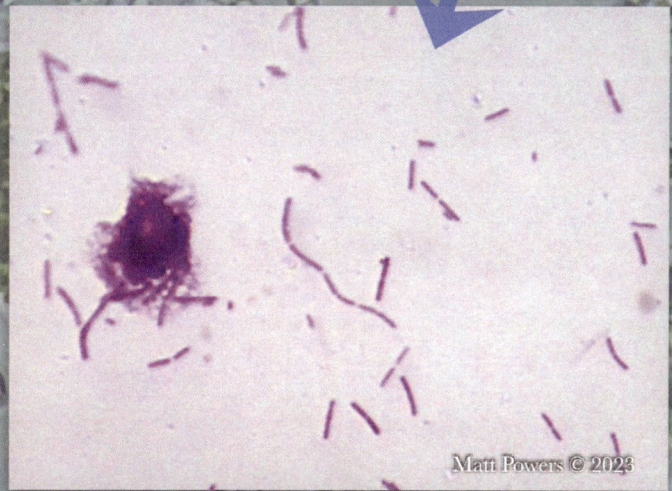

AZOTOBACTER
Gram-Negative Diazotrophic Aerobe

FAMILY: Pseudomonadaceae

ORDER: Pseudomonadales

PHYLUM: Pseudomonadota

FORM: Bacilli – Cocci (sometimes filamentous)

SIZE: 1.6 – 2.7 μm in diameter but usually 2 μm, 3 – 10 μm long but usually 4 μm (*A.Vinelandii* is typically 1.6 – 2.5 μm by 3 – 5 μm & range from rods to coccoid in expression).

APPEARANCE: Ranging from fat-rods to ovoid to spherical, motile (*A.vinelandii* especially) with peritrichous flagella or non-motile. Thick walled cysts can be 1 – 3 μm in diameter

ABUNDANCE: Found primarily in the soil but sometimes water

FOODS: Plant exudates, N_2, nitrates, urea, ammonium, sugars, alcohols, & salts of organic acids

PH & TEMPERATURE PREFERENCES: 7 – 7.5 pH but can range minimum 4.8 pH to maximum 8.5 pH, 30°C

REPRODUCTION: Asexual Reproduction (Binary Fission) & HGT

USES: N-Fixer (Ammonium), PGPR, Phytohormones, Siderophores

A
B
LAB-CERTIFIED LIVE CULTURE
1000X
1000X
HIGHLY MOTILE RODS
STALKED BUNCHES
Matt Powers © 2023

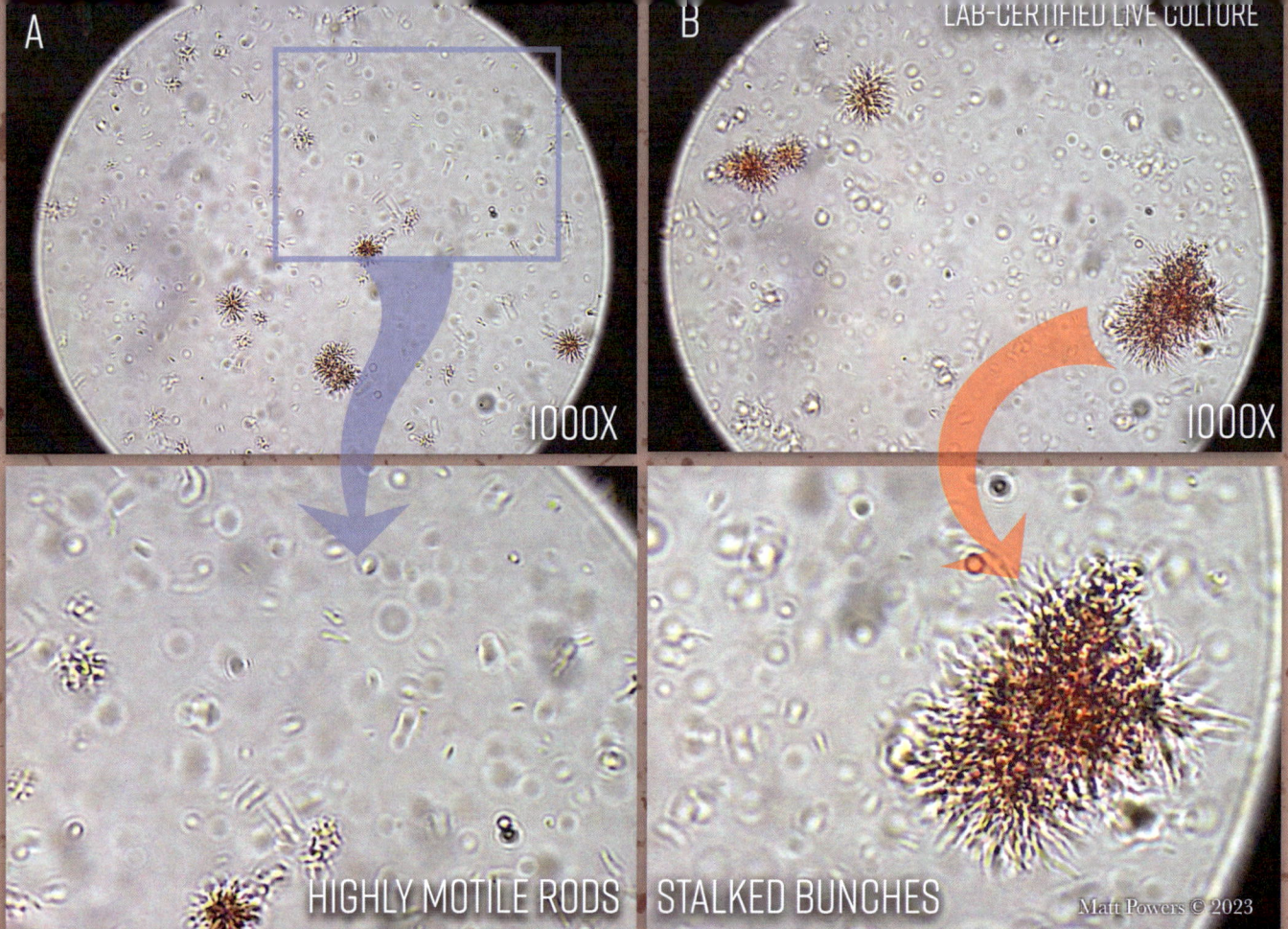

RHODOPSEUDOMONAS PALUSTRIS
Gram-negative Purple Non-Sulfur Bacteria

FAMILY: Nitrobacteraceae

ORDER: Alphaproteobacteria

PHYLUM: Pseudomonadota

FORM: Bacilli

SIZE: 0.3 – 2 μm wide, 2 – 10 μm (typically 0.6 – 0.9 x 1.2 – 2.0)

APPEARANCE: Rod-shaped, highly motile, swarming singular forms as well as stalked, nonmotile bunches that look like 360° star fish – both are formed with every asymmetrical cell division

ABUNDANCE: Found in soil, water, manure lagoons, marshes, swamps, and extreme environments of all kinds as well as in healthy compost and IMO preps

FOODS: Prefers molasses but also consumes organic matter, all sugars (carbon), light, CO_2, and inorganic compounds

PH & TEMPERATURE PREFERENCES: pH 6.9 and 37.9°C, but it can grow across pH 5.5 – 8.5 and 20 – 40°C

REPRODUCTION: Asymmetric Cell Division (Binary Fission) & Budding

USES: Biocontrol, Bioremediation, Algae Remediation, Facilitating Microbe in Effective Microbes® consortium, Biostimulant, Plant Growth Promoting Rhizobacteria, N-Fixation in low oxygen environments (or with B.subtilis), & potential Endophyte

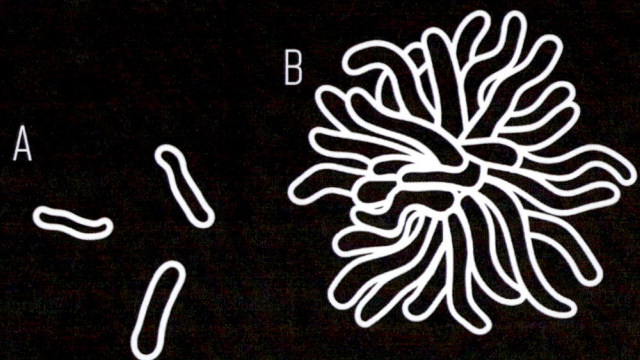

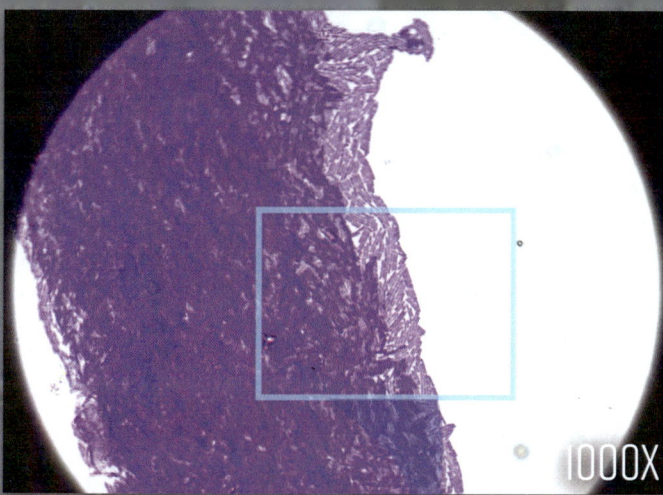

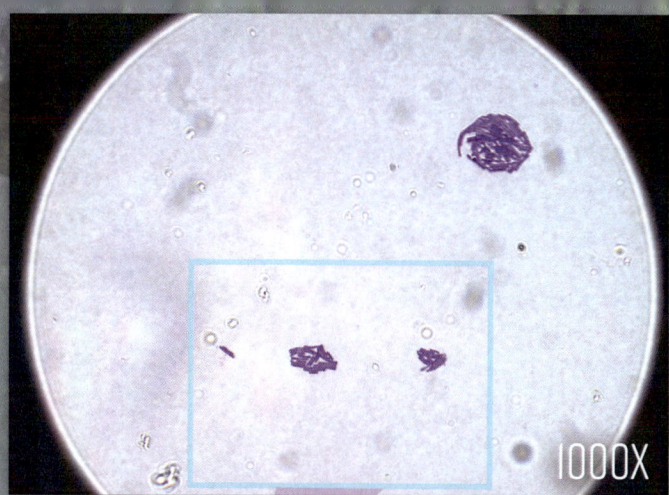

TAKE A CLOSER LOOK

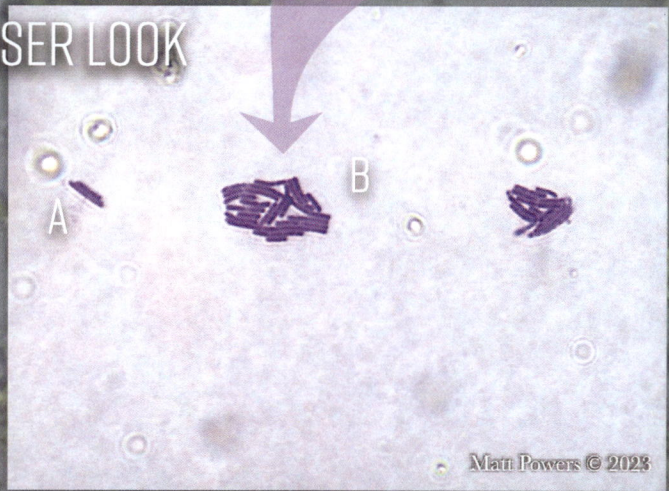

1000X

1000X

Matt Powers © 2023

ACETOBACTER ACETI

Gram-Negative Obligate Aerobe

FAMILY: Acetobacteraceae

ORDER: Rhodospirillales

PHYLUM: Pseudomonadota

FORM: Bacilli – Cocci , non-spore-forming

SIZE: 0.6 – 0.9 μm wide, 1 – 4 μm length

APPEARANCE: Rods that can fatten up to ovoid, motile and non-motile, with and without peritrichous flagella

ABUNDANCE: Found in fruits, flowers, vinegar, kefir, and fermented foods

FOODS: Ethanol, plant sap, and sugars

PH & TEMPERATURE PREFERENCES: 4 – 6 pH, 29 – 30°C but up to 37°C in thermotolerant species

REPRODUCTION: Asexual Reproduction (Binary Fission) & HGT

USES: IAA, Vinegar Production, & PGPR

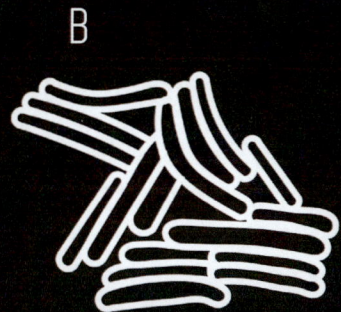

400X

1000X

1000X

SHORT FORM

LONG FORM

Matt Powers © 2023

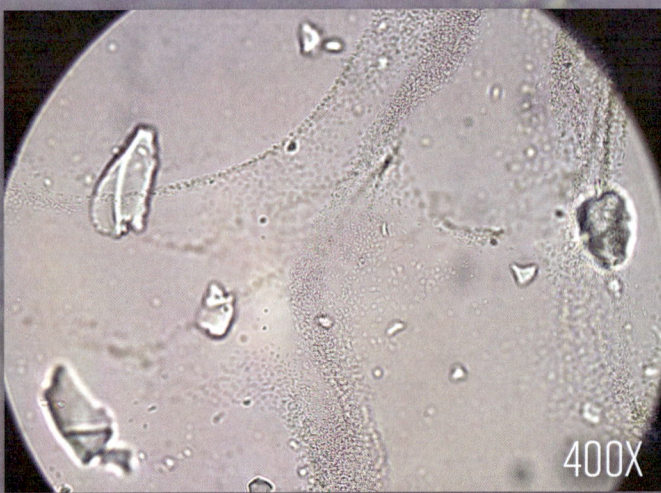

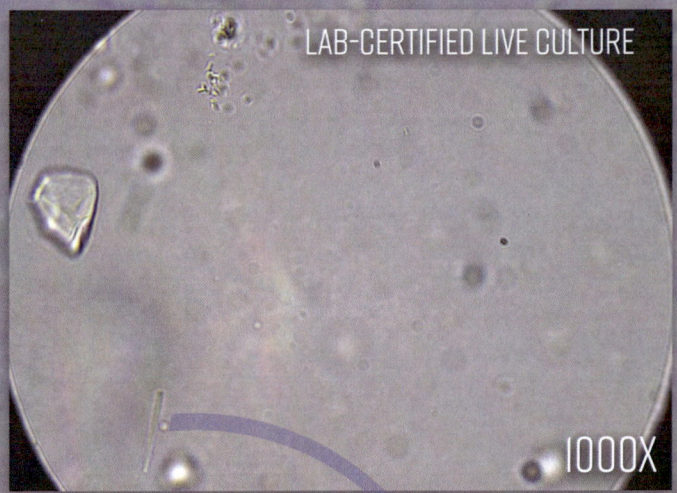

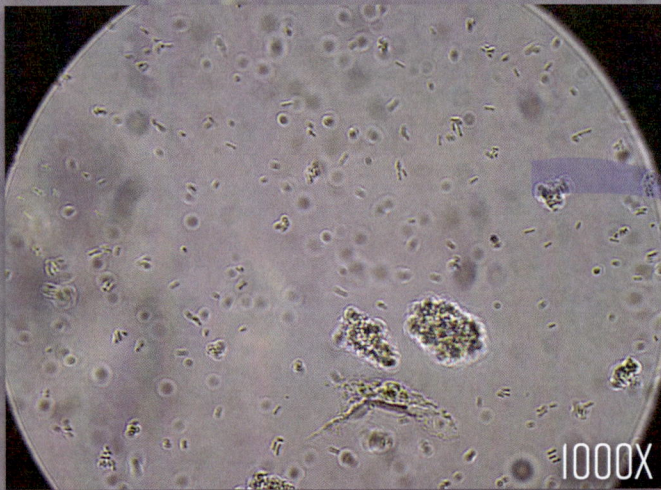

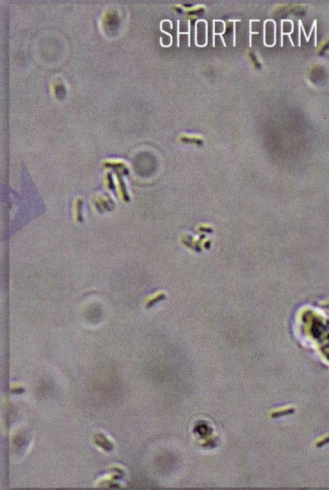

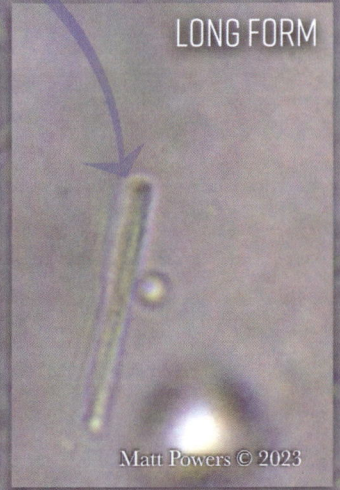

LACTOBACILLUS

Gram-Positive Aerotolerant Anaerobe

FAMILY: Lactobacillaceae

ORDER: Lactobacillales

PHYLUM: Bacillota

FORM: Bacilli, non-spore-forming, can form chains

SIZE: 0.5 – 0.9 µm diameter, 1.5 – 9 µm long

APPEARANCE: Rods that vary in size, motility, and chain length and formation

ABUNDANCE: Found in yogurt, cheese, wine, human digestion, the mouth, and other parts of the body

FOODS: Lactic sugars, plant sugars, & sugars

PH & TEMPERATURE PREFERENCES: 5.5 – 7.5 pH, 28 – 37°C (different strains prefer opposite sides of the spectrum)

REPRODUCTION: Asymmetric Cell Division (Binary Fission) & HGT

USES: Biocontrol, Controlled Composting, LAB production, and a key member of the EM® consortium.

AZOSPIRILLUM

Gram-Negative to Gram Variable Nitrogen-Fixing Aerobe

FAMILY: Azospirillaceae

ORDER: Rhodospirillales

PHLYUM: Pseudomonadota

FORM: Bacilli

SIZE: 0.6 – 1.7 μm diameter, 2.1 – 3.8 μm long, non-spore-forming

APPEARANCE: Wiggling, highly motile plump rods that can be slightly curved or straight with polar and lateral flagellum – has demonstrated pleomorphism in alkaline conditions or severe stress with non-motile and irregular shaped forms

ABUNDANCE: Thought to have originated in aquatic environments and then later made its way onto land, Azospirillum is found all over the world in the rhizospheric soils of a large diversity of plants

FOODS: Plant exudates: amino acids, sugars, & sugar acids

PH & TEMP PREFERENCES: 30 – 41 °C & 5.5 – 7.5 pH are optimal ranges, but Azospirillum can tolerate 5 – 42 °C & pH 5 – 9

REPRODUCTION: Asexual Reproduction (Binary Fission) & HGT

USES: PGPR, IAA, Gibberellins, ISR, & N-Fixation

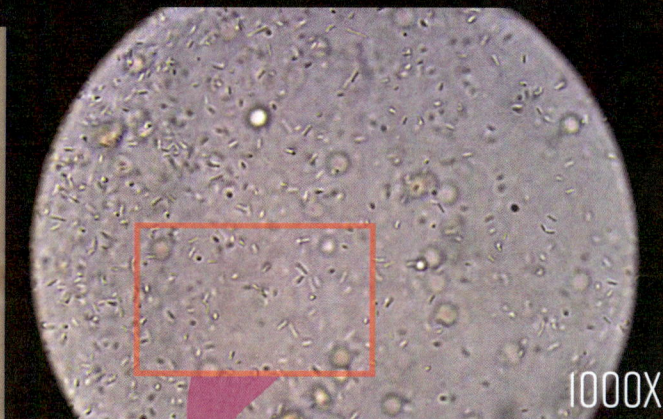

1000X

IMPERCEPTIBLE AT 100X

100X

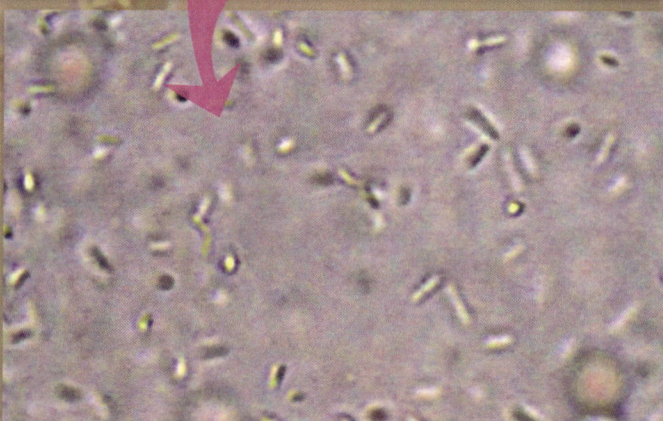

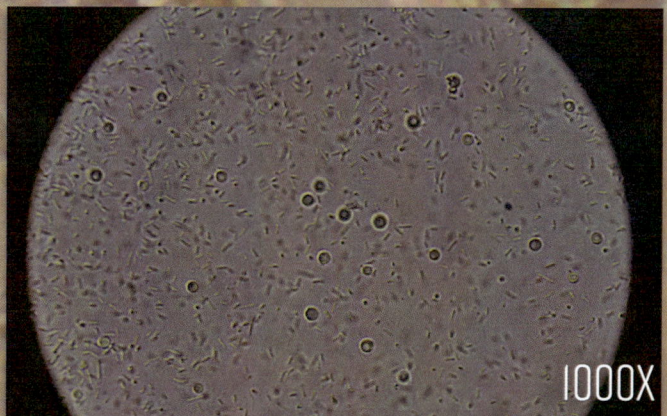

1000X

LAB-CERTIFIED LIVE CULTURE

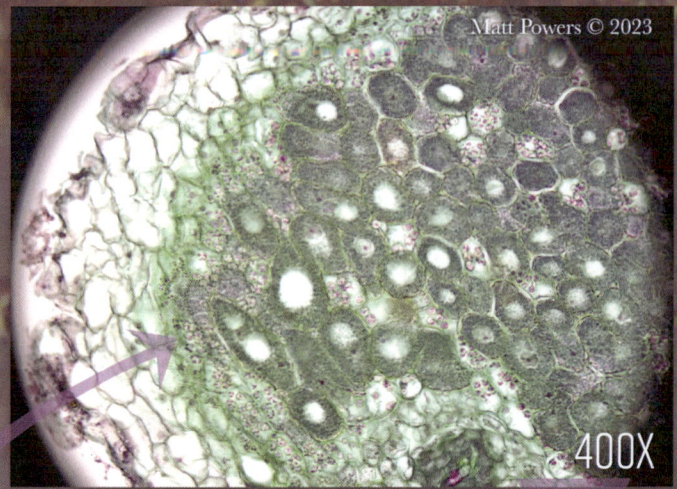

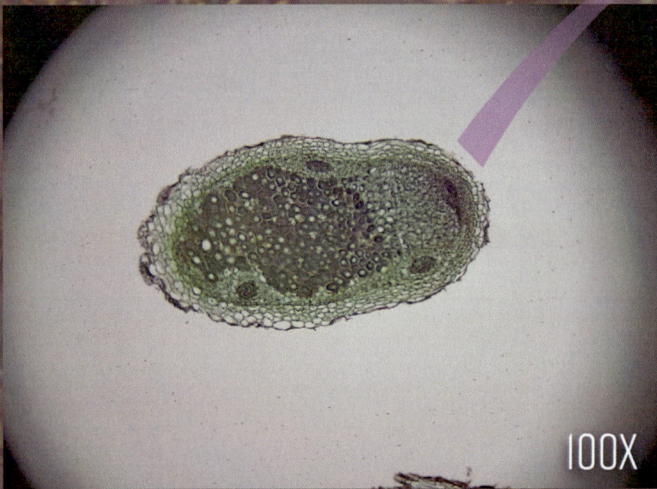

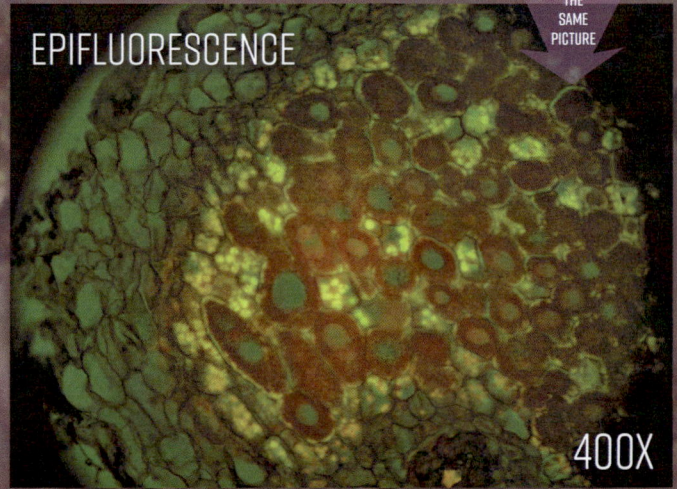

400X

EPIFLUORESCENCE

THE SAME PICTURE

100X

400X

RHIZOBIUM

Gram-Negative Diazotrophic Aerobe

FAMILY: Rhizobiaceae

ORDER: Rhizobiales

PHYLUM: Pseudomonadota

FORM: Bacilli, Nodule forming

SIZE: 0.5 – 2 μm long, 0.1 – 0.8 μm in diameter with nodules 1 – 3 mm in size

APPEARANCE: Rod-shaped, usually flagellated, but often y-shaped or irregular once inside the plant and nodule. It can look like bradyrhizobium and other bacilli at 1000x. Always use the context and the nodule type to differentiate between them.

ABUNDANCE: Found in association with legumes and many nodule forming nitrogen-fixers but also in the digestion and manures of animals feeding on those same plants as well as in compost using those same manures.

FOODS: Plant exudates (malate)

PH & TEMPERATURE PREFERENCES: Prefers Alkaline but can grow in pH 4 – 9, 25 – 30°C

REPRODUCTION: Asexual Reproduction (Binary Fission)

USES: PGPR, N-fixation, IAA, some solubilize P, & some release ACC deaminase

RHIZOBIUM NODULE

LAB-VERIFIED SAMPLE OF A CLOVER ROOT NODULE

VESICLES

CONTAINING RHIZOBIUM

1000X

EPIFLUORESCENCE

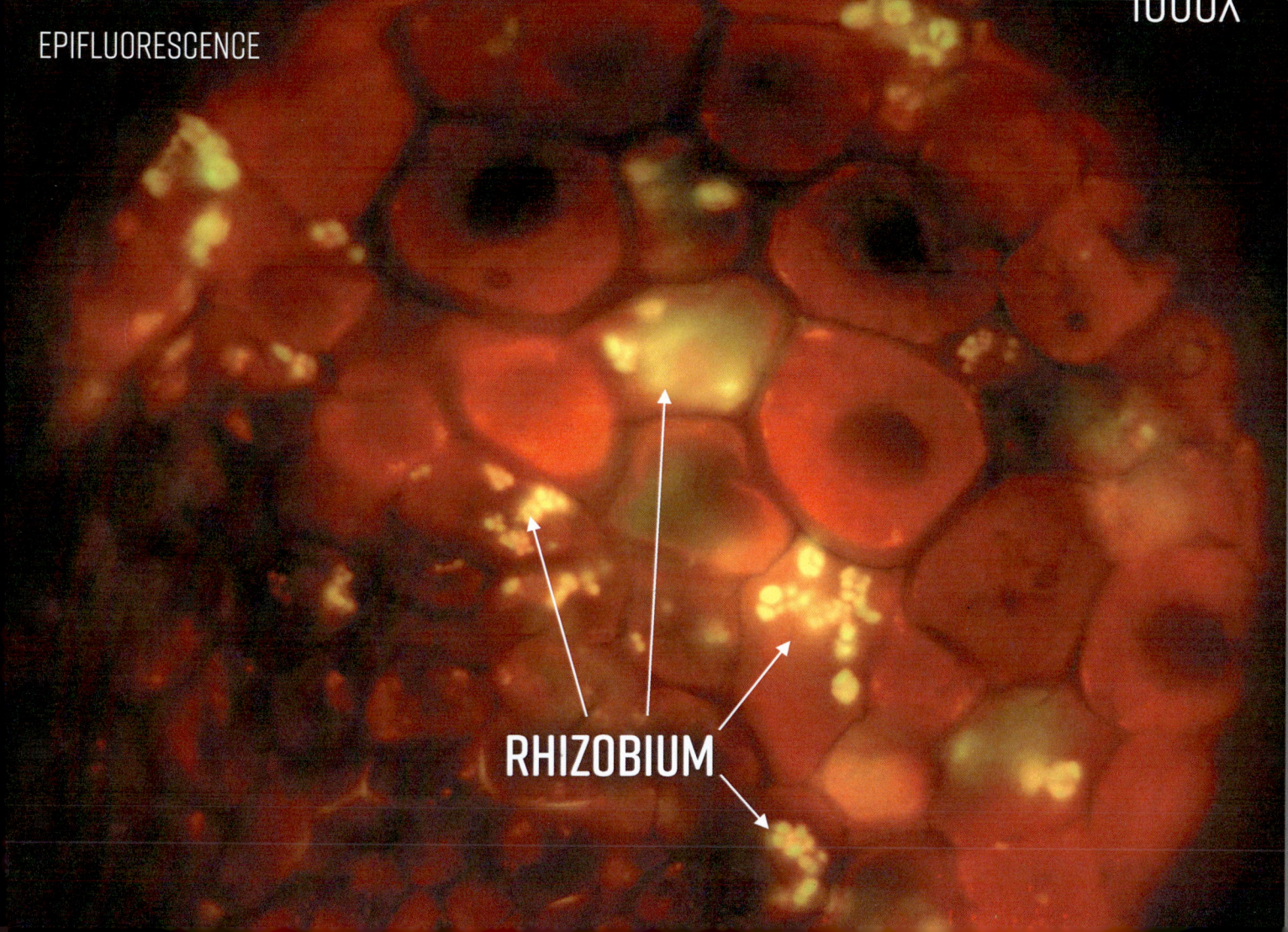

RHIZOBIUM

HERBASPIRILLUM
Gram-Negative Endophytic Diazotroph

FAMILY: Oxalobacteraceae

ORDER: Burkholderiales

PHLYUM: Pseudomonadota

FORM: Vibrio – Spirilla

SIZE: 0.6 – 0.7 µm diameter, 1.5 – 5 µm long

APPEARANCE: Curved rods or spirals with motile with polar flagella

ABUNDANCE: Found ubiquitously as an endophyte in plants – especially *H. seropedicae* is found in plants with high sugar content like sugar cane, pineapple, corn, and sorghum, but also rice suggesting it is much more abundantly spread. It is found at very low concentrations in compost (~0.03%) likely implying that it breaks down in the process.

FOODS: Plant exudates & sugars

PH & TEMP PREFERENCES: pH 5.3 – 8, 30 – 34°C

REPRODUCTION: Asexual Reproduction (Binary Fission) & HGT

USES: Endophytic, N-Fixation, PGPR

GLUCONOBACTER
Gram-Negative Obligate Aerobe

FAMILY: Acetobacteraceae

ORDER: Rhodospirillales

PHYLUM: Pseudomonadota

FORM: Almost Bacilli – Cocci , non-spore-forming

SIZE: 0.5 – 1 µm diameter, 2.6 – 4.2 µm long

APPEARANCE: Round to ovoid with sometimes polar flagella, motile and non-motile (*G. cerevisiae* is non-motile)

ABUNDANCE: Found in plants, saps, flowers, fruits, bees, and bee hives

FOODS: Plant exudates, sugars, and ethanol

PH & TEMPERATURE PREFERENCES: 25–30°C, 5.5–6 pH though most strains can grow at 3.5 pH

REPRODUCTION: Asexual Reproduction (Binary Fission) & HGT

USES: PGPR, IAA, N-Fixation, Endophytic, Vinegar Production, Phytohormones, P solubilization, Siderophores, Biocontrol, Z Solubilization, Composting, Vitamin C, & Promotion of *C.Elegans* Nematodes in Compost

SYMBIOBACTERIUM

Gram-positive Facultative Anaerobe

FAMILY: Coriobacteriaceae

ORDER: Coriobacteriales

PHYLUM: Actinobacteria

FORM: Pleomorphic & Filamentous, mostly non-spore forming

SIZE: 0.7 – 0.8 diameter and 2.7 – 7.7 µm length when bacilli form, 5 – 100 µm in length when filamentous

APPEARANCE: Rod-shaped, alone and sometimes in pairs, sometimes found in biofilms

ABUNDANCE: Less commonly found in soil but found in human and animal gut biomes

FOODS: Primarily carbohydrates with a few species that can break down cellulose and lignin

PH & TEMPERATURE PREFERENCES: 4.5 – 7.5 pH but ideal 5.0 – 6.0 pH, 20 – 37°C

REPRODUCTION: Primarily asexual reproduction but some do form endospores

USES: Bioremediation for heavy metals and toxins and cellulase production

PAENIBACILLUS POLYMYXA

Gram-Negative Diazotrophic Facultative Anaerobe

FAMILY: Paenibacillaceae

ORDER: Bacillales

PHYLUM: Bacillota

FORM: Bacilli, endospore forming

SIZE: ~0.6 µm wide, ~3 µm long

APPEARANCE: Rod-shaped with peritrichous flagella, motile

ABUNDANCE: Found in soil, rhizosphere, and marine sediment

FOODS: Plant exudates

PH & TEMPERATURE PREFERENCES: 6.5 – 7 pH, 30°C though it can resist heats up to at least 50°C and tolerate pH's as low as pH 4.

REPRODUCTION: Ellipsoidal endospores

USES: PGPR, N-Fixation, & Biocontrol

It's also unfortunately necessary to delineate what is best for the soil and what is best for humans at some point – some plants are fortified by toxic levels of aluminum, so it benefits them, but not anything consuming them: it's defensive tactics are too severe. The same is likely the case on the microbial side. We have Burkholderia as a prominent example of this: the majority of it in compost is *Burkholderia cepacia*. Immune-compromised individuals and those diagnosed with cystic fibrosis can be infected with this, and it's resistant to most antibiotics, so it's a real danger for them, so if you have either of these issues, DO NOT INHALE COMPOST OR SOIL FUMES. While I'm no doctor, I can read the names of pathogens and calculate their percentages all the same using the DNA sequencing outputs.

The more we look into our world, the more we find microbes of all sorts inside and outside of us. We are covered with microbes and many are ones that could cause us harm IF they were at high enough numbers and if our body were weakened and open to infection at the same time, but it baffles many to find the few microbes we know (and fear) to be nearly ubiquitous.

ONLINE RESOURCES:

- *Bergey's Manual of Systematic of Archaea and Bacteria* (BMSAB)
  https://onlinelibrary.wiley.com/doi/book/10.1002/9781118960608
- The MicrobeWiki https://microbewiki.kenyon.edu/index.php/MicrobeWiki
- R-Soil Database https://www.R-SoilDatabase.com

ARCHAEA

Archaea were once thought to be bacteria since they are indistinguishable from bacteria in light microscopy, but they are an entirely different branch of the tree of microbial life. Archaea are a bit of a mystery as they haven't been isolated for the most part in the lab – they have been detected and differentiated using genetic testing primarily. You can find them at a rate of 0.5%–1.2% of your compost, so one out of every 100–200 bacteria could really be an archaea, and its size is in the most common range, so it's like trying to find a needle made of hay inside a hay stack. They are amazing microbes that feed on a vast array of foods from metal ions to carbon to light to ammonia and more! These microbes are everywhere in nature but are most apparent in extremophile settings.

ARCHAEA FOUND IN JOHNSON-SU COMPOST

NITROSARCHAEUM
Archaea found in Johnson-Su Compost

FAMILY: Nitrosopumilaceae

ORDER: Nitrosopumilales

PHYLUM: Thaumarchaeota

FORM: Bacilli

SIZE: 0.3 – 0.5 μm in diameter, 0.6 – 1 μm in length

APPEARANCE: Rod-shaped, non-motile

ABUNDANCE: Found in soil, oceans, rivers, lakes, and marine sediment – considered to be the dominant archaea representing 5% of all prokaryotes found & detected at 0.6% in Johnson-Su compost

FOODS: Ammonia oxidizer

REPRODUCTION: Binary fission

CANDIDATUS NITROSOTENUIS
Aerobic Archaea found in Johnson-Su Compost

FAMILY: Nitrosopumilaceae

ORDER: Nitrosopumilales

PHYLUM: Thaumarchaeota

FORM: Bacilli

SIZE: ~0.4 μm in diameter, 0.6 to 3.6 μm in length

APPEARANCE: Rod-shaped with possible flagella, motile & non-motile

ABUNDANCE: Found in soil globally as well as hot springs, freshwater, and activated sludge - *there is no pure culture of it in existence.*

FOODS: Ammonia oxidizer

REPRODUCTION: Binary fission

MYCORRHIZAE = 'FUNGUS ROOTS'
THE SYMBIOSIS OF MYCORRHIZAL FUNGI & PLANT ROOT

Fungi

Fungi are eukaryotic microbes that are significantly larger than bacteria though there are a few larger-sized bacteria and several prominent filamentous bacteria, so there's some overlap between these groups morphologically that can be initially confusing. The fungi we'll be seeing under the microscope are primarily filamentous but some are spherical and ovoid like yeasts and spores. Yeasts are ubiquitous in the air, soil, and water – many are also endophytic. Mycorrhizal fungi, a filamentous fungi, are smaller than typical saprophytic fungi. Some mycorrhizal fungi are obligate rhizopheric residents but others can persist independently of roots. While bacteria can be motile and non-motile, fungi are non-motile (though depending on how you classify fungi and who you are talking to, there are some molds with motile spores though most folks are firm on fungi being non-motile and want the molds with motile spores to be separately categorized). Speaking of categorizing, the levels of taxonomy can be confusing – I tend to jump between phylum, species, and genus when I'm writing and talking about things. The other layers are vital for organizing all the complexity, no doubt, but often it's those midlevel layers that shift and change as new information arrives. The name of a useful fungi like *Glomus intraradices* may change its name to *Rhizophagus irregularis* but it's still the same useful species we are all familiar with and it's still in the *Glomeromycota* phylum.

Most published identification guides are non-phylogenic and considered highly artificial: that means they are inaccurate when tested genetically. Almost all the texts I've found rely upon the non-phylogenic, "traditional" keys and methods; almost everyone is pausing and waiting for the DNA methods to improve. The entire mycological world is holding its breath in this way, BUT despite these limitations, we can still categorize things by genus and sometimes have a strong yet still tentatively conclusive identification at more precise taxonomic levels.

WHAT TO LOOK FOR?

When it comes to fungi, it depends on our context most of all. Are we looking at compost? If so, we are looking at saprophytic fungi not mycorrhizal fungi – that's a different ballgame altogether. Elaine Ingham identifies good composting fungi as those with uniform hyphal thickness, ideally darker in color, with well-defined septa, and ideally as long and thick as possible. Chris Trump and Stephen Raisner have both pointed out that sporulation differs in appearance in comparison to mature hyphal strands, clear (or hyaline) hyphae are not at all always bad, AND mycorrhizal fungi do not share the same morphological features as saprophytic fungi (Ascomycota and Basidiomycota). Mycorrhizal fungi have to be stained or viewed through epifluorescence lighting because they are generally transparent (hyaline) though as always, the soil environment itself can influence coloring (arid climates tend to have more reddish/orange hyphae and spores). Yellow vs brown coloring may have more to do with the types of humic acids present in the soil than genetically predisposed morphologies. Uniform is good, but there's flaring, constricting, and other variations to be seen in AMF (Glomeromycota) especially, so it's really important to have a baseline of comparison across fungal types and their expressions to ground oneself and then to dig deeper and morphologically identify species using keys, either online or in book format. Additionally, AMF fungi share the same hyphal diameter and hyaline coloring as many actinobacteria, so we have to know what we are looking at: is it soil from the rhizosphere? Is it compost? Can we fool ourselves easily in this space? YES – phytophthera has the same hyphal diameters and is hyaline as well, BUT luckily the behavior, location, pattern, and other indicators can tell us who is who even without DNA testing to double check.

Many specifics must be waded through to identify most species – the professional microbiological community uses these keys and develops their own specialized keys for their own studies and products. We can use these keys to get specific, but

The width does vary but it is not a wide variation, and the appearance could be due to damage, bunching, and/or distance variation affecting the image.

As the hyphal wall goes out of focus, it seems to thicken…

Septa are divisions between sections of hyphae

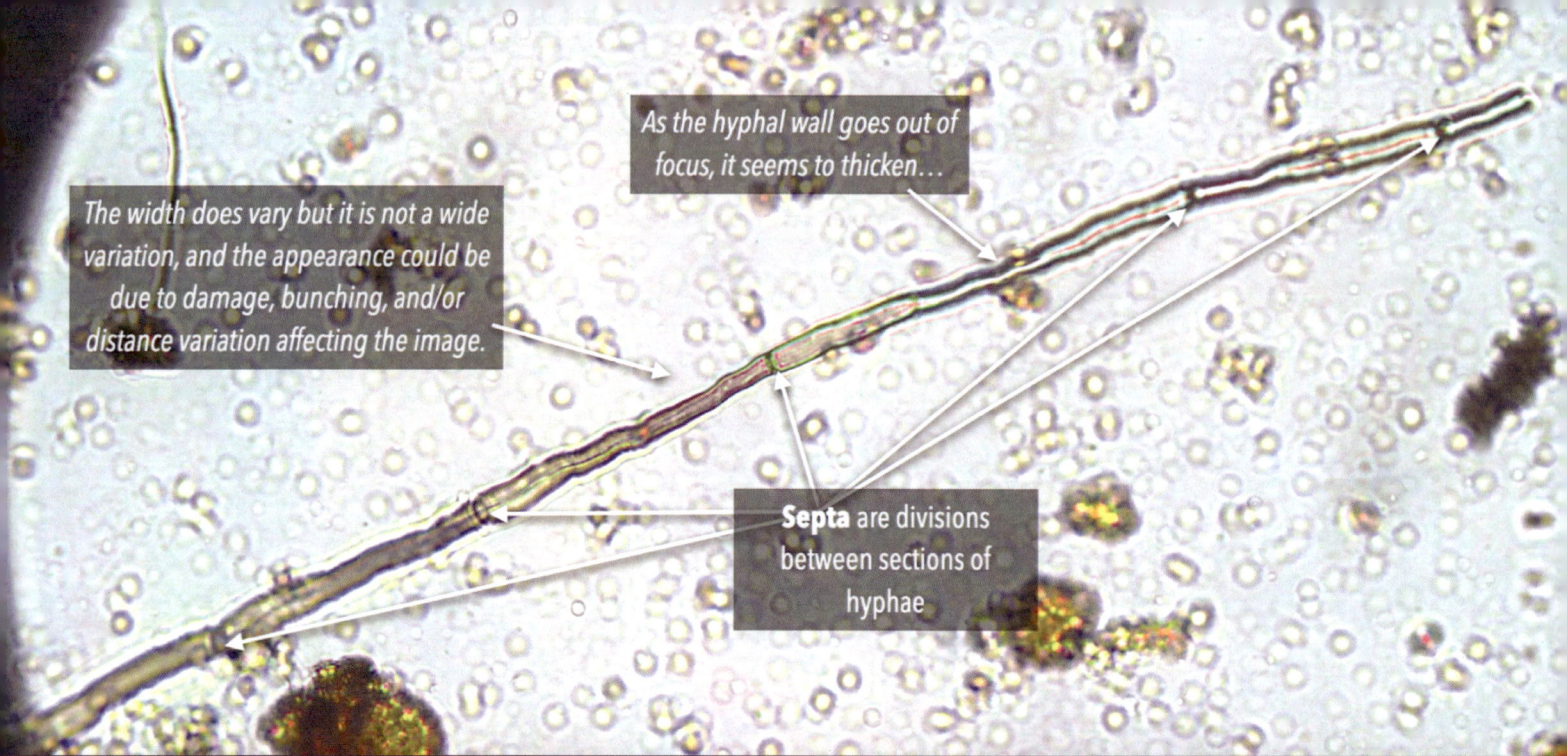

DECOMPOSING WOOD CHIP FRAGMENT

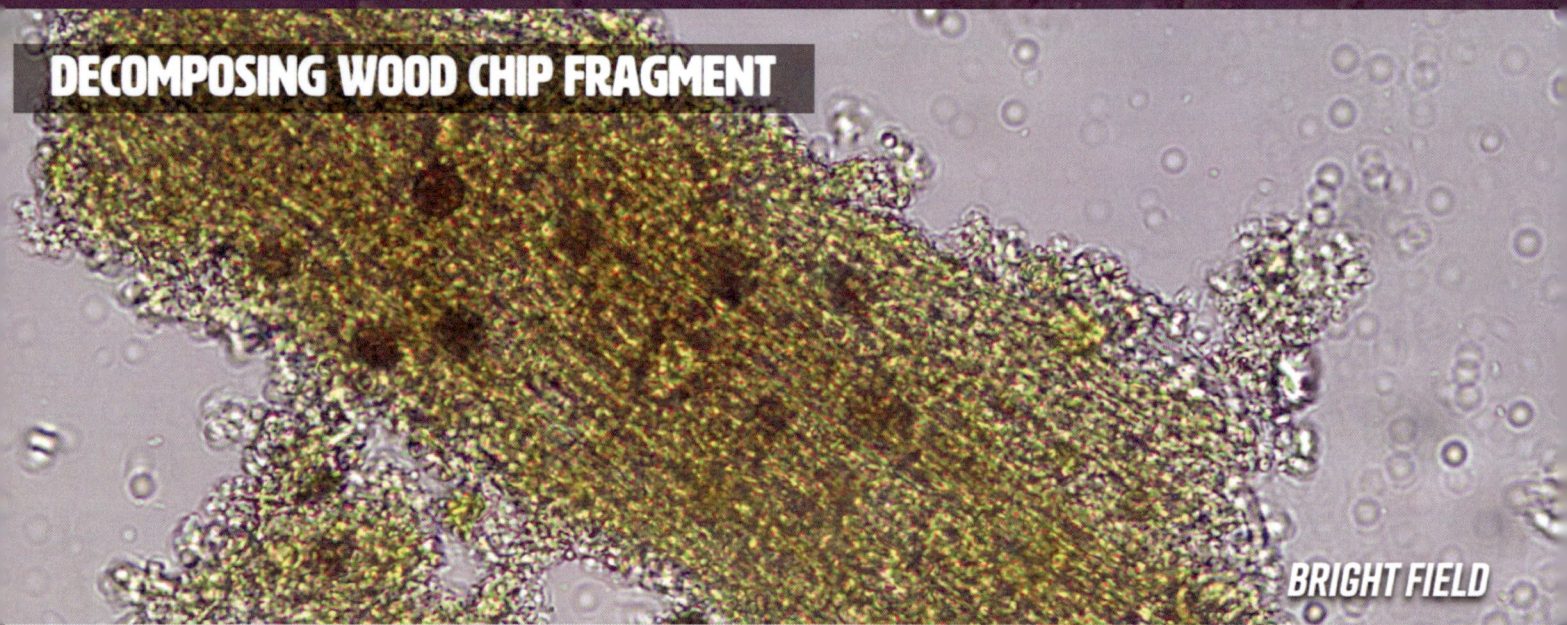

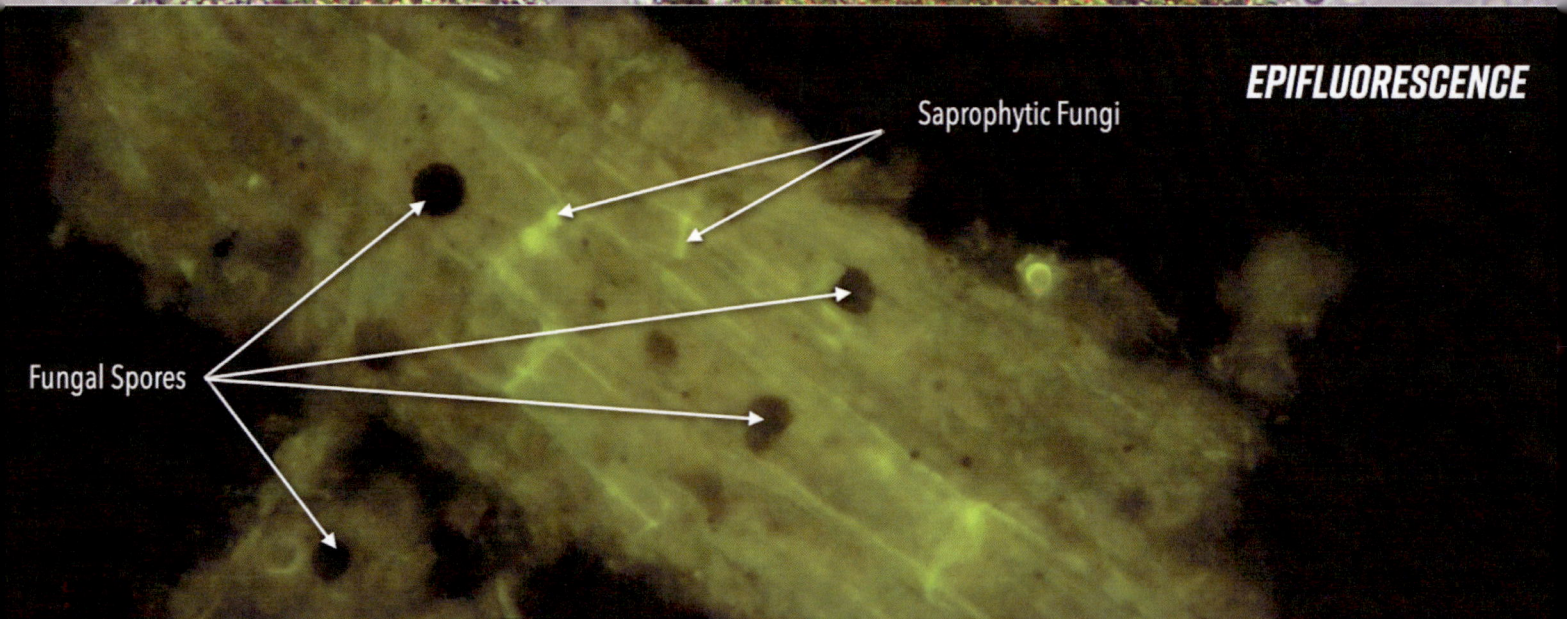

BRIGHT FIELD

EPIFLUORESCENCE

Saprophytic Fungi

Fungal Spores

FUNGI'S DIGESTIVE ENZYMES RELEASE IONS & FORM **CRYSTALS** ON THEMSELVES, ROOTS, & THE ENVIRONMENT AROUND THEM. THEY **GLOW** BECAUSE THEY CONTAIN PHOSPHORUS (AND USUALLY CALCIUM TOO).

THEY TRANSPORT PHOSPHORUS THROUGH THEIR HYPHAE LIKE A HOSE.

we can also identify most by their genus, family, order, and phylum at home with a microscope using this book and the online keys listed within these pages. And while the keys are very useful, the final word does come down to the fact that we have only identified 1-10% of all soil microbes (depending on which expert you're talking to the number shifts slightly – to me this indicates they really have no idea).

What is known is that the commonly found fungi are often not the desirable fungi we are seeking – this is why we have to so often buy or brew inoculants ourselves: they are not found at effective concentrations in nature. The most common soil fungus from the Zygomycota genus is *Mortierella* with *Mucor* and *Rhizopus* the next most ubiquitous. On or in damaged or sick plants, *Pythium* is most often found though others like *Phytophthera, Aphanomyces, Dictyuchus,* & other oomycetes can easily be found and observed. *Chaetomium* is the most common Ascomycotan fungus found in soil – Ascomycetes being the most commonly found fungi in the soil. Basidiomycota fungi are common in the temperate (woody) climates while rarer in the soil of islands and Asia, and the soil basidiomycota they have found often do not grow well on agar in a laboratory setting. Ectomycorrhizal fungi peak in their abundance in the temperate forest zones further away from the equator and even at high altitudes while arbuscular mycorrhizal fungi is more abundant in the tropical and warmer and lower altitude temperate zones. Deuteromycota fungi are even more commonly found than all the above already listed: *Alternaria* and *Penicillium*, for example, are always found in soil samples tested for fungi. While *Rhizopus* may be familiar to us as primary to Korean Natural Farming IMO-1 collection technique, the others are also helping cycle dead, decaying, damaged, or sick organic matter. When studying soil fungi, we cannot help find overlap with the study of spoilage fungi. It's successional – we have to decompose the organic matter before it can be reincorporated back into soil, and since the soil is essentially a large digestion vat for microbes to break down organic matter, we'll always find these microbes in situ. It's the next stage in succession we most often focus on: beneficial associations around plant roots and inside plant roots, on plant surfaces, in the phloem, and in the leaves though it should be noted that thermophilic decomposition is a 3rd context. The microbes in a hot compost pile are not the same as a the decomposers in natural soil conditions. Those populations shift through decomposition from primarily endophytes to a mix of saprophytes and endophytes at the end of the process. Garden soils therefore constitute a 4th context as they are a hybrid state of the compost selection process, the in situ native microbes, and the inoculants we bring in from outside our system. It is vital to know our context to know what to look for, test for, and how to interpret the answers.

40X

400X

00X

600X

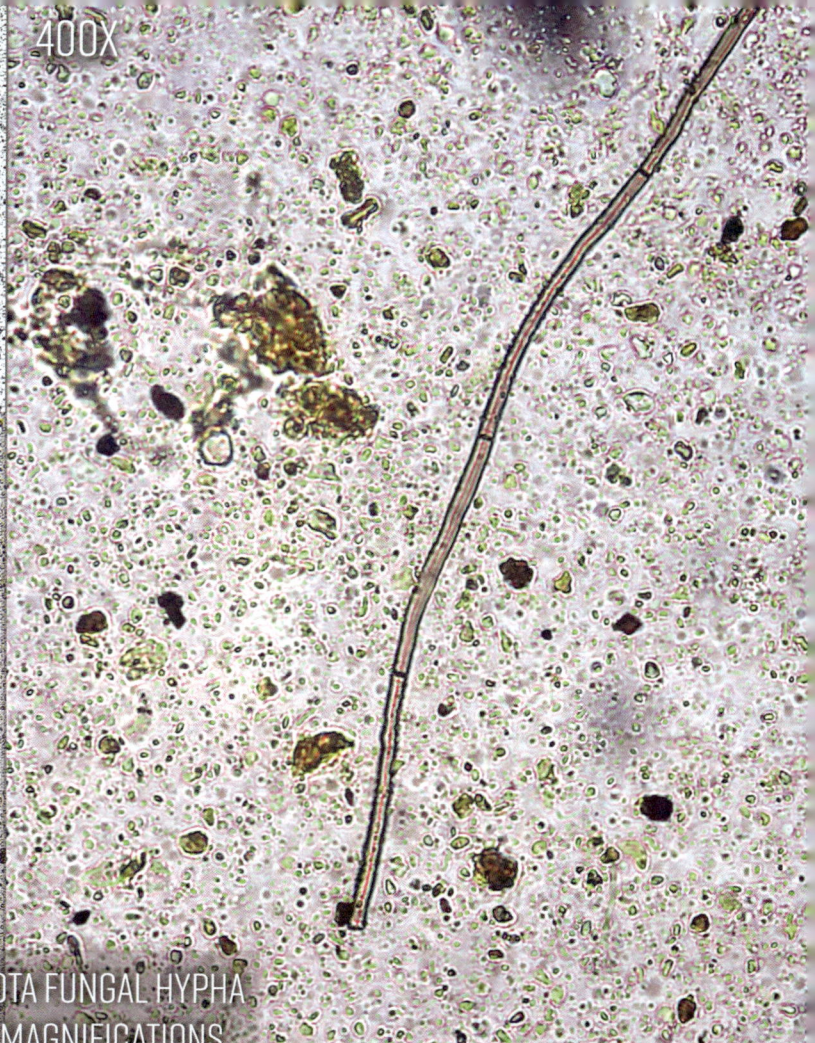

A BASIDIOMYCOTA FUNGAL HYPHA
@ DIFFERENT MAGNIFICATIONS
FROM JOHNSON-SU COMPOST

0 μm

For the microscopist, the essential question is: *What does it look like?* How far across the field does it stretch? Can you measure it? 40x field of view is usually ~400 micrometers wide (though my 4k camera chops mine up a bit). Take a picture with a grid overlay, hemocytometer, or a micron ruler. Document the magnification. Write down the color and the width as well though know that after a bit of practice, you'll be able to readily discern actinobacteria's relative thinness compared to a typical basidiomycota's fat diameter. Dr. Elaine Ingham has suggested that the smallest bacteria we can see at this resolution typically is one micrometer wide, but that's pure guesstimation and depending on the microbes (refer to the charts), hydration of those microbes, the pH, the oxygen levels, and your ability to see things with your microscope and eyes, you can arrive at very different conclusions. It is best practice to use a micron ruler or a grid overlay to calculate more precisely the area or width until you are comfortable recognizing the difference. While it is easy to fall into the bigger is better or darker is better, it's not so cut and dry. There's quite a spread of color expression and hyphal variation within fungi. Fungal hyphae can be brown, yellow, white, and hyaline (clear) and flared, constricted, uniform, looped, and in parallel track H's with spores of a variety of shapes and earth tones as well.

What do the spores look like? This is often the only way we can differentiate between species of AMF. Their spores can also vary, so it can be vital especially as you gain fluency to use the INVAM online keys, the R-Soil Database, and other fungal databases. Spores can be round, ovoid, irregular, pear-shaped, pointed, and a variety of colors: any combination of hyaline (clear), white, cream, yellow, and brown. Some soil fungi never seem to sporulate in a controlled environment, making their identification impossible.

Can We Trust Our Numbers?

Remember how fungi grow and spread: by spores. This means that all fungi present came from a spore and was spread by the release of a spore. We often think of spores spread by the wind but these are released in the soil environment and so do not float on the breeze: they spread a limited distance and then sporulate when the conditions are right. That means that everything fungal is spreading in pockets of expressions that move outward from a single point consuming the food it finds and leaving decomposition and fungal hyphae in its wake. That means your soil when examined under the microscope is NEVER a uniform expression: it can't be if it's no-till. It will most likely be low or high in fungal abundance, and not an even distribution nor a representation of the average ratios. That means that while we can prove fungal abundance, we can't prove out fungi being present even when we don't find it in our sampling. Because of this, it is vital to not fool ourselves with numerical thinking in this space. F:B ratios with a microscope have historically led to misinterpretation – I've seen this in the published journals and heard this personally from certified soil labs as well.

We need to examine fungal abundance and bacterial abundance, but we need to do it separately and not let being so close to the problem fool us by missing the bigger picture: it's far too time consuming for its lack of accuracy as a practice to be relied upon or perhaps even practiced. This is why Microbiometer has taken off as have PFLA testing – these are great for establishing generalized estimations of biomass and rough ratios (though the fungi #'s can be inaccurate). In this book, you will find a faster, more pragmatic and effective way of analyzing the soil that uses microscopy in tandem with other tests because we don't want to be tricked by arbitrary numbers – we want to understand the holistic context.

It's also important to note that measuring fungi *in general* is also not very useful: which fungi? Yeasts, mycorrhizae, wood saprophytes? The difference is everything!! Just saying *fungi* is too ambiguous to be functionally useful. This is why examining the soil and compost with a microscope for who is there in terms of fungi is far more valuable than trying to count or attach a number to its ratio in relation to bacteria – again, which bacteria?!

Pockets of expression dominate the natural world; why would soil be any different? *The endemic species that exist in one valley only. The peak predators that cover massive ranges.* We should expect to see things like this and keep our eyes open because the almost ceremonial work that so many lab scientists do, gathering telegraphed data, reinforces their own misconceptions of whatever they are working on. It's not just the scientists getting funding from big corporations with precise expectations – there's a tendency to fool oneself in science if one is not careful and there are folks stretching the truth to get better headlines for their journal articles or to appear ahead of one's peers. It's purely ego-driven, and we all suffer for it.

What about Colonies?

This is likely the most common way to examine fungi after microscopy – they grow it out on an agar petri dish. These colonies are morphologically identified using characteristics, but there are lookalikes even in the colony stage of expression, so even this area of science is having to adjust and recognize the impact of DNA testing.

This is the Characterization list from the *Pictorial Atlas of Soil and Seed Fungi*:

- "Color and tint on colony surface and reverse (Ref: Standard color charts, i.e., Rayner (1970), Ridgway (1912))

- Smell or fragrance of culture

- Surface structure: aerial hyphae (quantity), cottony, crustaceous, embedded, furrowed, homogenous or heterogeneous, powdery (floury), raised, resupinate, shrunken, sloppy, sticky, thin or thick, velvety, water soaked, yeastlike

- Pattern: arachnoid, flowery, radiate, zonate,

- Margin: irregular, smooth

- Growth: restricted, spreading

- Pigment exuded: color, watery

- Organs formed: fruiting structures, sporodochium, sclerotia, setae, stroma, synnema, rhizomorphs" (Pictorial Atlas of Soil and Seed Fungi)

RECOMMENDED REFERENCES WITH COLONY KEYS:

Pictorial Atlas of Soil and Seed Fungi by Tsuneo Watanabe. 3rd ed.

Identifying Fungi: A Clinical Laboratory Handbook by Guy St-Germain and Richard Summerbell. 2nd ed.

BUT WAIT, do soil fungi even grow on agar in a laboratory?

Most do not because the conditions are different. The petri dish grows the fungi that the new conditions favor most. Some fungi do not sporulate under any known conditions, and even some sporulating fungi we cannot agree as a mycological community on how to classify or differentiate them. Soil is a diverse and challenging environment that for fungi is seemingly limitless in most instances – the petri dish environment is very different and limited though still very useful. Just like with the microscope, it is very powerful when it is used in conjunction with other testing and is used as a tentative identifier. We, in general, want to avoid making definitive statements because it is very difficult and perhaps impossible to definitively identify anything.

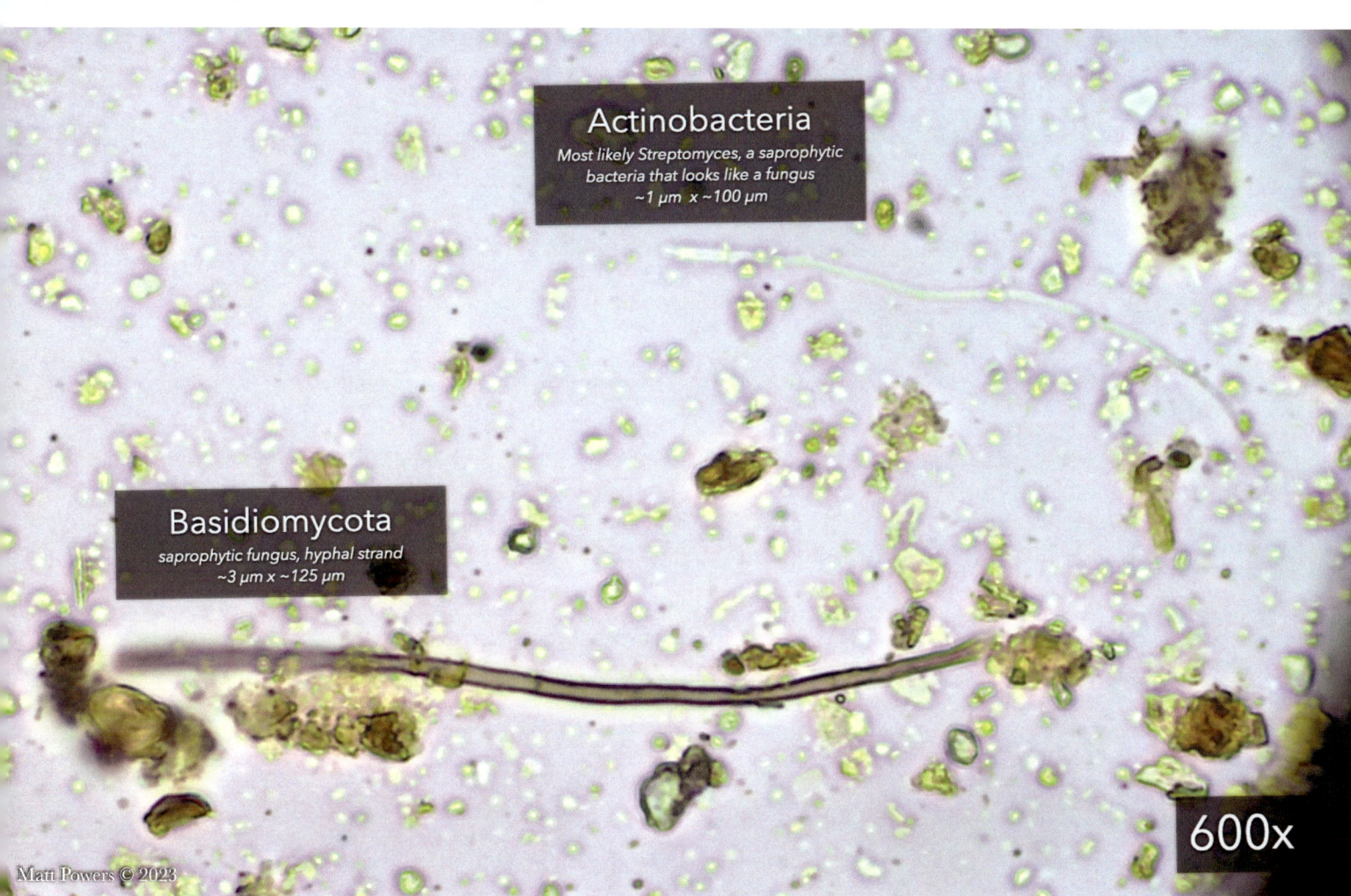

Actinobacteria
Most likely Streptomyces, a saprophytic bacteria that looks like a fungus
~1 μm x ~100 μm

Basidiomycota
saprophytic fungus, hyphal strand
~3 μm x ~125 μm

600x

Matt Powers © 2023

How many strands of fungi per drop?

In the fungal composts I'm testing, I'm seeing basidiomycota hyphae at a rate of 1 – 2 threads per FOV @ 400x – 600x with threads reaching across or nearly across the FOV. In terms of size, these are ~3 μm wide and anywhere from a small fragment in length to hundreds of microns in length. You'll see examples throughout this book – I try to include references to rulers or the magnification usually for size, but after a while, you'll know the difference. The skinnier the fungi is, the more likely it is to be clear and less likely it is basidiomycota. The skinnier fungi is either false fungi, mycorrhizal, or sporulating larger fungi (but even those tend to sporulate fat). It's also important to note that streptomyces is a plant growth promoting endophyte, so despite it being false fungi and lambasted among some circles, it is vital to plants, soil, compost, and nutrient cycling. It's also the 2nd most prominent bacteria in compost. It can absolutely be there in low numbers under the microscope, BUT it does inhibit fungal growth in its behavior. Having both implies a good balance and optionality for soil microbes and plant roots.

It should also be noted that a lot of this depends on your climate: temperate climates can see 20–50% of their soil being comprised of fungi!! This is why context is so important for interpreting our findings – what's considered high fungal counts in compost in arid and tropical climates will be considerably higher in temperate climates because there is already more fungi, more spores, and more fungal foods in the environment, soil, and compost ingredients. Knowing our context also guides our understanding of which fungi we are looking for, expecting, and perhaps needing to add to our compost , KNF preps, or soil: is it Basidiomycota we are looking for? Ascomycota? Glomeromycota? Zygomycota?

WHAT'S "GOOD" FUNGI?

We have to approach it first from which category of fungi: saprophytic, endophytic, pathogenic, or mycorrhizal? It is important to note that some studies suggest that approximately 17% of wood saprophytic fungi are also endophytes – Trametes versicolor (Turkey Tail) is a mushroom forming saprophyte (a white rot fungi) that is also an endophyte that boosts grain yields and likely works with many plants beneficially. A lot of students were trained to see any saprophytic fungi as beneficial, but obviously some are better than others, AND sometimes these endophytic saprophytes manifest as pathogens when the plant is weak or is calling out to be composted. Peter McCoy of *Radical Mycology* has referred to them as "vocal fungi" because they are trying to tell us something is wrong. We have to hold all these potentials and different pathways of expression in our minds as we evaluate each fungi from the category in which it belongs. White and brown rot fungi, endo and ectomycorrhizal fungi, yeasts, trichoderma, dark septate endophytes (DSEs), and pathogenic fungi must all be examined within their domains because they all morphologically express themselves differently. Sporulation vs mature hyphae must also be considered. While microscopy is still the best way to roughly ID fungi quickly, it is a morphological process of elimination and still not entirely definitive unless dealing with purified and verified samples or a DNA tested isolation from a wild sampling. There's continuous adaptation, evolution, horizontal gene transfer, epigenetics, and likely even more at work that we need to be aware of – we are going to find new things, miss things, mistake things, and confuse things. It's the nature of the space and manner in which we are currently able to test and document things. For instance, streptomyces is the next most prevalent microbe genus in the soil and in compost – that means, an actinobacteria is more present than all beneficial bacteria. You're likely going to see that reflected in the samples you view under the microscope – the best compost had those same ratios, so we can't just say that "false fungi" are bad. They are ever-present and, after e.coli in 1st, are the 2nd most abundant bacterial genus. Perhaps we should call them lookalikes to shed the connotations associated with "false" and especially "pathogenic". Does that mean phytophthera is good - NO IT DOES NOT! That's another lookalike fungi but it is a water mold that attacks and kills entire nurseries and oak forests. It's a terrible epidemic in some areas. This is why we need better IDing, testing, and sharing of information – there's nuance to this space.

To quote a 40-year leading expert, published scientist, and respected author, Tsuneo Watanabe: "We are now in confusion and at a loss between the traditional, morphological taxonomy and the modern, molecular taxonomy [DNA testing]." As an expert in this field, when examining fungi under the microscope, it is his hope to have samples "identified tentatively at least to the genus level whenever possible" (Watanabe, 2010). The page before this he blames the lack of taxonomical certainty for basidiomycota identification on the lack of "adequate microscope observation". This means that even seasoned experts cannot always ID samples to genus level especially for basidiomycota but at times we can ID the genus with the microscope – this implies that the species level can be impossible to ID. Also note his usage of the word *tentatively*: definitive identification is not possible. Keep this in mind when you are frustrated and stuck. This is precisely why I created the R-Soil Database – it's the only way around or to work through all these limitations. Once we develop the proven correlations across tests, across time, and from soil health to fruit and leaf nutrition, biology, and energy, we will be able to have more definitive guidelines around time and place, strain and consortium, timing and season, redox and

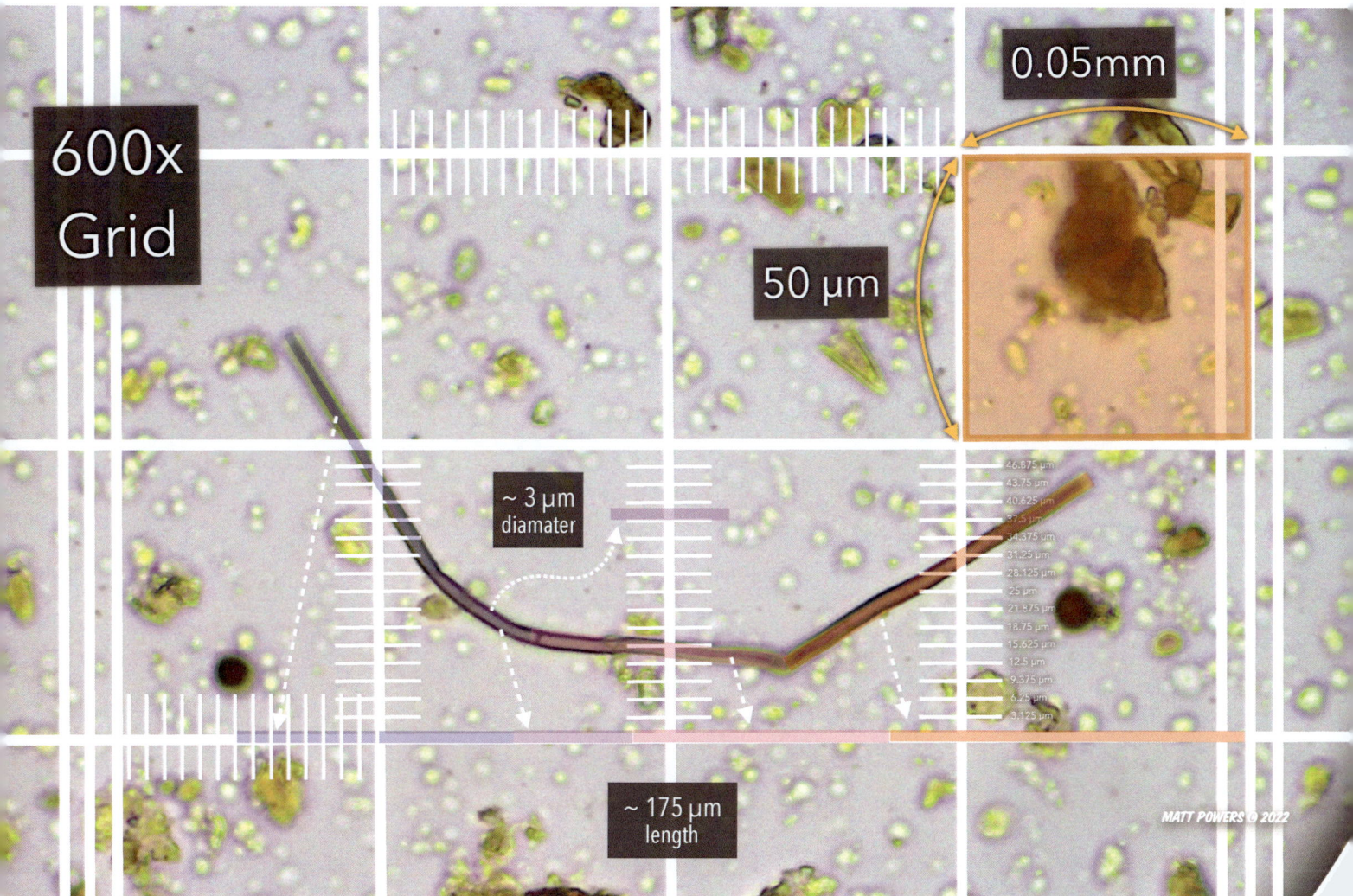

600x Grid

0.05mm

50 µm

~ 3 µm diamater

~ 175 µm length

46.875 µm
43.75 µm
40.625 µm
37.5 µm
34.375 µm
31.25 µm
28.125 µm
25 µm
21.875 µm
18.75 µm
15.625 µm
12.5 µm
9.375 µm
6.25 µm
3.125 µm

moisture. The microbial behaviors around these simple dynamics have never been possible until now. That being said, we can ID quite a lot with the microscope.

WHO ARE THE LOOKALIKE FUNGI?

1. *Is it thin and colorless?*
2. *Does the thickness change from thin to thick?*
3. *Is it Tattered? Breaking up?*
4. *Is it chaotic, irregular, and/or bulbous?*
5. *What's the context? Is it a sick plant? Is it anaerobic soil?*

With identification we can't just take one feature and declare its identity – we have to look at several factors and features. If it is a combination of factors like the above 1–3, 1–4, or even 1–2+4, we can usually identify it as beneficial or not.

Current prevailing thought would have us say: it's actinobacteria or an oomycete most likely – but which one makes all the difference! They are both considered "false" fungi. Oomycetes are plant pathogens that indicate water logging, sick plants, reduced and alkaline–neutral pH/Eh, and anaerobic conditions while actinobacteria in large numbers indicate alkaline and oxidized soils though they are present everywhere and do not indicate detrimental conditions in lower or even moderate concentrations.

Fungi form strands, so anything tattered or breaking up is organic matter, false fungi, or even microarthropod remains depending on your piles composition. It's also important to recognize that there really is no "bad" fungi – they each have a role or are telling us something. The paradigms of good and evil don't play out well in the microbiological space as many microbes shift roles over time and share many forms like Trichoderma also being a wood saprophyte via its teleomorph. Plus, streptomyces being one of the most prevalent bacteria in soil and compost (not to mention it being responsible for over 75% of worldwide penicillin production) means we're going to see it and other actinobacteria in our viewings.

GENERAL INDICATORS OF BENEFICIAL SAPROPHYTIC FUNGI

- Well-Defined Septa & Uniform Width (or Nearly So) – Beneficial Fungi Can Also Be Damaged By Microbes, Slide Covers, & Soil Solution Preparation. Sporulating fungi can vary in their expression widely as well.

- Wide Hyphae with Thick Walls – Thicker the Cell Walls, the Older the Fungi, but this also varies by type

- Golden, Tan, Reddish, Brown – usually the darker the more beneficial but saprophytic fungi can be clear when sporulating and mycorrhizal fungi can be hyaline, so knowing our context is critical to properly assess things.

- No shredding, wisps, fragments, clear breaks

- Does not flare NEON RED ORANGE when put under epifluorescence lighting (that would indicate microplastic)

- Does not have to glow with phosphorus under epifluorescence lighting to be beneficial (most do not)

GENERAL INDICATORS OF "FALSE" FUNGI

These "bad" fungi aren't really fungi but look like fungi & compete with it

- Non-Uniform Septa or Lacking Septa

- Wispy, Chaotic, Snarls, or Broken Fungi (Norcardiodes genus actinobacteria form mycelium and it breaks up into cocci and bacilli shapes, and it's often in the Top 10 most common microbes found in the soil and compost I've tested AND it's disease-causing for those with compromised immune systems).

- Exceptionally thin Hyphae (though it can be young, low oxygen conditions, or a specifically thin strain)

- Hyaline (crystal/clear) Hyphae (though it can be a young hyphae, a non-saprophyte, or a less common species of fungi)

Dr. Elaine has said that in compost typically actinobacteria are 1–1.5 μm, Oomycetes 1.5–2.0 μm, Ascomycetes 2-2.5 μm, and Basidiomycetes greater than 3 μm. If we follow the logic of 'bigger is better', then we'd reason that Basidiomycota are preferable yet rusts are in that phylum. Most mycorrhizal fungi are the same diameter hyphae and color ranges as oomycetes and actinobacteria. In fact, the overlap leaves us relying upon other measures and context to understand our fungi more than their diameter and color. In the past yeasts have also been seen as indicators of problems, but they are ubiquitous and often used in biofertilizer fermentations and found in plants endophytically, so to rule them out makes no sense. If they are dominant in a sample, it signals a problem if we are examining our soil, but they should be abundant in most of our biofertilizers (many microbes need yeast or bacilli species in context to even reproduce).

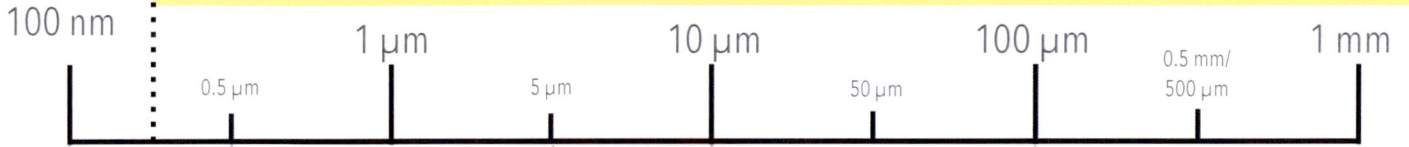

ELECTRON MICROSCOPY

EYE

LIGHT MICROSCOPY

100 nm | 0.5 µm | 1 µm | 5 µm | 10 µm | 50 µm | 100 µm | 0.5 mm/500 µm | 1 mm

YEASTS
DIAMETER

ARBUSCULAR MYCHORRHIZAL FUNGI (AMF)
HYPHAE DIAMETER — SPORE DIAMETER

Matt Powers © 2022

TRICHODERMA
X
CONIDIA — HYPHAE DIAMETER

ACTINOBACTERIA
HYPHAL DIAMETER

AMF

RHIZOPHAGUS IRREGULARIS
HYPHAE — SPORES

RHIZOPHAGUS CLARUS
HYPHAE — SPORES

RHIZOPHAGUS AGGREGATUS
HYPHAE — SPORES

GIGASPORA MARGARITA
HYPHAE — SPORES

PARAGLOMUS BRASILIANUM
HYPHAE — SPORES

FUNNELIFORMIS MOSSEAE
HYPHAE — SPORES

CLAROIDEOGLOMUS ETUNICATUM
HYPHAE — SPORES

SEPTOGLOMUS DESERTICOLA
HYPHAE — SPORES

WHITE ROT

PHELLINUS VIETNAMENSIS
A VIETNAMESE WHITE ROT
HYPHAE — W X L SPORES

PHELLINUS IGNIARIUS
GLOBAL WHITE ROT
X — SPORES — HYPHAE

TRAMETES VERSICOLOR
SPORE WIDTH — HYPHAE — SPORE LENGTH

PHYTOPHTHORA
DIAMETER — OOSPORES

AVERAGE SOIL HYPHAE
DIAMETER

0.25 µm
The Abbe Diffraction Limit

WHAT ABOUT YEASTS

Yeasts are ubiquitous in the natural world. They are typically much larger than the bacteria you'll be looking at; his is how you can easily differentiate between the purple non-sulfur bacteria and beer yeast in your biofertilizer brew or EM extension. They are saprophytes that can be endophytes, so an over abundance of them indicates decomposition, fermentation, and, likely, alcohol production. The Oriental Herbal Nutrient (OHN) Korean Natural Farming (KNF) prep uses alcohol as a preservative, but it likely has other effects: triggering an immunological response at the very least. Effective microbes and all yeast-based biofertilizer fermentations likely have some alcohol content as well.

MYCORRHIZA VS SAPROPHYTES VS ENDOPHYTES

Mycorrhiza	Saprophytes	Endophytes
Spore Forming	Spore & Non-Spore Forming	Primarily Non-Spore Forming *(there are some endophytes in grass leaves that form spores that spread on the wind)*
Found in Roots & the Rhizosphere	Found in Compost & the Rhizosphere	Found inside Plants, Roots, Compost, & the Rhizosphere
Thin, Clear, Opaque, Yellowish, Brownish	Thick–Thin, Clear–Many Colors	Round/Ovoid
AMF, Ectomycorrhizal Fungi, Ericoid	Yeasts, White Rot, Brown Rot, Soft Rot, Trichoderma	Saccharomyces cerevisiae
Glomeromycota	Basidiomycota & Ascomycota	Ascomycota

SEPTA

Septa are the divisions in the hyphae – Basidiomycotan septa are plugged while Ascomycota has a central pore, so they have less movement of bacteria within them than Zygomycota which lacks these divisions, but this rather new concept of bacteria being inside fungal hyphae has evolutionary roots and this pattern plays out in rhizophagy as well. It's theorized that B12 is provided to fungi via endohyphal bacteria! That's because like in rhizophagy, the host consumes the biology inside it. In Zygomycota, and in other fungi lacking septa, the bacteria moves freely within them and can leak out.

"Fungi internalize bacteria the way that plants do. The more bacteria internalized - the less pathogenic is the fungus. If the fungi can get nutrients from bacteria--they do not need to attack plants to get nutrients - and so they don't. The septa play

roles in managing the bacteria in hyphae. " - James F. White, scientist, professor, author, and contributor to

Transformative Paleobotany

SPORES

It's important to note that spore color and hyphae color range likely due to the mineral and humic content of the soil environment. This is why more oxidized soils in the more arid regions have redder or more orange colored spores: the oxidation of iron turns soils red and orange, and those pigments, hallmarks of their environment, carry over into the spores. We also can see this in the ranges of color and the split around yellow/tan (cream can be on a spectrum with either) in the spread and range of species. The colors are environmental and thus it opens up the potential for them to be fallible indicators of species as specialized and unique soil situations exist all over the world. Most rules of thumb are developed across points of commonality (where it's easiest to farm and where we've farmed traditionally), so we can easily miss exceptions to the rule. Also, mycologists in the forest have results that are rarely compared with the results found in agricultural fields despite them all being part of this larger continuum of expression.

AMF SPORES BY COLOR

HYALINE	WHITE	CREAM	GOLDEN	YELLOWISH	TAN	YELLOWISH BROWN	BROWN	REDDISH/ ORANGE BROWN
R. AGGREGATUS	R. AGGREGATUS	R. AGGREGATUS	R. AGGREGATUS	R. AGGREGATUS	R. IRREGULARIS	R. IRREGULARIS	C. ETUNICATUM	C. ETUNICATUM
P. BRASILIANUM	R. IRREGULARIS	R. IRREGULARIS	R. CLARUS	R. CLARUS	R. CLARUS	R. CLARUS	S. DESERTICOLA	S. DESERTICOLA
	R. CLARUS	R. CLARUS	F. MOSSEA	F. MOSSEA		F. MOSSEA		F. MOSSEA
	G. MARGARITA	P. BRASILIANUM	G. MARGARITA	G. MARGARITA				
		G. MARGARITA						

To a degree we can ID spores to their type (mold, basidio, alternaria, etc.), but down to the species can be challenging as many different types of arbuscular mycorrhizal fungi have similar looking spores. If you are looking at an inoculant mix, it can be difficult to ID the different strains as they are all the same family and phylum – their spores look alike and can vary in similar ways. Most identification practices these days rely upon DNA or molecular identification, but we can use these morphological indicators to ID a mycorrhizal spore. AMF spores are just one part of a larger picture. There are mold, rust, mildew, saprophyte, and pathogen spore to consider as well. AMF spores have 3 layers to their structure – these are often used to identify and differentiate AMF spores. Taking all spores into context won't fit in a book – that's why the mycological and microbiological communities rely upon online databases like INVAM which is superb for this exact task.

Ectomycorrhizal Fungi

While partnering with only around 5 – 8% of plant species, specifically cold temperate trees that are of value to their ecosystems and the timber industry for the most part, the ectomycorrhizal fungi constitute a small group of fungi, approximately 5% of all fungal species. You'll see these in association with trees like Douglas fir, alder, pine, manzanita, beech, and chestnut and at higher altitudes and latitudes. These are important to forestry workers, native nurseries, and timber operations, not for agricultural purposes or for composting. With native nurseries, test inoculation rates as you would for AMF but know that a good inoculation rate for EMF is below 50% while AMF is above 50%.

WHAT ABOUT ERICOID MYCORRHIZAE?

Ericoid mycorrhizal fungi (ERMF) like *Pezizellaericae* is even more rare than EMF preferring azaleas, cranberries, heather, and blueberries to partner with, all members of the Ericaceae family. ERMF prefer nutrient poor, acidic soils as do the plants – because their environments are rather austere, plants do not require heavy inoculation applications to achieve symbiosis: this fungus is adaptive and effective! It is not clear what the ideal inoculation rates for ERMF are yet. As images begin to compile in the R-Soil Database, we will begin to see the range of healthy plants emerge.

DAMAGED FUNGI?

Low oxygen, fungal predators, or just too much shaking with too much sharp sand – all these can lead to damaged fungi. Some predators like fungal feeding nematodes puncture the hyphae and suck out the insides! Or is it plastic? Turn on your epifluorescence lamp and see!

NEMATODE KILLING FUNGUS?

Are you seeing hoops of 3 cells of hyphae? Those are nematode-trapping fungus, but if there are no nematodes around, they'll look and behave just like regular saprophytic fungi. They kill root feeders primarily, and we know this because it releases a smell that is akin to what roots smell like to attract them. There's also parasitic fungi that kills root-feeding Nematodes – they form spores inside them,

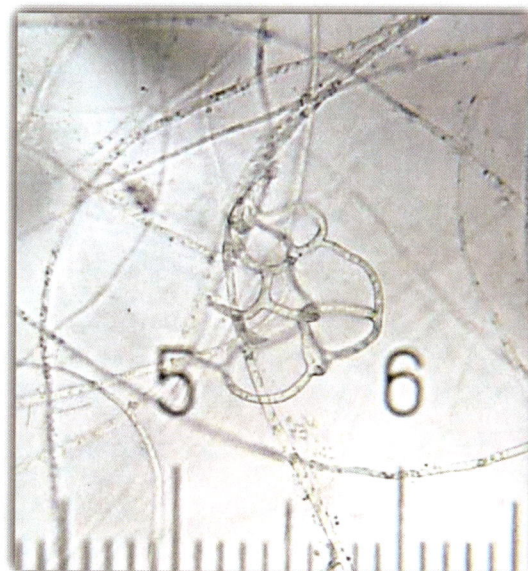

A nematophagous fungus of the genus Arthrobotrys

100X

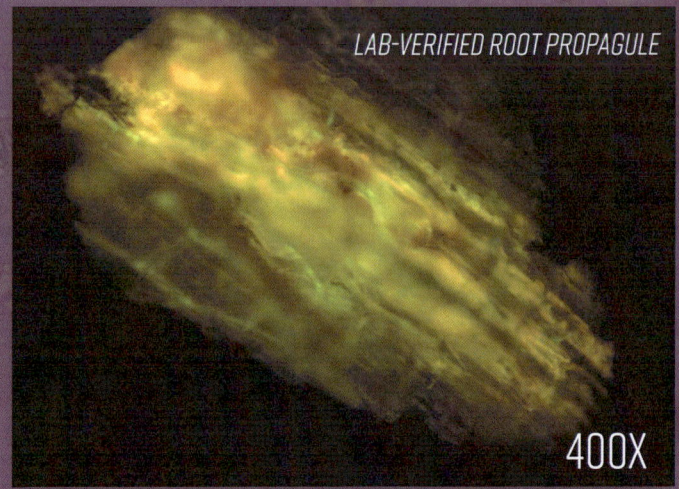

400X

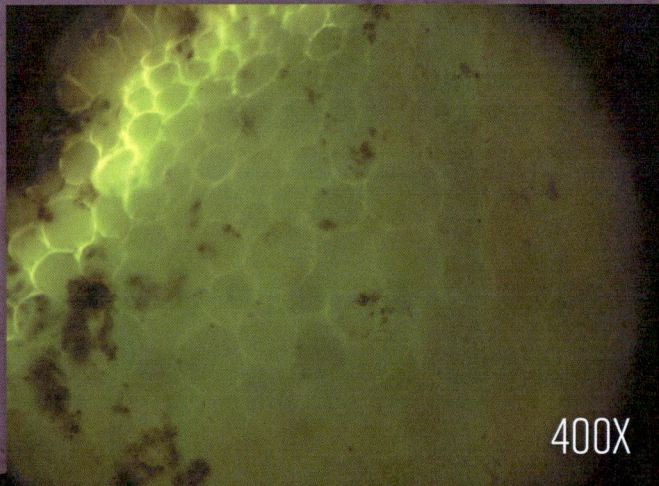

400X

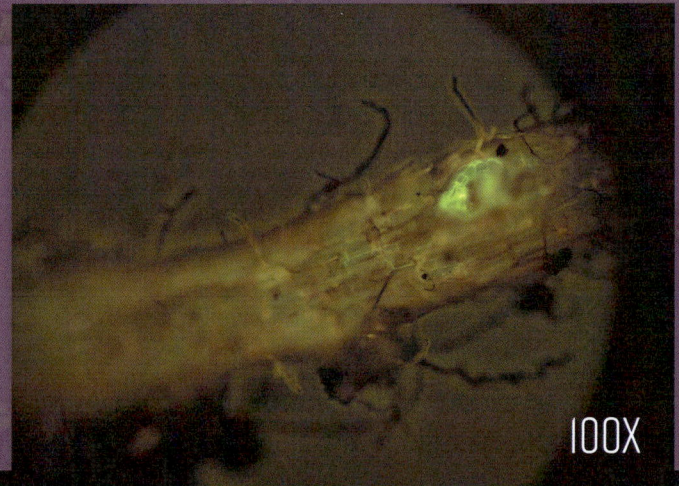

100X

ARBUSCULAR MYCORRHIZAL (AM) FUNGI

Endomycorrhizal Obligate Symbionts i.e. Vesicular Arbuscular Mycorrhizae (VAM)

FAMILIES: Glomeraceaes & Claroideoglomeraceae

GENUS: Glomus, Funneliformis, Rhizophagus, Sclerocystis, Septoglomus, & Claroideoglomus

ORDERS: Glomerales, Diversisporales, Archaeosporales, & Paraglomerales

PHYLUM: Glomeromycota

FORM: Filamentous, Endomycorrhizae, Arbuscular Mycorrhizal Fungi (AMF), spore-forming

SIZE: 12 – 18 μm diameter hyphae, 40 – 300+ μm diameter spores

APPEARANCE: Infects root cells and forms arbor-like hyphae inside the cell walls of root cells but does not penetrate the cytoplasm – some develop vesicular spores on the inside (*intraradices*) while others develop on the outside of roots

ABUNDANCE: Considered ever-present in almost all ecosystems in association with over 90% of plant roots

FOODS: Plant internal sugars

PH & TEMPERATURE PREFERENCES: Partners with plants all over the world in diverse ecosystems, but in the lab setting, these are the ranges: 4 – 8.2 pH, 10°C – 25°C

REPRODUCTION: Spores, root fragments, and hyphae fragments

USES: PGPM, Biocontrol, Enhanced Plant Immunity, P uptake, & Drought Tolerance

400X

AN ASPARAGUS ROOT INOCULATED WITH
ARBUSCULAR MYCORRHIZAL FUNGI (AMF)

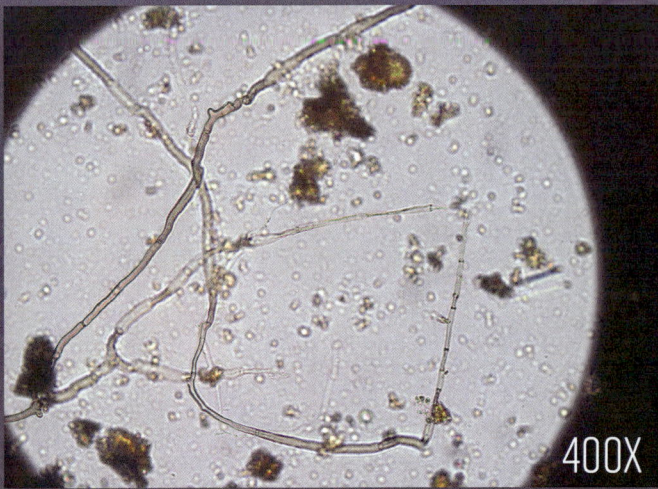
400X

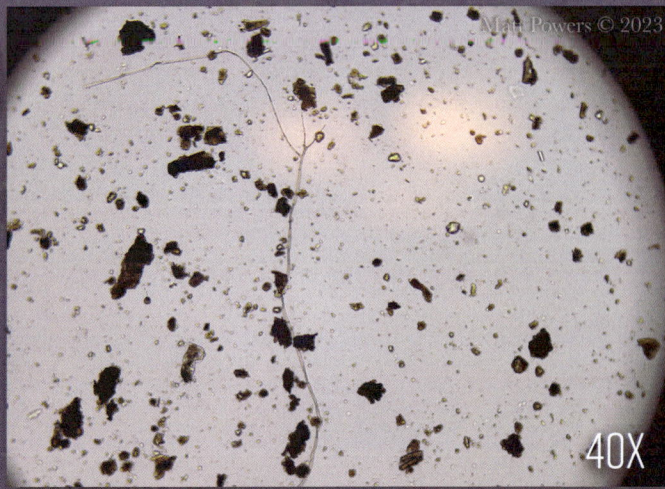
40X

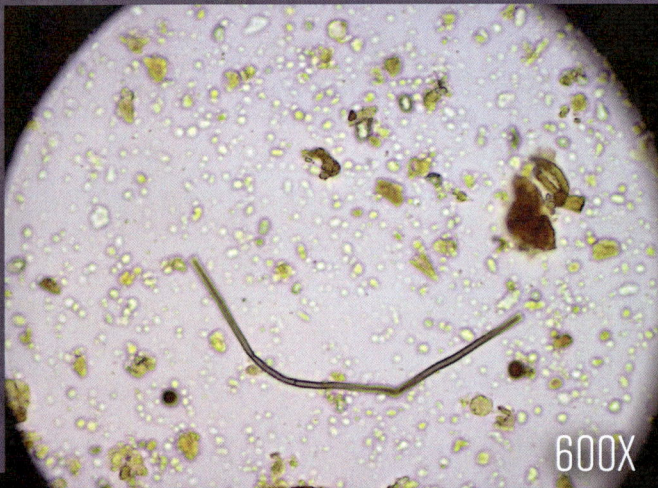
600X

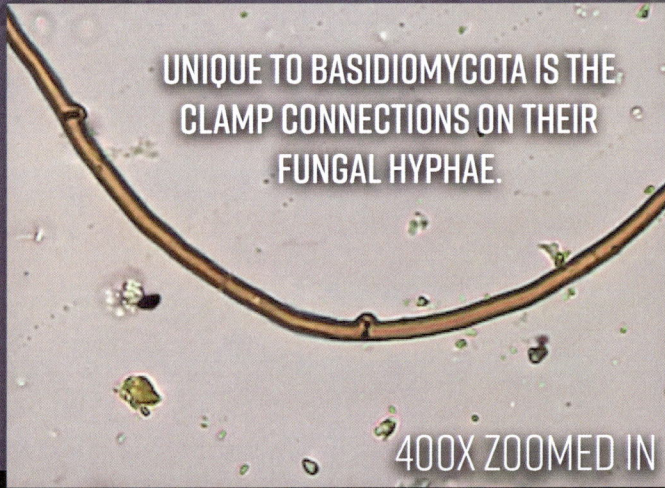
UNIQUE TO BASIDIOMYCOTA IS THE CLAMP CONNECTIONS ON THEIR FUNGAL HYPHAE.

400X ZOOMED IN

BASIDIOMYCOTA SPECIES

Predominantly Saprophytic Gilled Mushroom-Forming Fungi

PHYLUM: Basidiomycota

FORM: Mycelial though some are cocci like yeasts, spore-forming & non-spore-forming, but primarily gilled mushroom-forming mycelial networks with microscopic club-shaped fruiting bodies or *basidia* found hanging off their gills

SIZE: Generally 3 – 10 µm diameter hyphae: ranging from microscopic to visible with the naked eye as they form mycelial mats, and mushrooms range in size broadly

APPEARANCE: Ranging in expression from yeasts to hyphae forming, spore & non-spore forming, highly diverse in nature – those found in compost typically have clear to opaque brown, honey, or tan hyphae

ABUNDANCE: Comprising 1/3 of all known fungal species including most mushrooms, yeasts, smuts, rusts, white rot, brown rot, and other plant and human pathogens = common everywhere in the world. 90% of basidiomycota are white rot.

FOODS: Lignin, organic matter, and sugars

PH & TEMPERATURE PREFERENCES: A broad range that covers all the ecosystems with living plants & active decomposition

REPRODUCTION: Primarily spore forming (basidiospore), but some can only reproduce asexually

USES: Decomposition & Composting

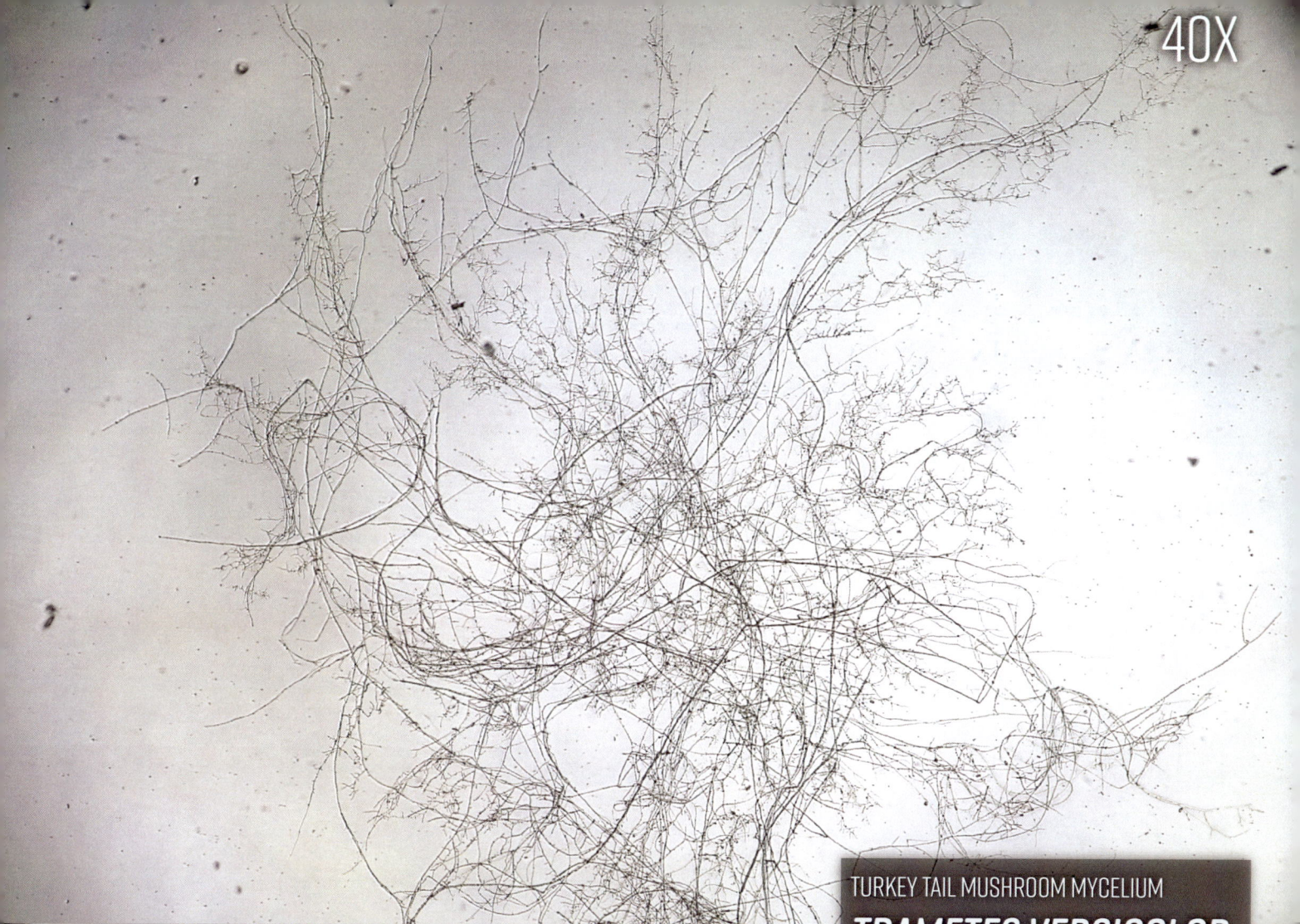

40X

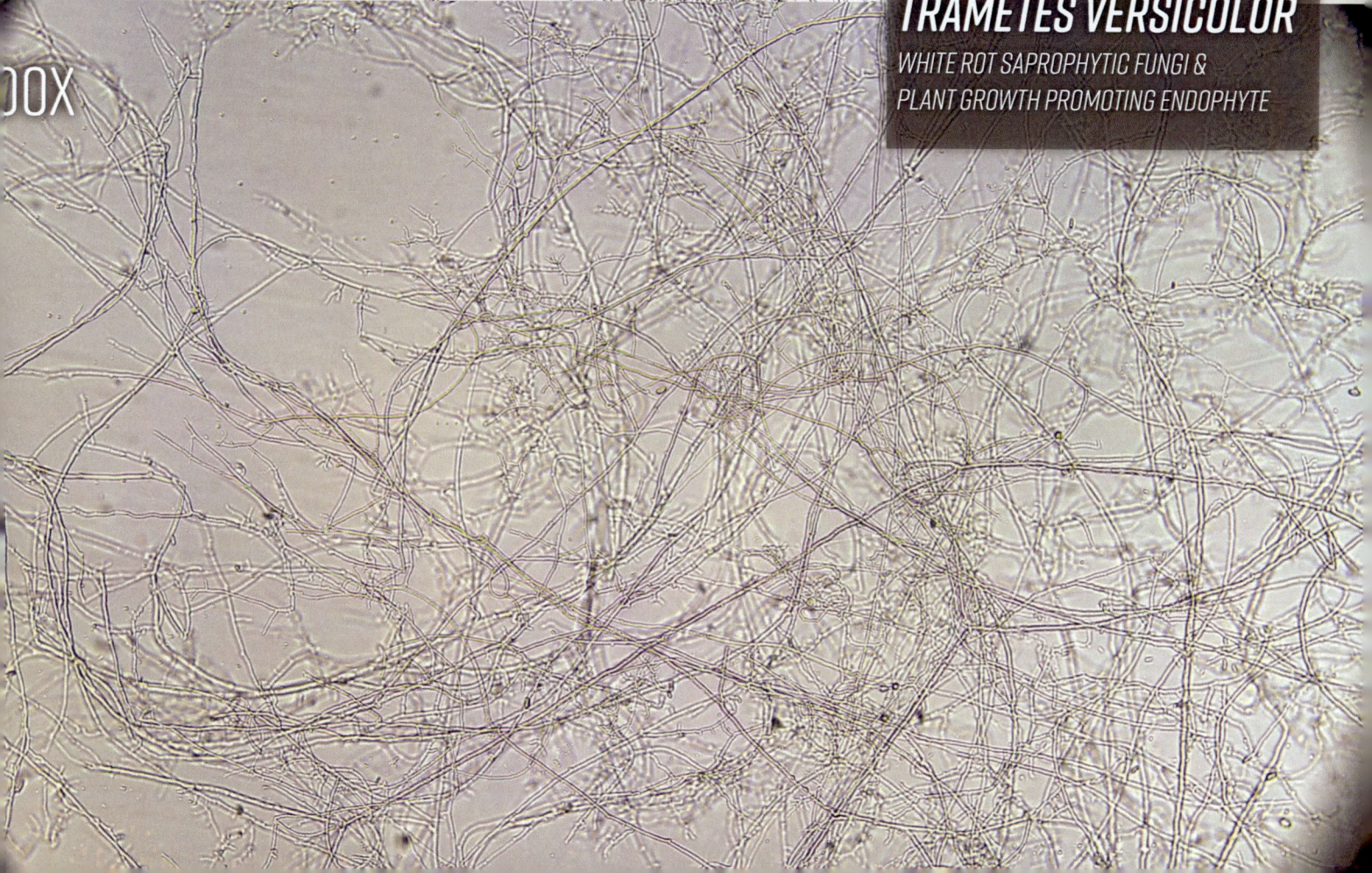

00X

TURKEY TAIL MUSHROOM MYCELIUM

TRAMETES VERSICOLOR

WHITE ROT SAPROPHYTIC FUNGI &
PLANT GROWTH PROMOTING ENDOPHYTE

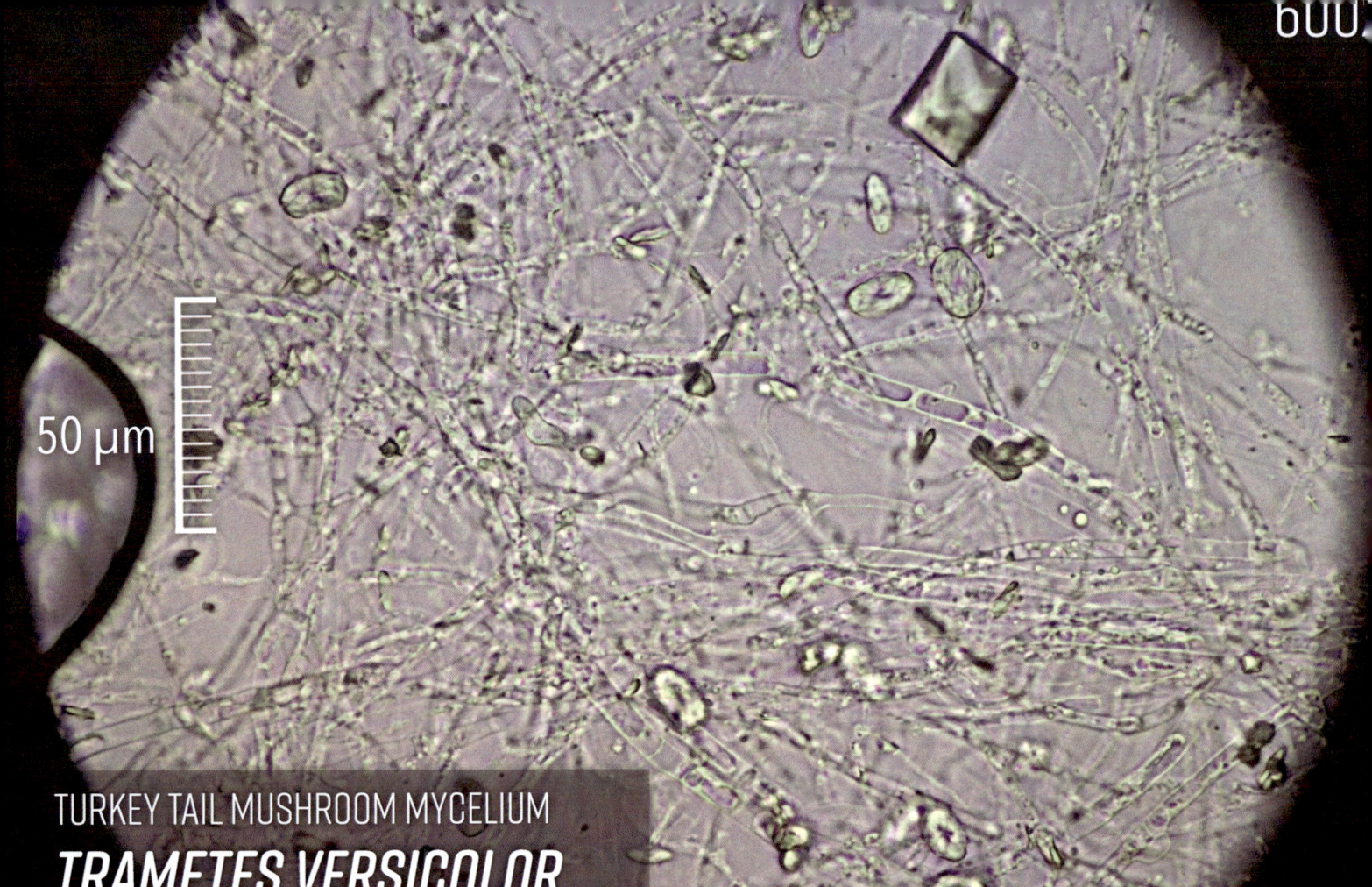

50 μm

600X

TURKEY TAIL MUSHROOM MYCELIUM
TRAMETES VERSICOLOR
WHITE ROT SAPROPHYTIC FUNGI &
PLANT GROWTH PROMOTING ENDOPHYTE

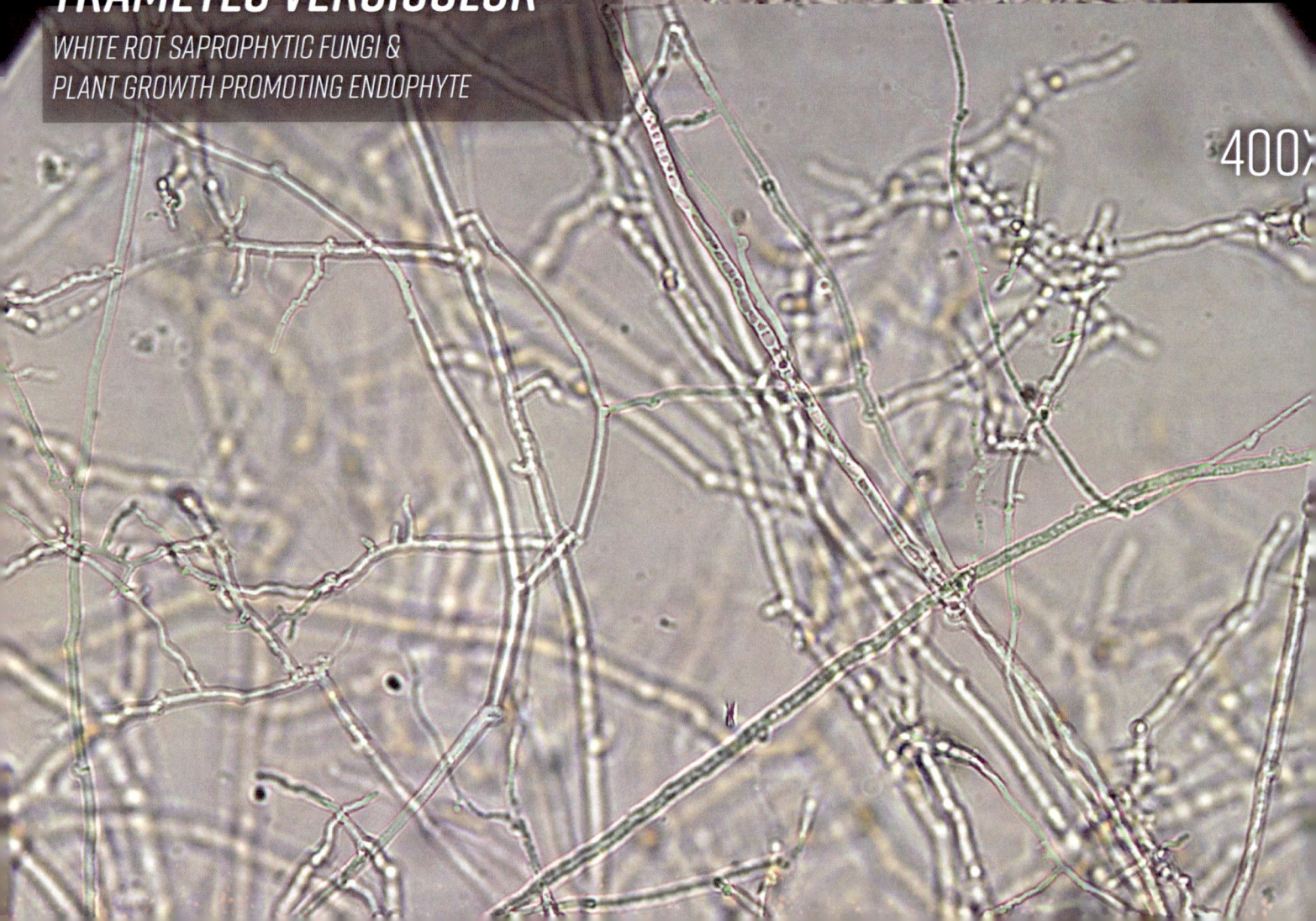

400X

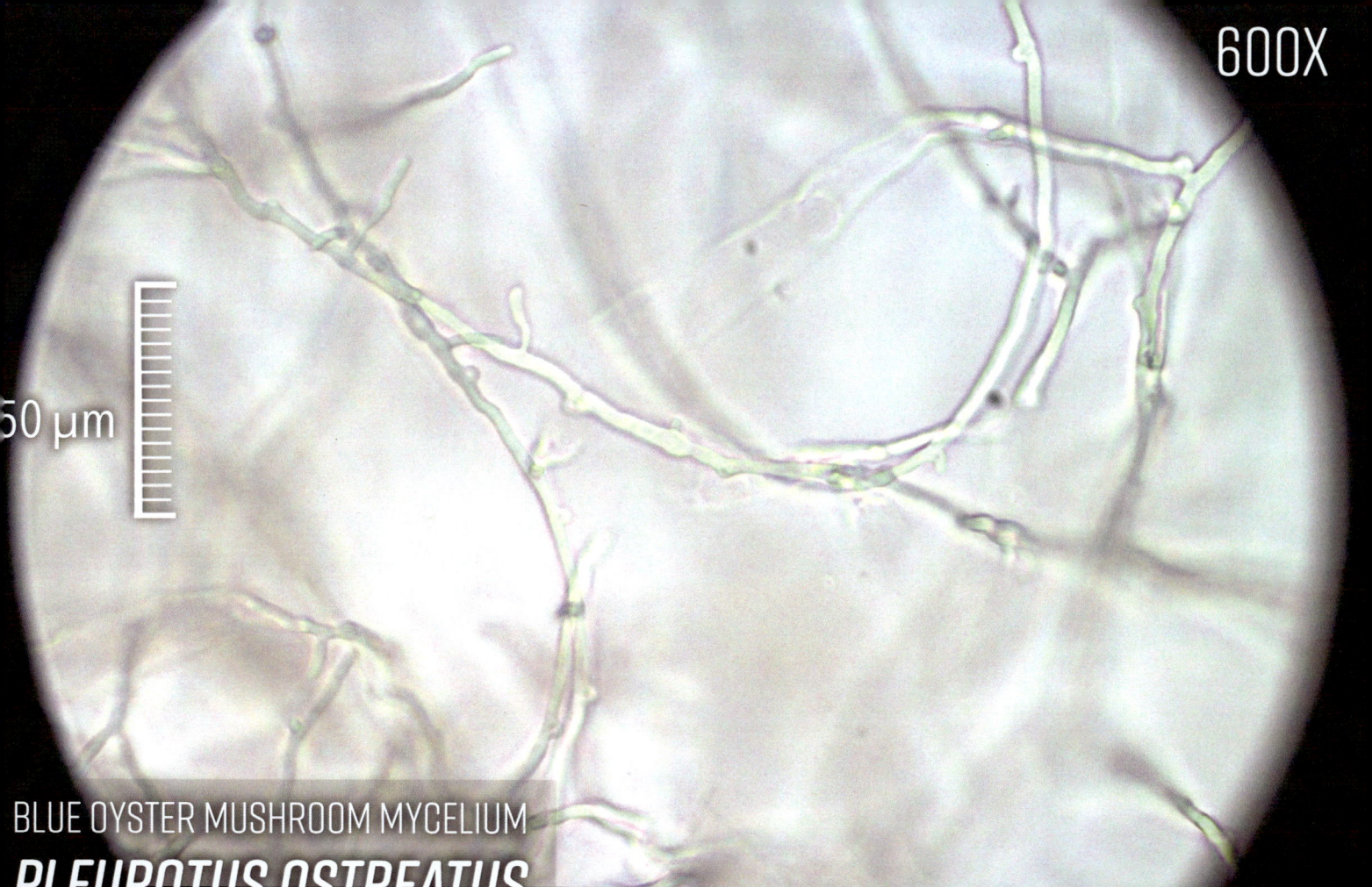

50 μm

BLUE OYSTER MUSHROOM MYCELIUM
PLEUROTUS OSTREATUS
WHITE ROT SAPROPHYTIC FUNGI

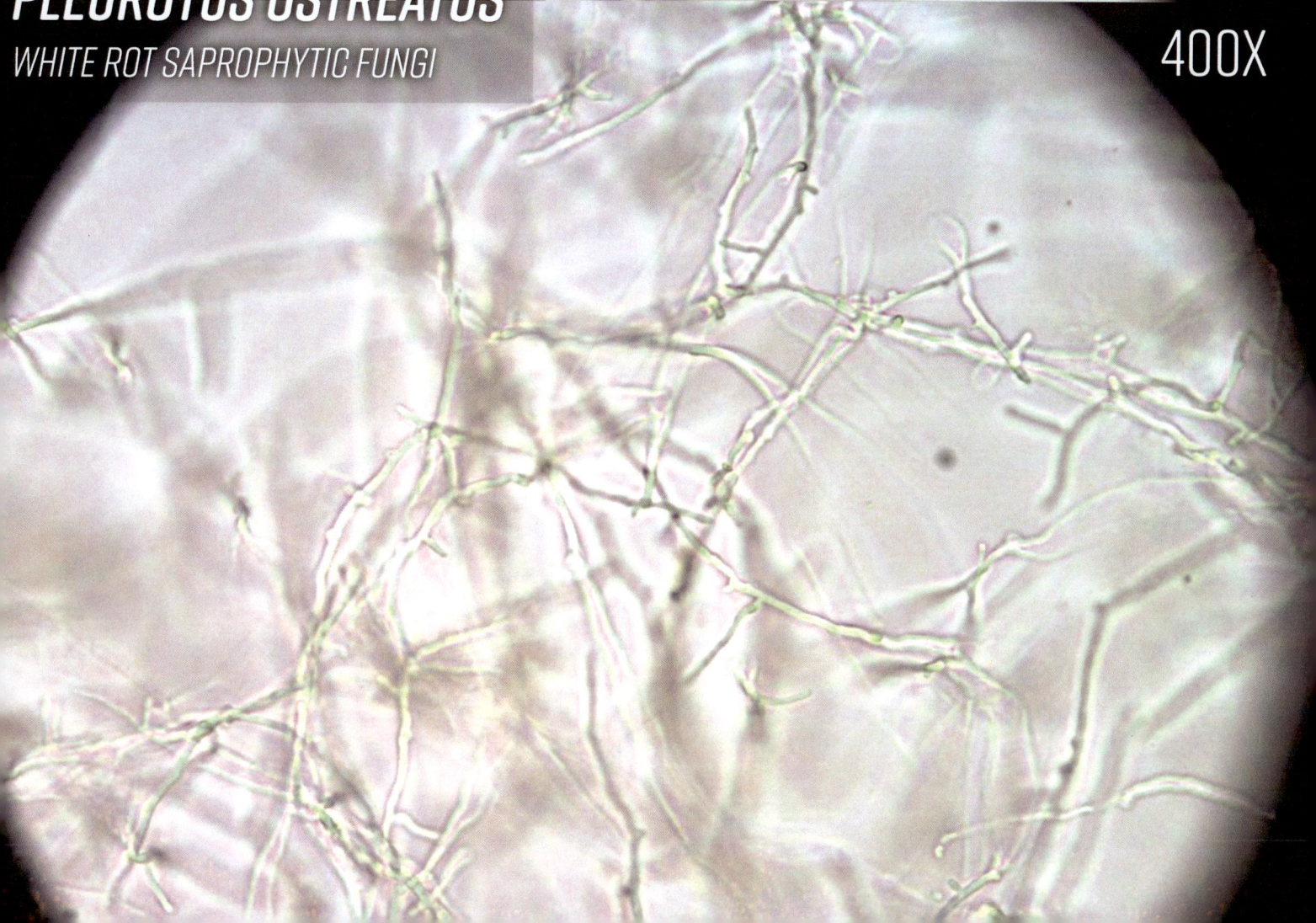

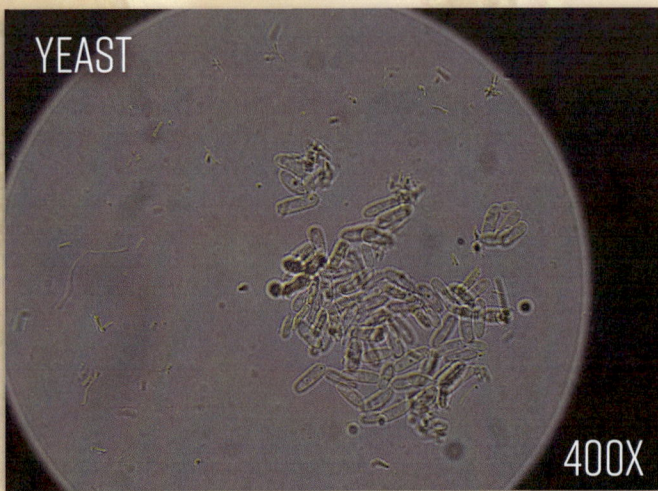

YEAST
400X

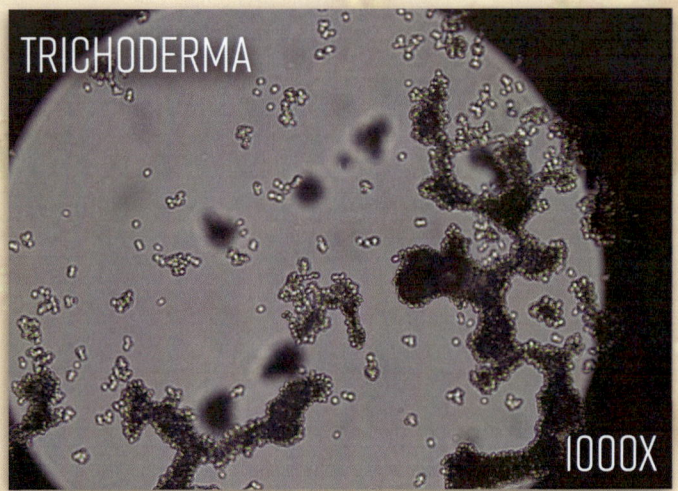

TRICHODERMA
1000X

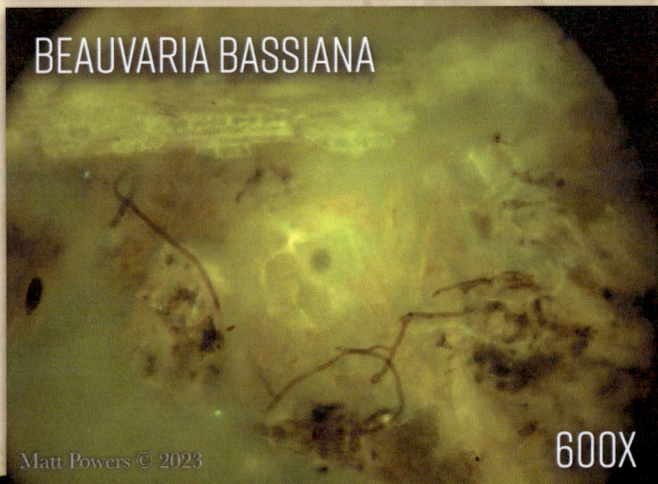

BEAUVARIA BASSIANA
600X

Matt Powers © 2023

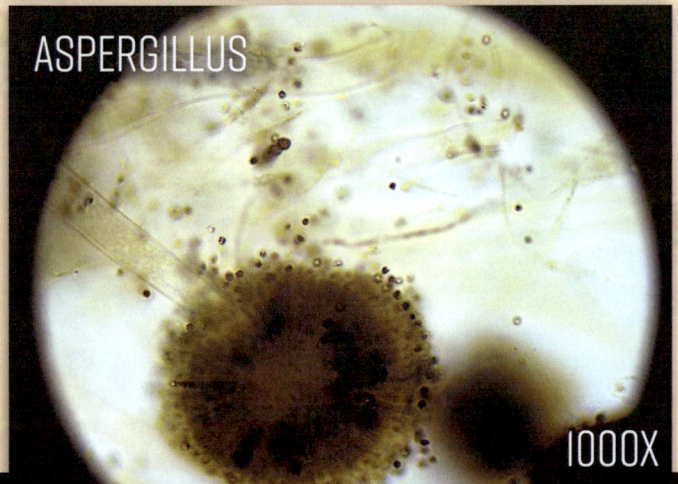

ASPERGILLUS
1000X

ASCOMYCOTA SPECIES

Saprophytic Fungi including pathogenic "vocal" fungi

PHYLUM: Ascomycota

FORM: Filamentous – cocci, spore-forming and non-spore forming

SIZE: Highly variable but all microscopic

APPEARANCE: Filamentous – cocci

ABUNDANCE: It is the largest phylum in the fungal kingdom – these fungi are everywhere: brewer's & baker's yeast, truffles, morels, lichens, pathogens, powdery mildew, and many more!!

FOODS: Organic matter, simple sugars, fungi, and more

PH & TEMPERATURE PREFERENCES: A broad range that covers all the ecosystems with living plants & active decomposition

REPRODUCTION: Asexual, spore-forming most commonly but also budding

USES: Biocontrol, Decomposition, Immunologically Stimulating, & Endophytes

1000X

1000X

EPIFLUORESCENCE

1000X

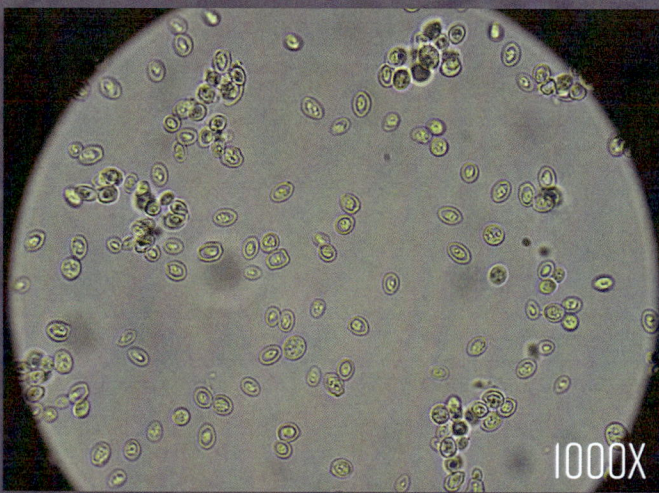

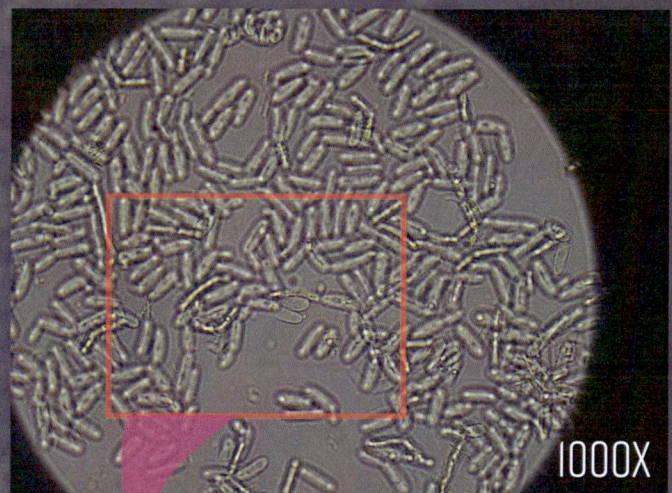

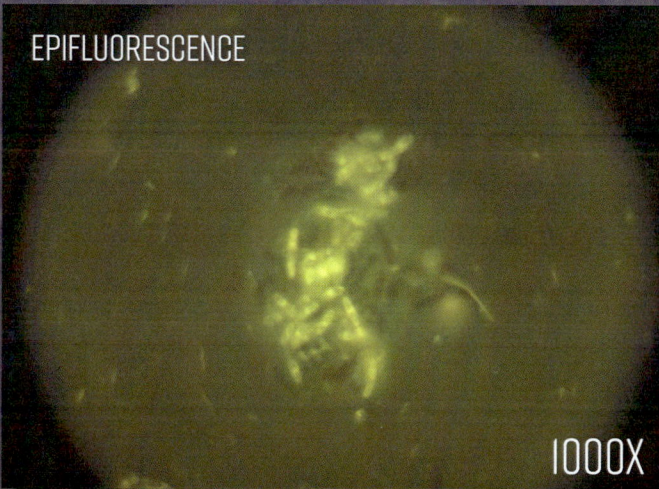

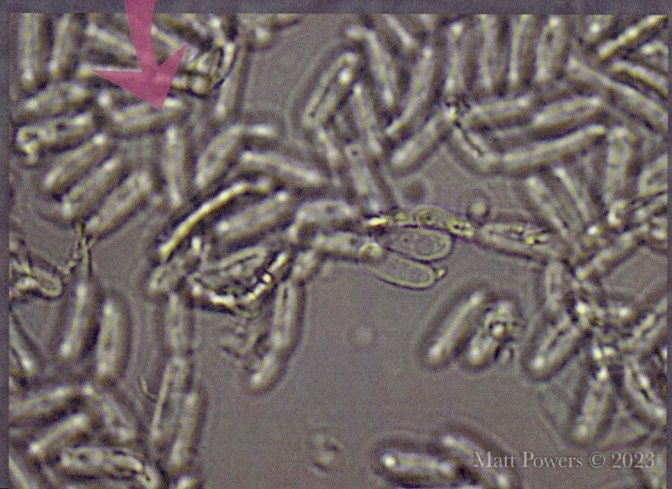

Matt Powers © 2023

SACCHAROMYCES CEREVISIAE

Gram-positive Aerobe - the Beer & Bread Yeast

A

B

C

FAMILY: Saccharomycetaceae

ORDER: Saccharomycetales

PHYLUM: Ascomycota

FORM: Single-celled fungi

SIZE: 5 –1 0 µm wide

APPEARANCE: Ovoid to round, ranging in size depending on moisture and nutrient availability

LOCATION: Fresh and salt water, soil, compost, vermicompost, inside the plant endophytically & the rhizosphere

ABUNDANCE: Found in the bark of all oak trees, as an endophyte in nearly all plants, on the skin of grapes, and in most fermentations

FOODS: Organic matter, but primarily sugars (carbon)

PH & TEMPERATURE PREFERENCES: Alkaline (pH 8 – 9), but they are found even at pH 5 –11, 15 – 40°C

REPRODUCTION: Budding

USES: PGPR & Endophytically

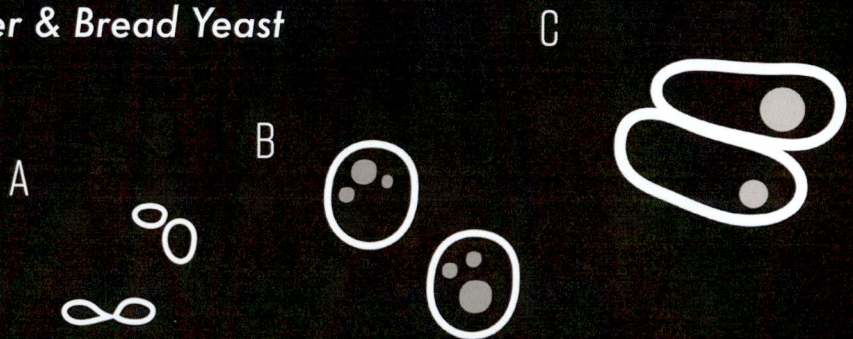

TRICHODERMA
Saprophytic Fungi

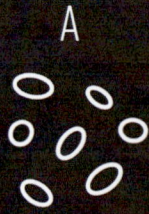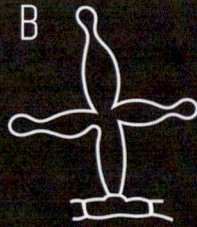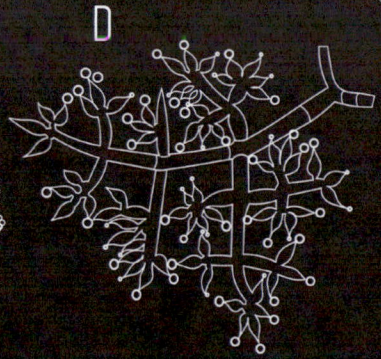

FAMILY: Hypocreaceae

ORDER: Hypocreales

PHLYUM: Ascomycota

FORM: Branching Mycelia (near 90° on primary & secondary) with ovoid conidia (spore sacs)

SIZE: 5–10 μm diameter hyphae, 3–5 diameter conidia

APPEARANCE: septate & transparent hyphae with tufts of transparent mostly 90˙ branches of conidiophores and those are branched with bunches of transparent phialides which are fatter on the bottoms usually with sticky bunches of ellipsoidal conidia on top. We commonly recognize it as the deep green mold on bread, decomposing citrus, and old coffee grinds.

ABUNDANCE: the most abundant culturable fungi in soil globally – also ubiquitous in water

FOODS: organic matter and especially fungi

PH & TEMP PREFERENCES: pH 5.5–7.5, 25–30°C

REPRODUCTION: Budding & Asexual (Conidiospores)

USES: Biocontrol, Fungicide, & Enabling AMF to Inoculate Brassicas

TELEOMORPHS: Hypocrea, a wood saprophyte that can be yellow/orange/brown/black

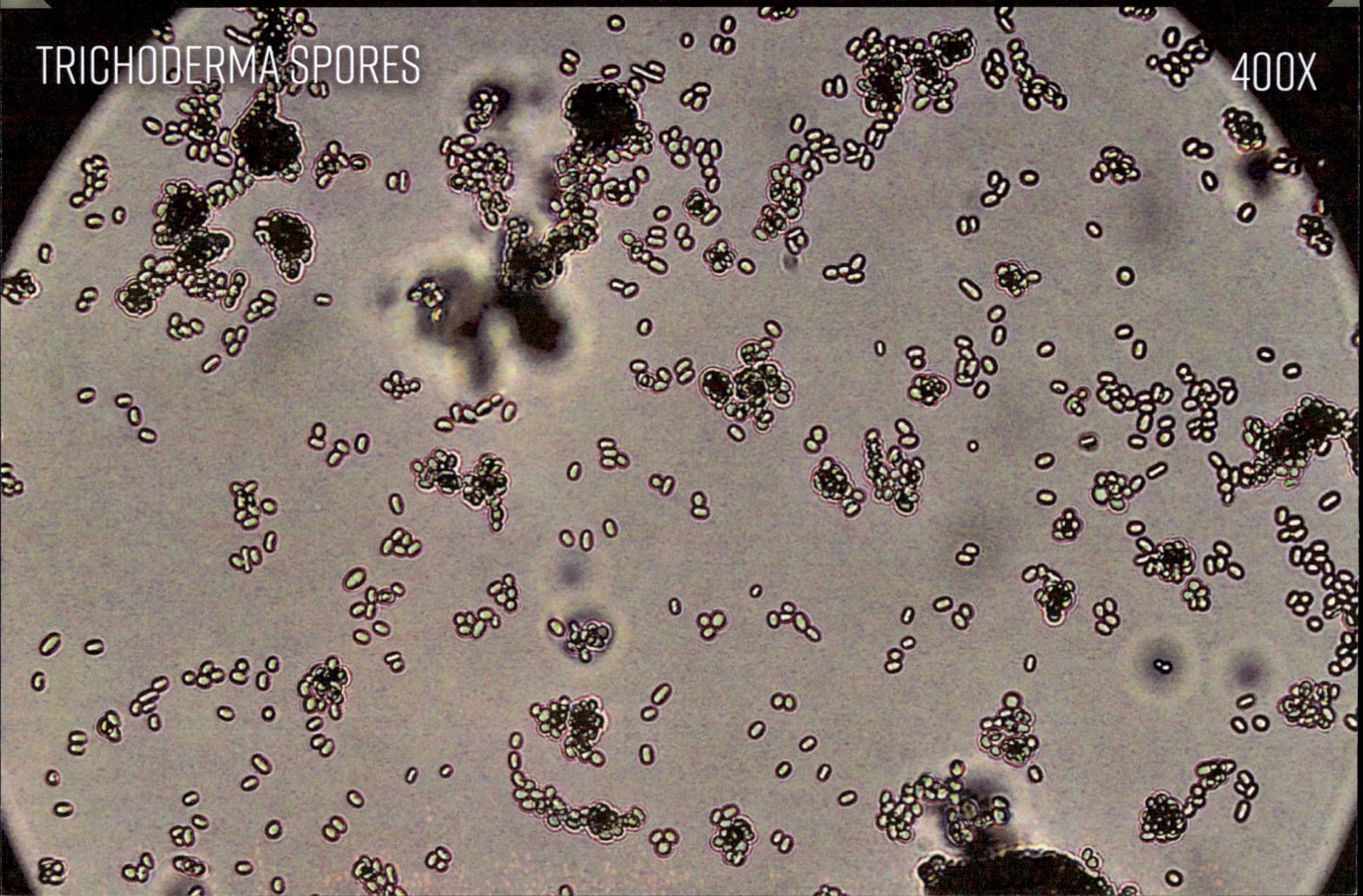

TRICHODERMA SPORES 400X

BEAUVARIA BASSIANA

A useful entomopathogenic white mold fungi that digests insect chitin

FAMILY: Cordycipitaceae

ORDER: Hypocreales

PHLYUM: Ascomycota

FORM: Filamentous fungi, hyphae + spores

SIZE: Typically 3 μm diameter hyphae but can have swellings up to 17 – 15 μm, over 250 – 400 μm long

APPEARANCE: Mature hyphae are anywhere from white to reddish-brown to tannish-brown, spores are ~1 μm in diameter. To the naked eye, the white mold is only visible when it overtakes an insect host, focusing on the joints and then turning the entire insect into a mass of white hyphae. Airborne spores look like typical airborne mold spores.

ABUNDANCE: Widely used as a way to control or eradicate insect pests, this fungi is found throughout the world's soils – it is part of the nutrient and decomposition cycles for all insects.

FOODS: Chitin

PH & TEMP PREFERENCES: Ideally 5 – 10 pH but can tolerate 2 – 12 pH, 25 – 27°C ideal temperature

REPRODUCTION: Asexual (but has *Cordyceps bassiana* as a teleomorph which does reproduce sexually).

USES: Biological Control of Insects & Decomposition

600X

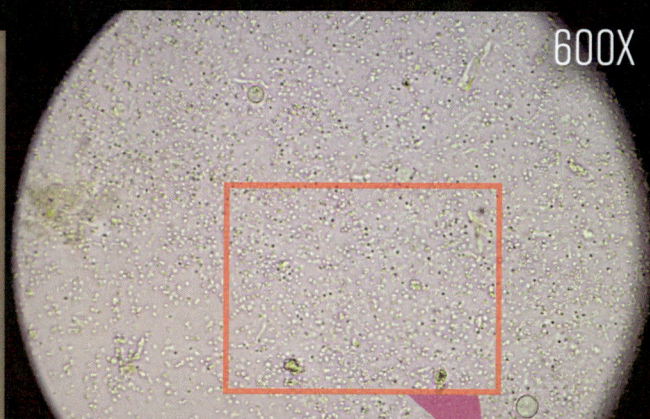

600X

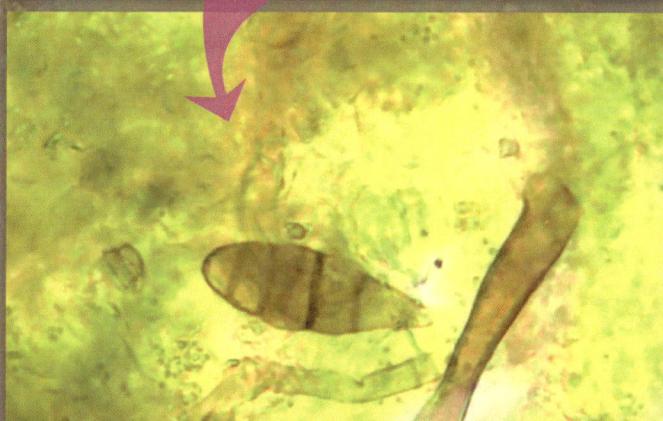

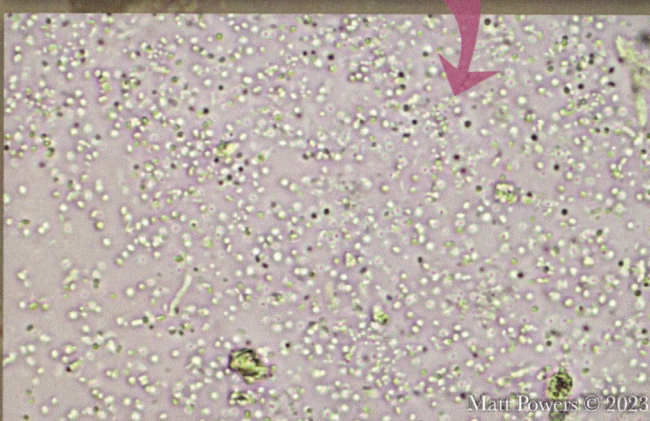

Matt Powers © 2023

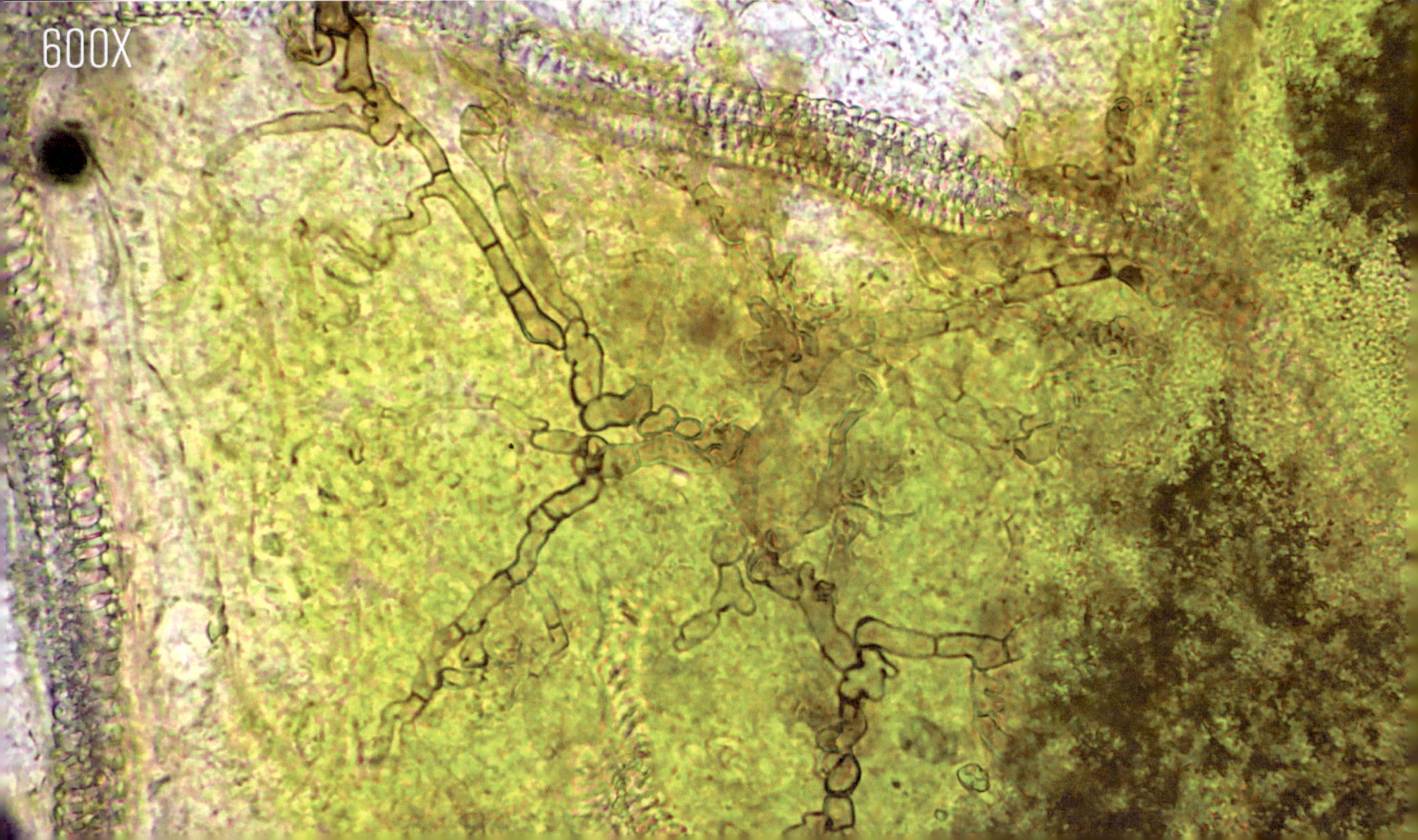

600X

CAN YOU SEE THE FUNGI
WRAPPING AROUND
THE INSECT HAIR?

MATT POWERS © 2022

600X

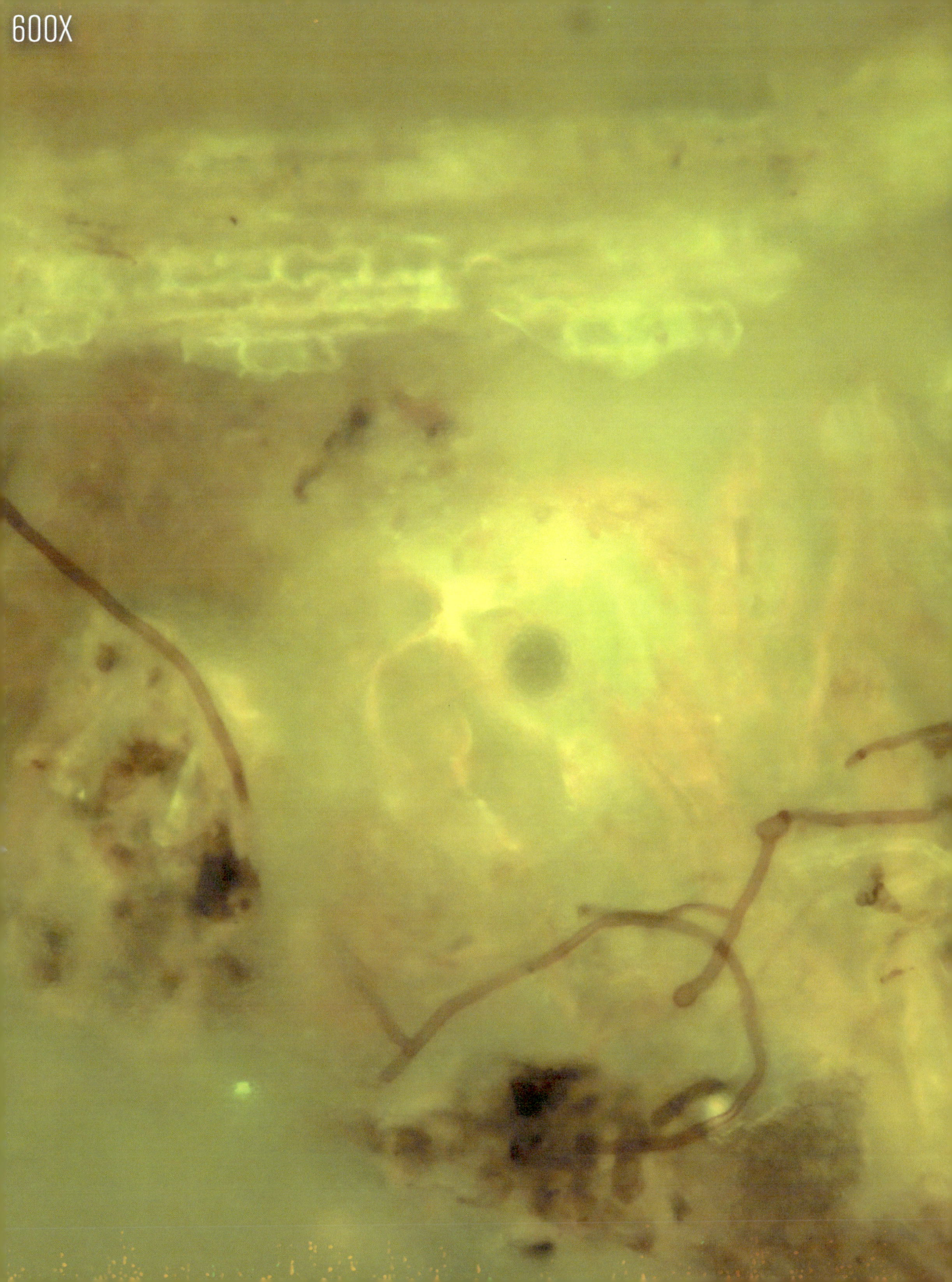

killing them, and then spreading out from the dead worm.

MORPHOLOGICAL GUIDES:

- *Illustrated Genera of Imperfect Fungi.* 4th Ed. Barnett, H.L. & Hunter, B.B. 1998.

- *The Identification of Fungi: An Illustrated Introduction with Keys, Glossary, and Guide to Literature.* Dugan, F.M. 2015.

ONLINE RESOURCES:

- **INVAM - The International Collection of (Vesicular) Arbuscular Mycorrhizal Fungi,** University of Kansas. This is the BEST online identification database - it's very thorough with pictures and measurements. https://invam.ku.edu

- **IBG – the International Bank for the Glomeromycota.** http://i-beg.eu
- **Handbook of Arbuscular Mycorrhizal Fungi.** Souza, T. (2015). doi:10.1007/978-3-319-24850-9 https://sci-hub.se/10.1007/978-3-319-24850-9 (Contains AMF Identification Key.)

- **E-FLORA BC: ELECTRONIC ATLAS OF THE FLORA OF BRITISH COLUMBIA** https://ibis.geog.ubc.ca/biodiversity/eflora/
- **Funginomi** https://funginomi.com/
- **IDphy: molecular and morphological identification of Phytophthera based on the types** https://idtools.org/id/phytophthora/factsheet.php?name=7956
- **CABI Digital Library** https://www.cabidigitallibrary.org
- **EOL.org**

The following profiles are to give you references but also to show you the amount of overlap between these samples and prior shown samples.

The Bad Guys

ANTAGONISTIC & PATHOGENIC BACTERIA & FUNGI

There's a long list of microbes that are concerning, but most you'll never see or encounter. Some are at such small numbers that they are harmless. Rhizoctonia's order, Cantharellales, Phythophthera's order, Peronosporales, and Rust's order, Pucciniales, all were completely absent from all compost and EM samples I examined. The only concerning basidiomycota species found in all my compost samples was Malassezia restricta which is linked to human dandruff but that was seen at a concentration of 0.001% (12 reads out of 706,720 total). The only viruses I've detected were ones that affect e.coli and facilitate HGT.

BUT, I did detect some things that are concerning:

- **Salmonella** is on average ~0.3% in composts tested and much lower in EM
- **Shigella** is on average ~0.09% in composts tested and not present in EM
- **Burkholderia cepacia** complex was at 0.06% concentration – *it can cause a fatal infection.*

If you have concerns about pathogens, you can find some answers fast with the microscope BUT there are always pathogens that look identical to harmless or even beneficial fungi and bacteria, so DNA testing is how we double check our work and how we make more definitive identifications possible – that's why this book is part II in a trilogy.

SALMONELLA
Gram-Negative Facultative Anaerobic Bacteria

FAMILY: Enterobacteriaceae

ORDER: Enterobacterales

PHLYUM: Pseudomonadota

FORM: Bacilli, non-spore-forming

SIZE: 0.7–1.5 μm diameter, 2–5 μm long

APPEARANCE: rods with peritrichous flagella, usually motile

ABUNDANCE: found primarily in intestines of animals and in animal manures but can also be found in composts, specifically *Salmonella enterica* at ~0.3% making it the 8th most abundant bacteria in exemplary thermophilic compost.

FOODS: Organic Matter (Chemotroph)

PH & TEMP PREFERENCES: 7.4 pH, 37°C

REPRODUCTION: Asexual Reproduction (Binary Fission)

DANGERS: Virulent & prevalent at low levels in compost due to manure usage

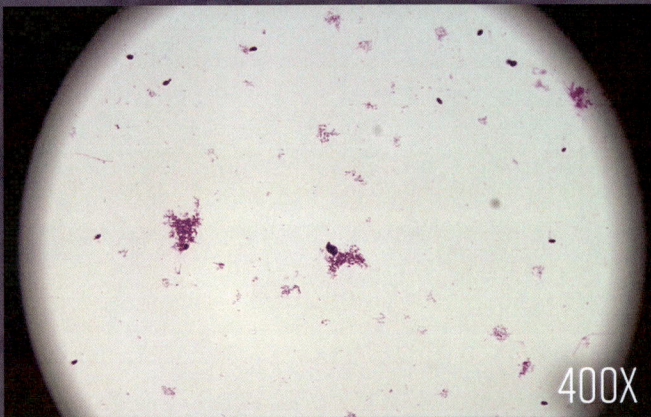

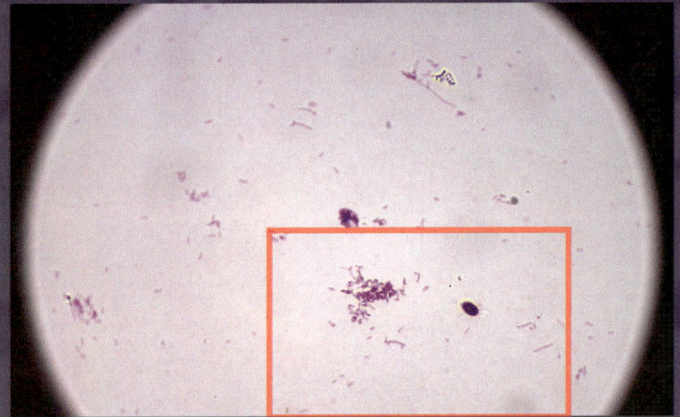

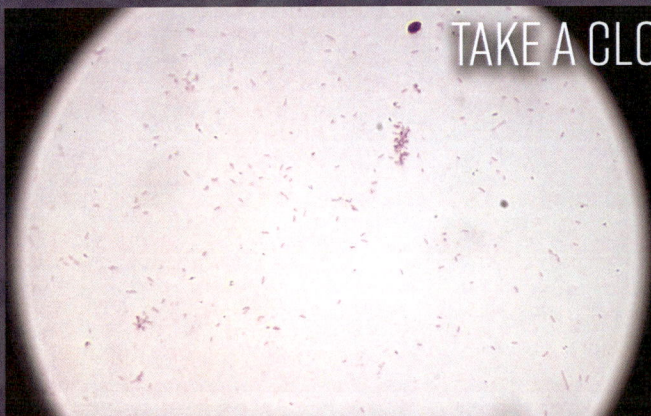

400X

TAKE A CLOSER LOOK

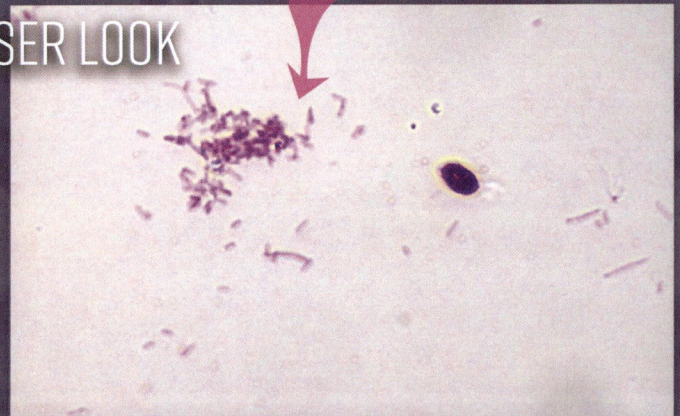

ASPERGILLUS
Pathogenic & Saprophytic Fungi

FAMILY: Trichocomaceae

ORDER: Eurotiales

PHLYUM: Ascomycota

FORM: filamentous, spore-forming (though some teleomorphs exist)

SIZE: 2–4.5 μm diameter with highly variable lengths of hyphae with globose and subglobose spores 2–3.5 μm, conidia 2.5–3 μm with conidiophores 120–800 μm in length

APPEARANCE: wildly diverse hyphae in length, width, & expression with long and thick conidiophore (stalks) with large vesicles (stalk heads), metula & phialides (extensions off the stalk heads), and conidia (spores) on top of those.

ABUNDANCE: it is the most common form of contamination of farm products in the world (in the field & in storage)

FOODS: they are known to grow in nutrient depleted environments - as long as there's moisture, but they prefer sugars of all kinds and carbohydrates. They are found in association with most commonly with starchy foods,

PH & TEMP PREFERENCES: 7 pH but a few prefer pH 8 or even 9 (like *A. Niger*), 21–28°C for most species but 37°C allows for differentiation of *Aspergillus* under the microscope and 50°C activates *A. fumigatus*.

REPRODUCTION: Asexual Spores

DANGERS: Avoid Inhalation – causes severe disease and in some cases produces aflatoxins causing all sorts of health problems. *A. fumigatus* grows well at 50°C, so composting can promote them – DON'T BREATH IN COMPOST FUMES!!

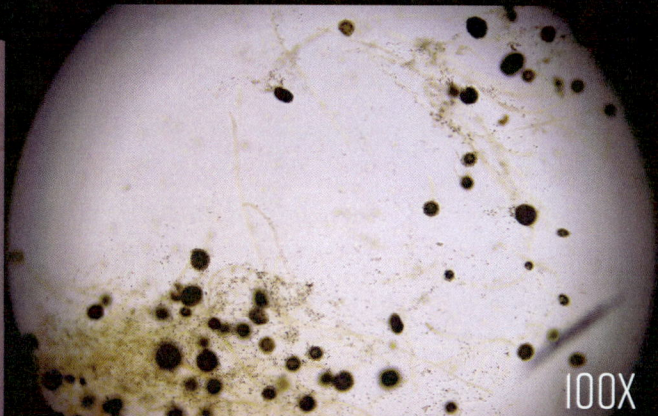

100X

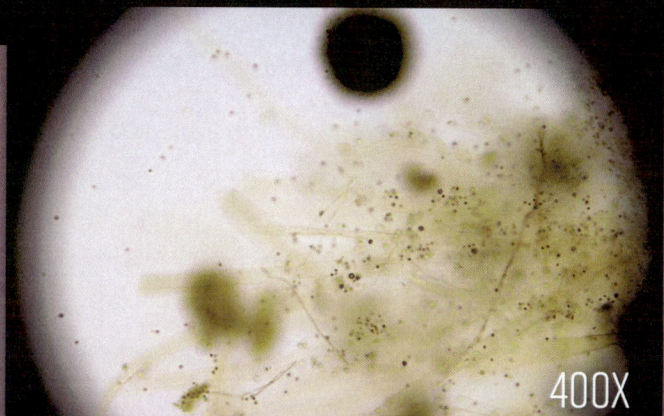

400X

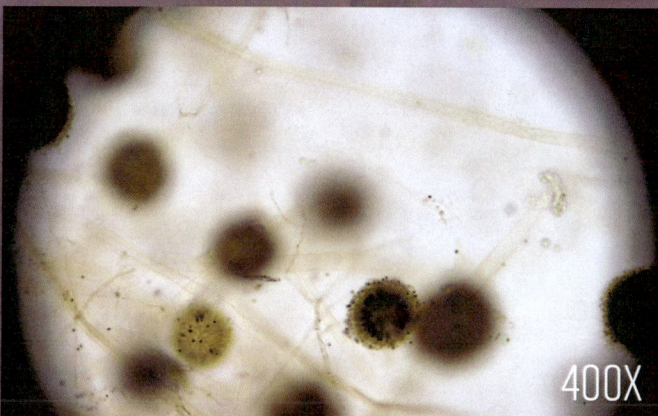

400X

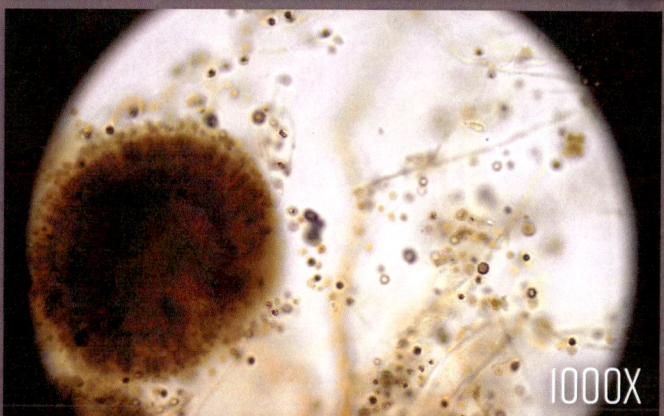

1000X

Protozoa

Protozoa are unicellular organisms found in soils as well as aquatic environments. They are part of the Protista kingdom, so you may hear folks refer to them as Protists though that would include protozoa in with algae and multicellular organisms. For our purposes, we need to keep all those separate, so we focus on protozoa specifically.

Protozoa are the first layer of trophic nutrient cycling – while nutrients are cycling among and between soil bacteria and fungi, the nutrients are not released to plants at the same rate and levels (symbiotic exchanges are the focus instead at that first stage). If you are seeing protozoa, the nutrients trapped in your bacteria and fungi are being released to the plant roots in a plant bioavailable form (they have digestive systems and excrete their waste through their cell walls). Bacteria and fungi can trigger plant roots to do a variety of things like nodulate, but their main focus is not to give up their nutrient banks unless they are part of the rhizophagy cycle, and in that case, the roots directly digest nutrients leaking from the microbes or they consume the entire microbe. Keeping track of each of these nutrient pathways is vital for evaluating the nutrient cycles in our soil or compost and how they relate to our growing plants.

AMOEBAE

Amoebae are single-celled organisms found in soil, freshwater, and marine habitats. They range in expression dramatically but each are in essence an oozing, transparent blob though some have tests (shells) they shelter in like hermit crabs, but their shells are made from minerals in the soil. Amoebae typically exhibit a pseudopod, like an arm or foot they form, which they use to explore the world around them. Since they are transparent, we can see what they are feeding on quite easily. Most of what the testate amoebae I have viewed are consuming large amounts of fungal spores. Many of these are mold spores – this would mean they are acting in their roles as biocontrol. Amoebae can form cysts to protect themselves when the environment is no longer favorable – they form a thick wall sometimes with multiple layers when they do, but they

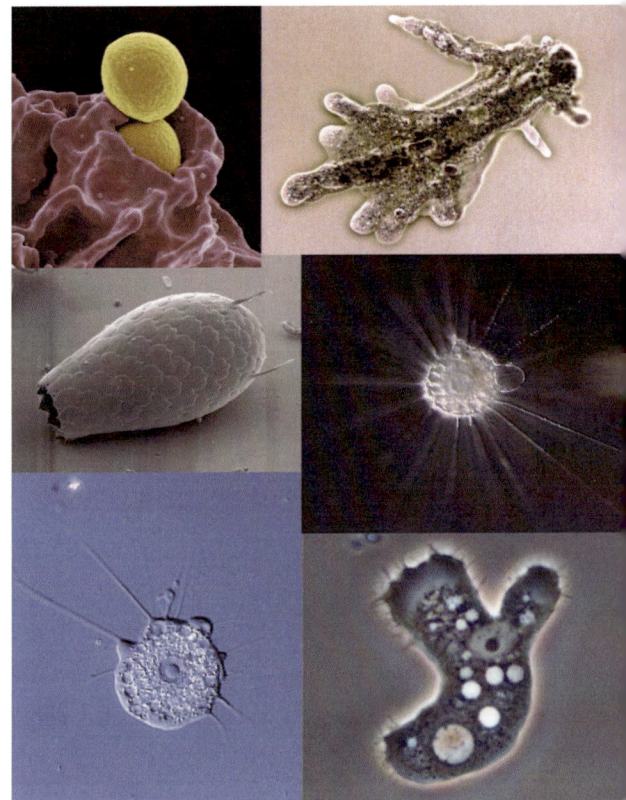

Respectively: NIAID, Cymothoa exigua, ja:User:NEON / User:NEON_ja, Jacob Lorenzo-Morales, Naveed A. Khan and Julia Walochnik, ja:User:NEON / User:NEON_ja, ja:User:NEON / User:NEON_ja, CC BY-SA 4.0 <https://creativecommons.org/licenses/by-sa/4.0>, via Wikimedia Commons

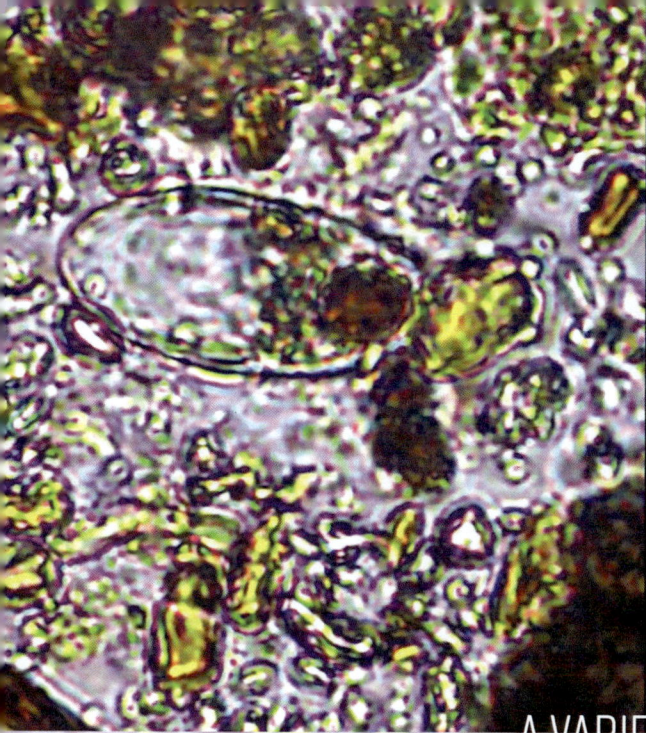

A VARIETY OF TESTATE AMOEBAE

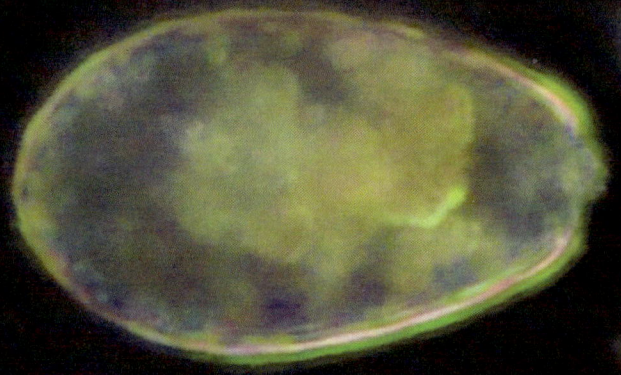

MOST LIKELY EUGLYPHA

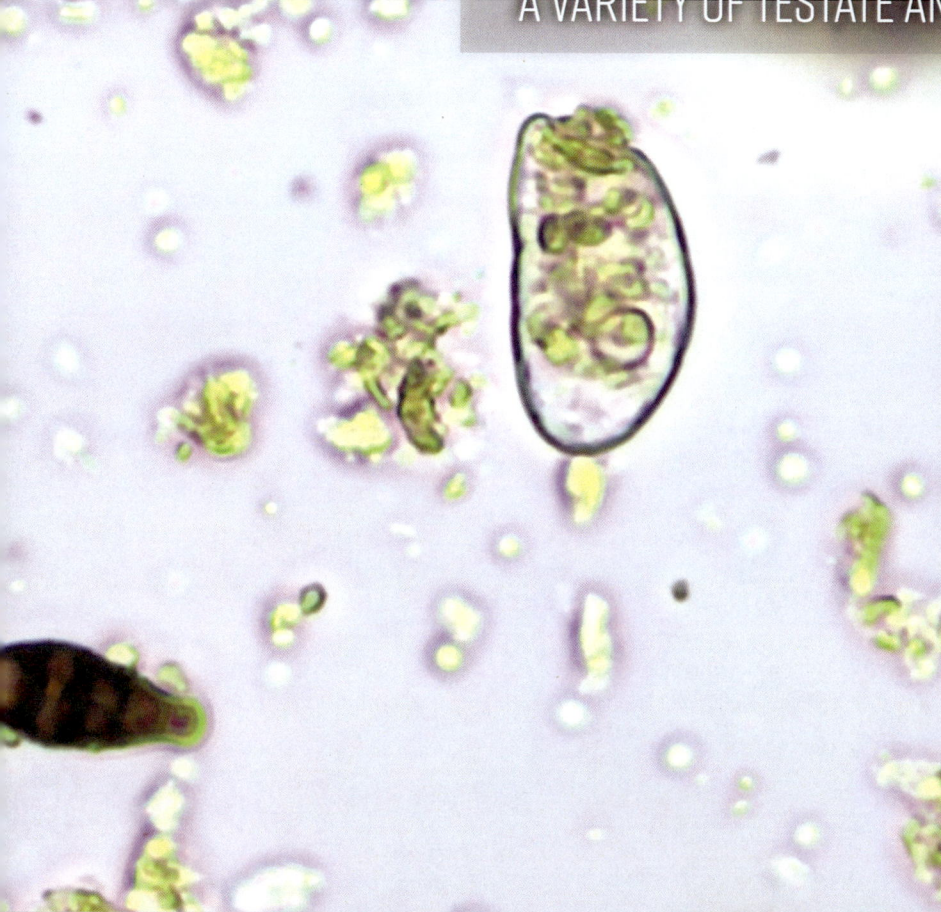

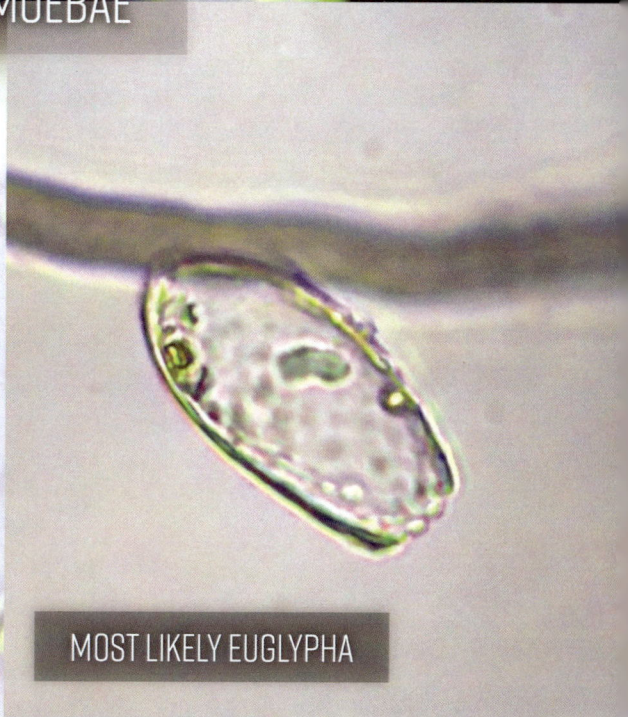

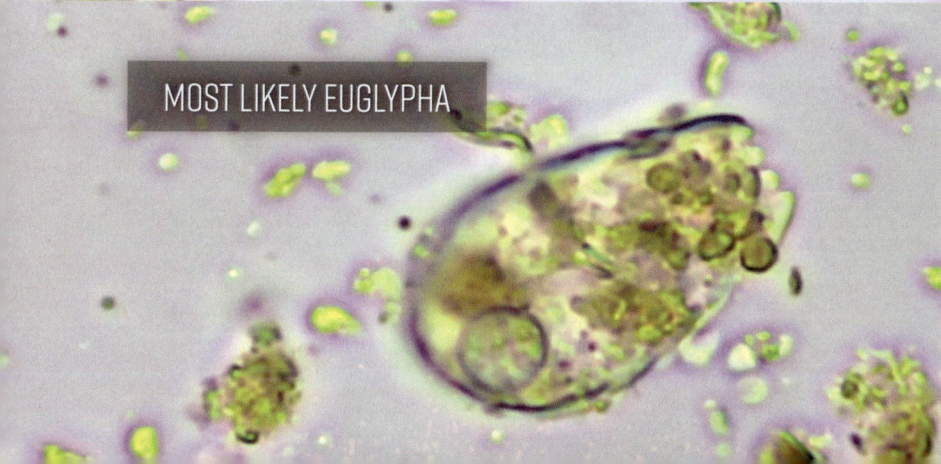

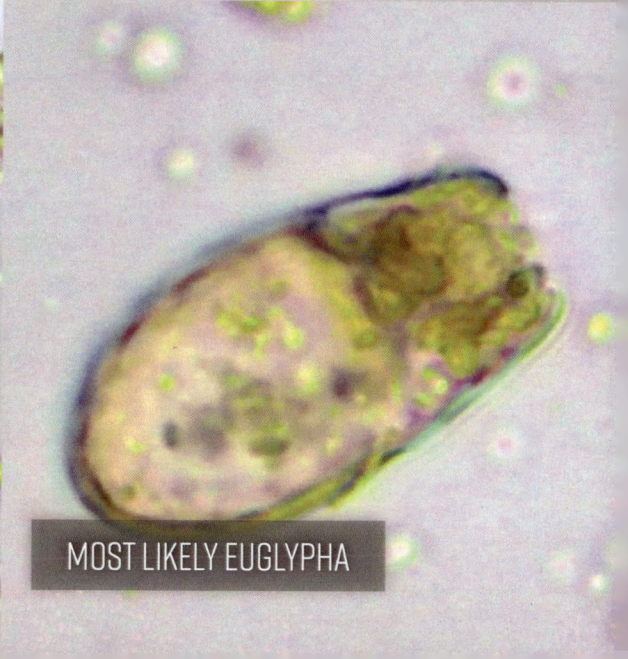

MOST LIKELY EUGLYPHA

MOST LIKELY EUGLYPHA

MOST LIKELY EUGLYPHA

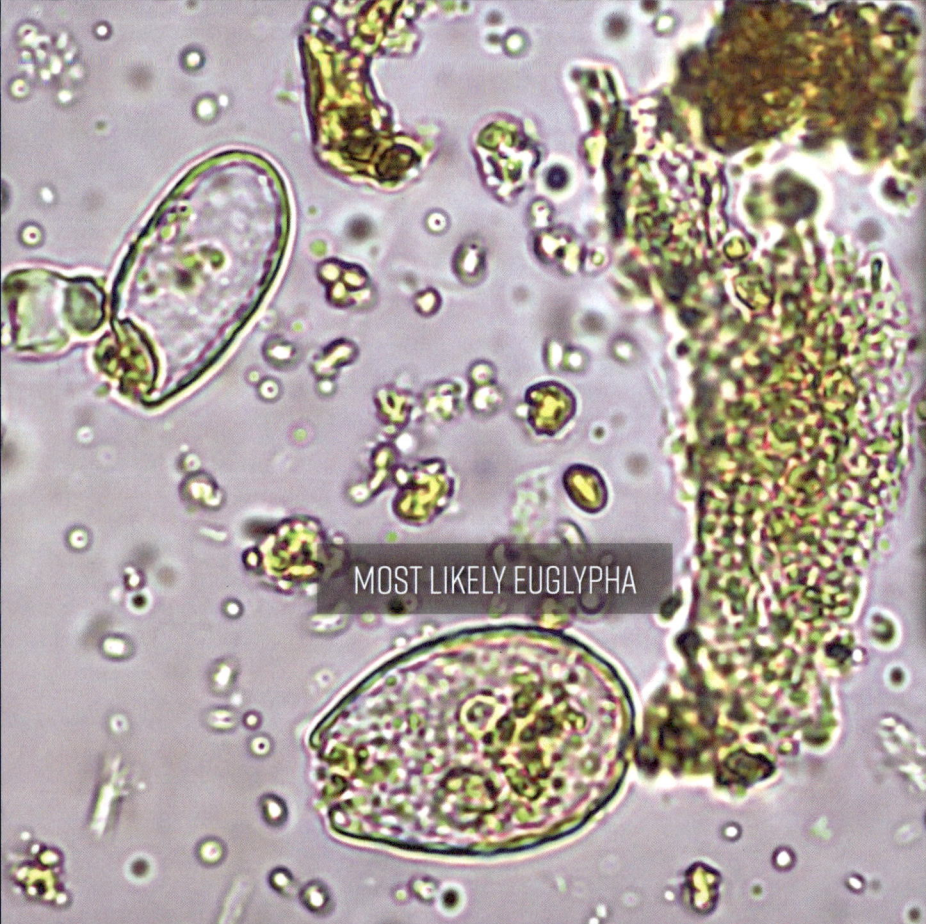

MOST LIKELY EUGLYPHA

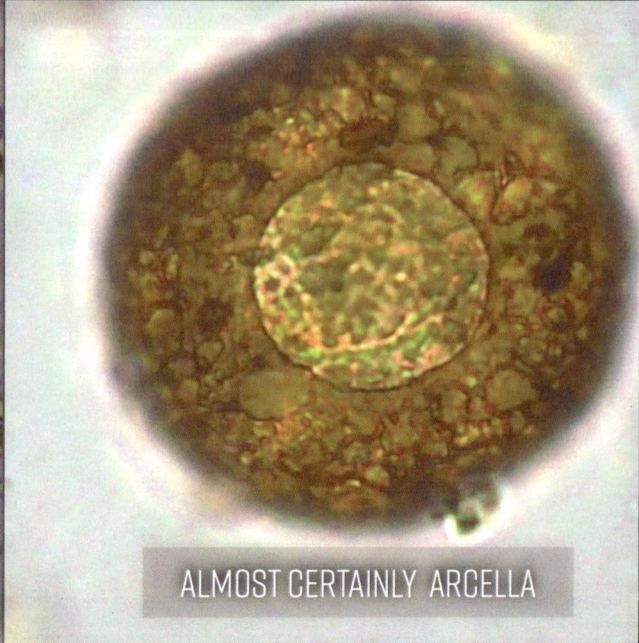

ALMOST CERTAINLY ARCELLA

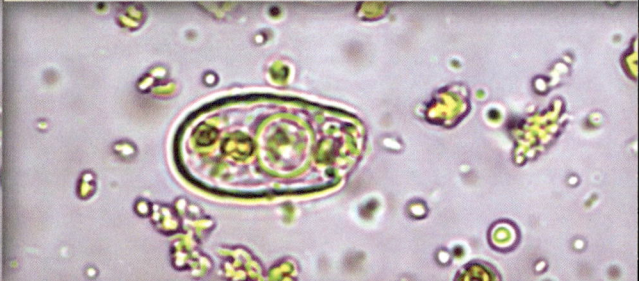

A VARIETY OF TESTATE AMOEBAE

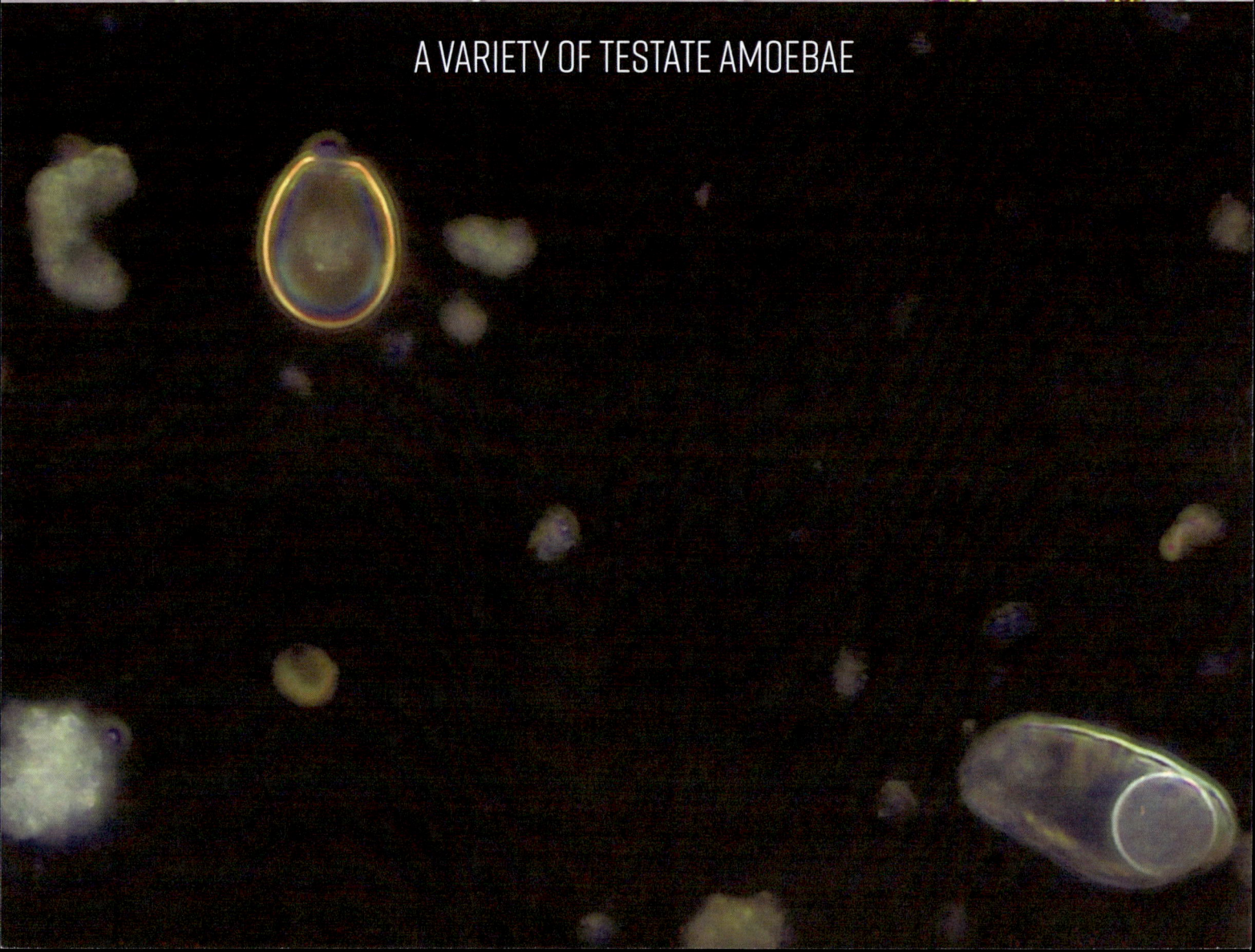

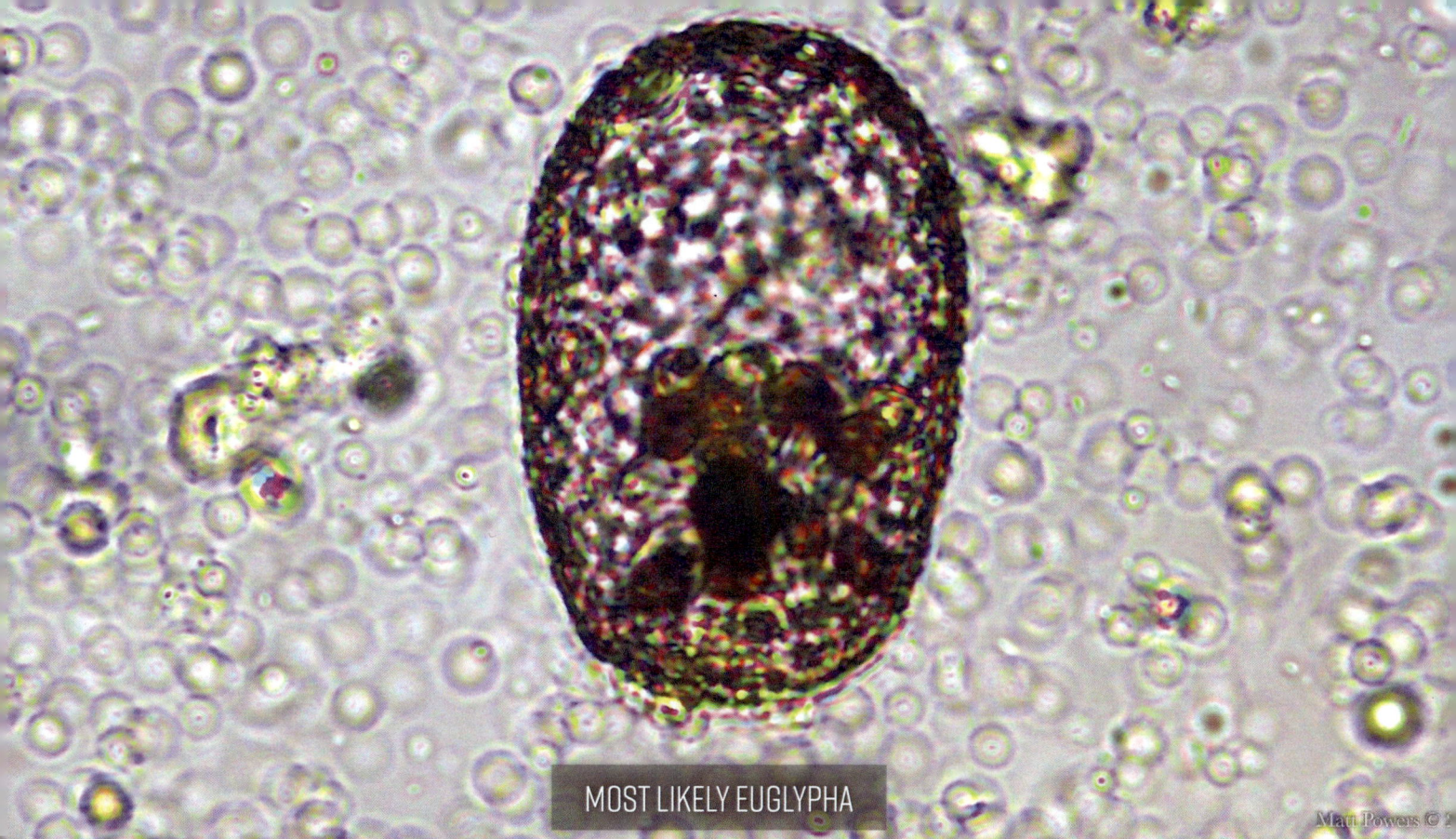

MOST LIKELY EUGLYPHA

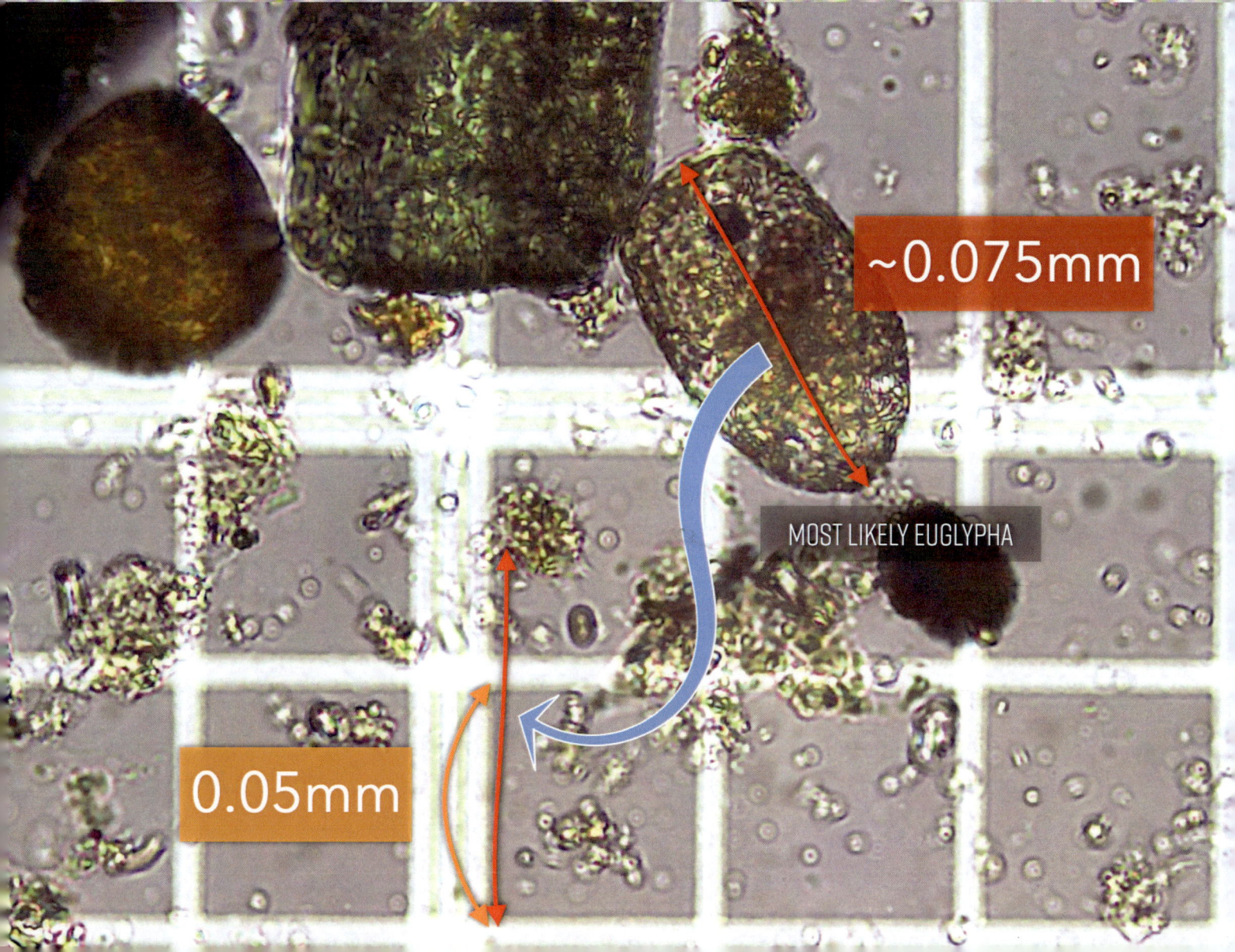

~0.075mm

MOST LIKELY EUGLYPHA

0.05mm

maintain much of their appearance internally. If you have cysts of microbes, that means that the soil environment has an issue that needs to be addressed: something is out of balance.

If you've never heard of brain-eating amoeba, I'm here to tell they exist. They are free-living, and this is yet another reason why I don't recommend tasting, licking, smelling, or inhaling fumes from compost or soil. I've met too many people who've gotten sick from exotic soil microbes at this point to ignore this potentiality – primarily hands-on workers with soil, not soil scientists, and on top of personal experiences with folks like that, compost and garden soils are making headlines with their pathogens at least in the UK currently being linked to illness. It's important to note that what benefits the soil and some soil processes may not be ideal for us to ingest, breath, or eat, and that's okay. It is just the way it is – despite a long held desire in the community to ingest, bath in, or taste compost products, it is not advisable.

- **Naked Amoebae** – these amorphous, ghostlike organisms are very hard to see, so you have to look carefully. They are more motile than testate amoebae – they ooze and pull themselves through soil. They are harder to find in thermophilic composts than static composts especially when made in conjunction with a mother compost – over time testate amoebae are selected for both because of the high moisture content and to deal with the regular pH/Eh swings. Having naked amoebae implies that your nutrients are cycling and spreading at a higher rate than testate amoebae would be able to at the same concentrations. Since their shape is constantly changing, sizing them is a challenge, but they can be anywhere from 10 – 50 μm in size, but they can stretch much further than that as well.
- **Testate Amoebae** – the shell is a test with an opening on one side (the top or side) – it's a portable shelter. They are found ubiquitously in hot compost, including Johnson-Su composts. They are found in nature in bogs, marshes, and wetlands as well as in marine environments – these environments help maintain their tests. They tend to attach to organic matter or fungal hyphae and do not travel through the soil medium as naked amoebae do. Testate amoeba can become quite large even over 100 μm in size.
- **Other Amoebae** – Vampyrellid amoebae look like a cross between an amoeba and a ciliate. They have an array of spiny pseudopodia. They are excellent cyclers of nutrients and prey on bacteria, fungi, and other protozoa.

Common Soil Amoebae:

Acanthamoeba (dangerous - naked), Hartmannella (concerning - naked), Vahlkampfia (possible biocontrol - naked), Naegleria (dangerous - naked), Chaos (great cycler - naked), Euglypha (beneficial soil cycler - testate), Centropyxis (beneficial soil cycler - testate),

Nebela (beneficial soil cycler - testate), Difflugia (an extremely common testate in composting, vigorous cycler - testate), and Arcella (rhizopheric amoebae that are key to the soil food web cycles - testate).

FLAGELLATES

These are unicellular organisms that look like a cross between amoebae and ciliates. They are extremely fast moving though awkwardly bumbling through their environments motored by 1-2 flagella, whip-like appendages located on opposite sides or both on one side of the flagellate body. These can be very hard to get in focus as they move so quickly. They can bunch up and look like a balloon on a string or stretch out and look like a ciliate with a tail

A Flagellate, Protozoa

and lacking their cilia (hairs) though it should be noted that we can differentiate them by size: flagellates are typically smaller than ciliates. Their cysts are also smaller than ciliate and amoeba cysts though they can be smaller or larger than amoebae. They range in size anywhere from 5 – 100 µm in length.

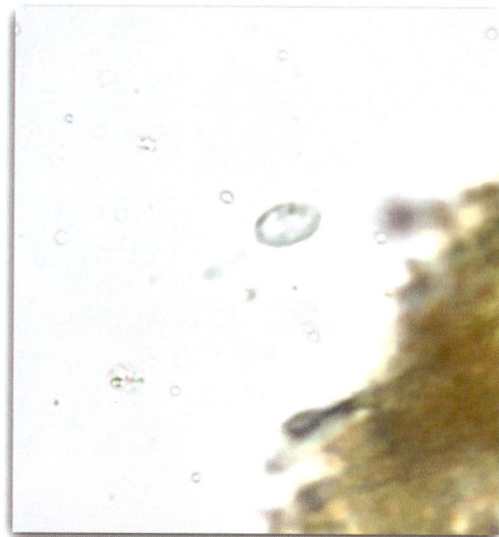

Are flagellates safe? Giardia and a host of other illnesses are associated with flagellates, so we don't want too many of these but a few zipping across our FOV every now and again are fine. They can move so fast that we can be unable to ID them other than by their speed and behavior: we know they're flagellates. If your FOV is full of them, that is a potentially dangerous sample. I've never seen more than an occasional meteoric flagellate passing by. They are found ubiquitously in soils and are vital to the soil nutrient cycle, but they can also be harmful, so wash your hands!

CILIATES

These mercurial unicellular organisms can be beautiful yet alien in their appearance – silvery, zooming, salmon-like, and covered in tiny hairs: their cilia. Their patterns and range of expression are diverse – they can be small with tufts of cilia or they can be large with refined expressions of cilia, but they all feed on bacteria (and some even feed on other microorganisms). They can reproduce asexually and sexually. They really like to come out to play when things are water

Small Ciliate vs Nematode vs Large Ciliate
All from the same sample & at 400x magnification

logged or in reduced oxygen conditions for the most part, but increasingly there is evidence that a good portion of ciliates are facultative anaerobes (they can tolerate higher oxygen levels) and some studies are showing that they may even be part of the transition from anaerobic to aerobic NOT the other way around (their zooming around actually aerates the soil!). I've seen some waterlogged samples exhibit zero ciliates even after days of being waterlogged, so they may not be hiding in your soil waiting for the waterlogged conditions either. Letting samples sit waterlogged can show us what can happen in our fields when they are left to sit waterlogged for too many days as well. The water logging can reveal encysted ciliates. Ciliate cysts are clear in the middle and have thin walls. They are larger than amoeba cysts and flagellate cysts. Ciliates can range in size from just a few microns to over a millimeter in length. Their behavior identifies them most clearly, so they are not hard to identify as a group. They are vital to the soil food web, but if we see them in high numbers that means something is wrong if we are trying to grow plants in this soil or amend farm or garden soils with that compost. Seeing 3 – 5 ciliates in a drop of 1:10 freshly diluted soil/compost:water is likely fine but reveals a vulnerability in the nutrient cycling system in the event of water logging UNLESS there's numerous types of other protozoa, nematodes, fungi, and bacteria at high densities. Ciliate numbers have often been correlated to higher soil organic matter levels, so their presence is an indicator of health in general, and having higher levels of cycling can only happen if we have diversity and abundance of all key microbes, but finding the sweet spot of coherence for your soil or compost is the key. It all depends on what's going in and what the purpose is for that compost or soil.

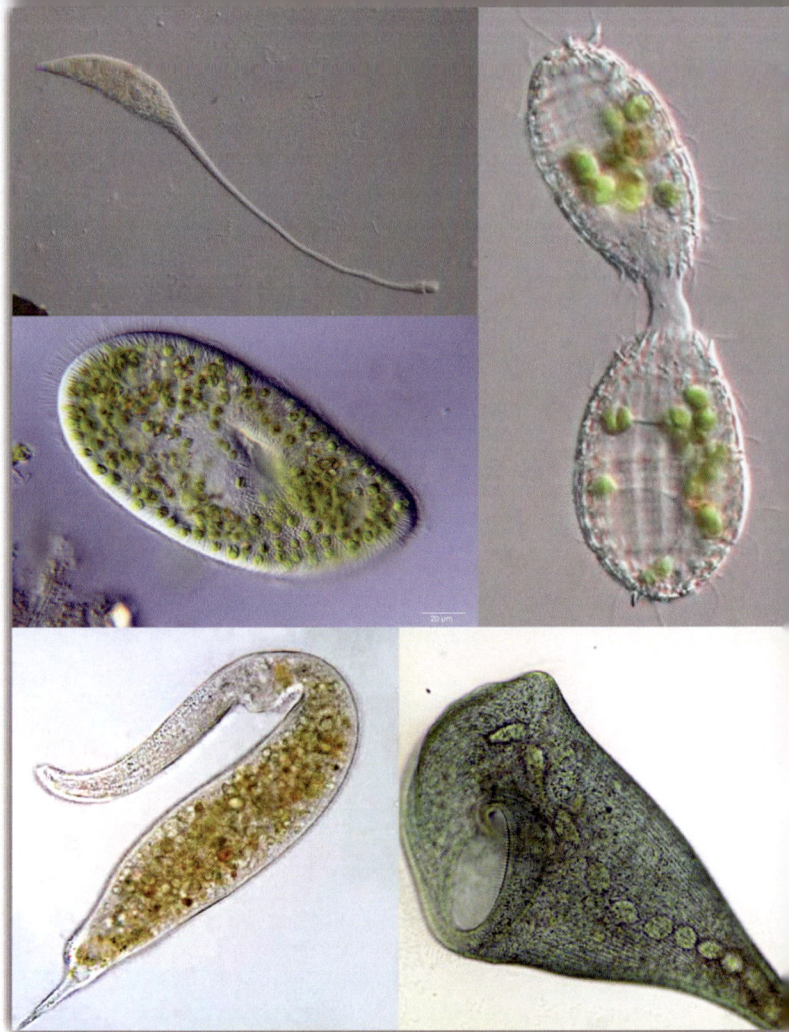

Respectively: Picturepest, Anatoly Mikhaltsov, Bernd Laber, Deuterostome, Flupke59, CC BY-SA 4.0 <https://creativecommons.org/licenses/by-sa/4.0>, via Wikimedia Commons

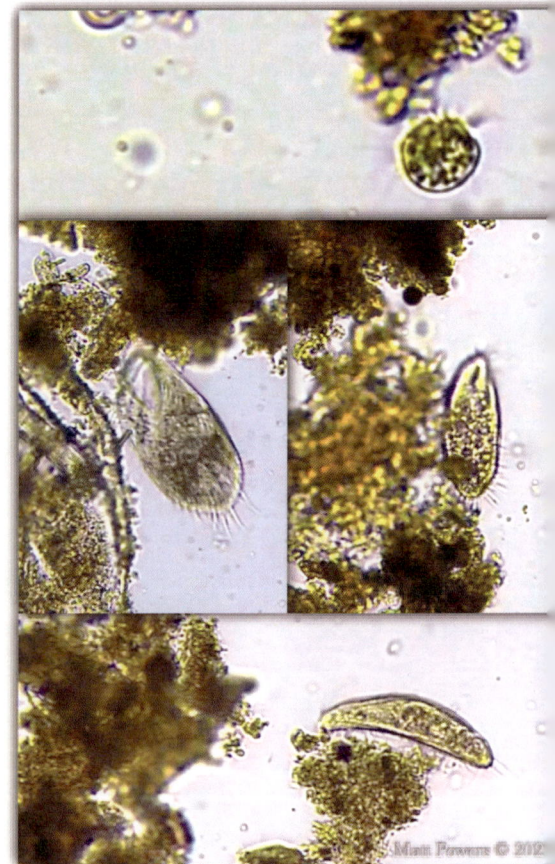

How do you know which ciliates you have?

Another ciliate key – it's all morphological keys unless you shift to DNA testing for this part, and THAT implies that you can catch ciliates for your DNA sampling. It's been done, but it's wild: they give them antibiotics to kill the bacteria inside them then they flush them repeatedly in drops of distilled water until it's really really clear and clean through and through, then they DNA sequence it.

BUT, do we need to know them individually?

At this point, we don't because we shouldn't ever see them in high numbers in healthy soils that are not waterlogged to the point of being anaerobic. There are stalked ciliates, and they too have the tell-tale signs of being a ciliate: cilia, and they are focused on feeding on bacteria. Their filtering and feeding habits look just like ocean filter feeders and shellfish.

Common Soil Ciliates: Colpoda, Tetrahymena, Paramecium, Spirostomum, Cyclidium, Blepharisma

Algae

If we are working in aquatic, wetland, or aquaculture systems, we will encounter algae. It is noticeably green and it is usually larger than fungal hyphae (but ectomycorrhizal forest fungi can be very long and thick. Algae can be found in soils at the surface layer or in areas where that surface layer was recently turned in to the soil.

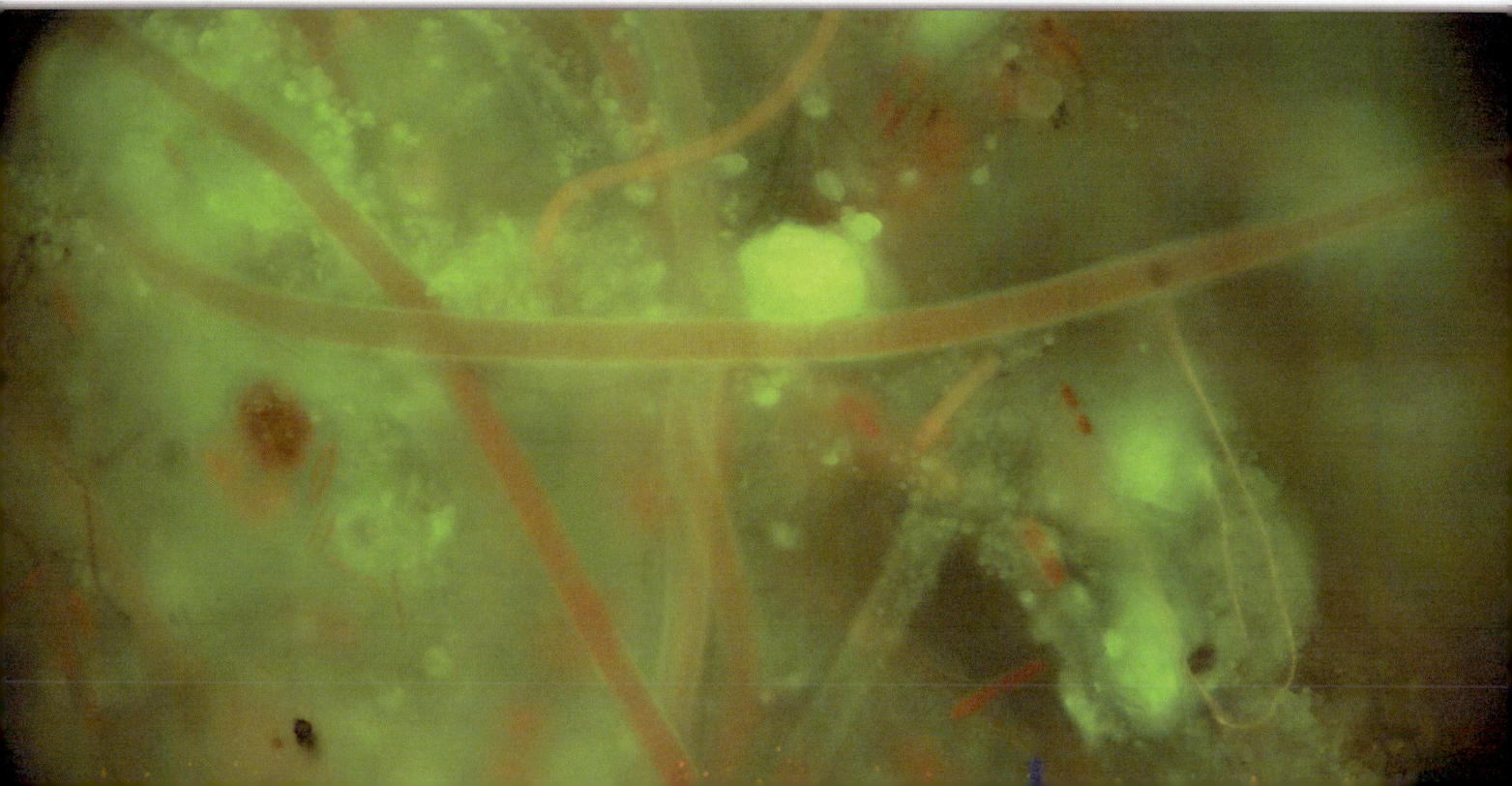

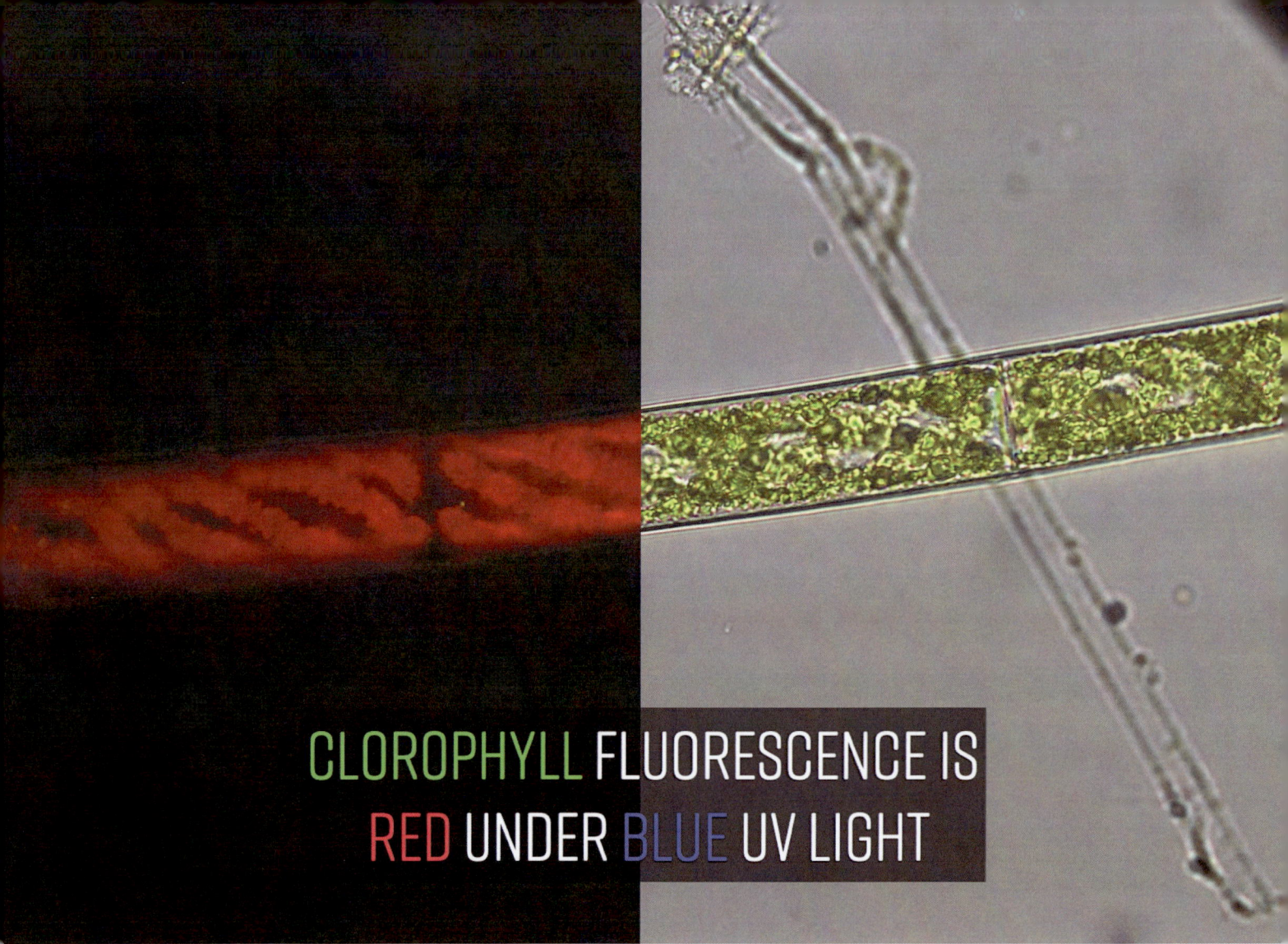

CLOROPHYLL FLUORESCENCE IS RED UNDER BLUE UV LIGHT

Algae and cyanobacteria look incredibly beautiful under magnification, but algae and other aquatic species can indicate water-logging and anaerobic conditions. Often in chains of elaborate design, they can easily dominate the entire field of view in a tangled mass. Algae and all photosynthetic cells and biology turn RED in the epifluorescence nm range that we are utilizing in this book, so we can quickly flash the samples we are viewing with epifluorescence lighting and see the fungi, the algae, and the phosphorus bearing minerals instantly.

Unless you are looking at pond water or composting in an area with a high water table, you likely are not going to encounter algae or the online sensations: tardigrades, or as they are more fondly known Water Bears. They are not signs of good soil or compost, so we should not be seeing them in our piles or soils.

Pollen & Spores

We can find spores throughout compost and in the topsoils everywhere in the world. Pollen is more likely to be at the surface level, but there are so many microbes making spores below the surface that we should always except to see spores in healthy soil and compost samples. Pollen tends to be 3–30x larger than our average fungal hyphae is thick and can often be symmetrical and/or bizarre looking. Bacteria can be on the surface of these specimens. Spores can greatly range in size from below 0.5 µm to nearly 500 µm. There are spores just beyond the limit of resolution for light microscopy, but in general we are talking about pollen, spores, and cysts all looking alike and covering a wide range of expression, size, and overlap between them. This is why water logging samples can help us imagine what happens in times of crisis in our

A Basidiospore – notice the purple green halo: that is an artifact of the lens and not really there.

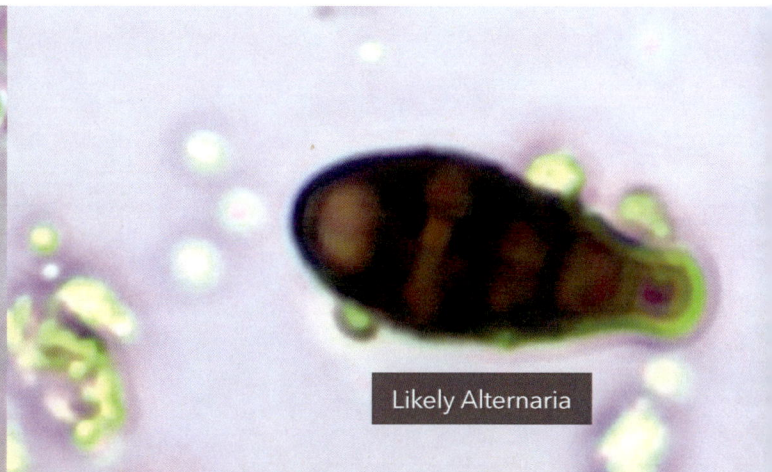

Likely Alternaria

Likely Alternaria

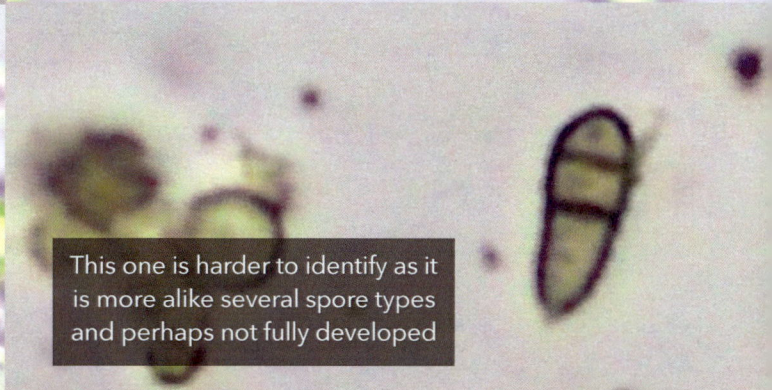

This one is harder to identify as it is more alike several spore types and perhaps not fully developed

A Conidiospore

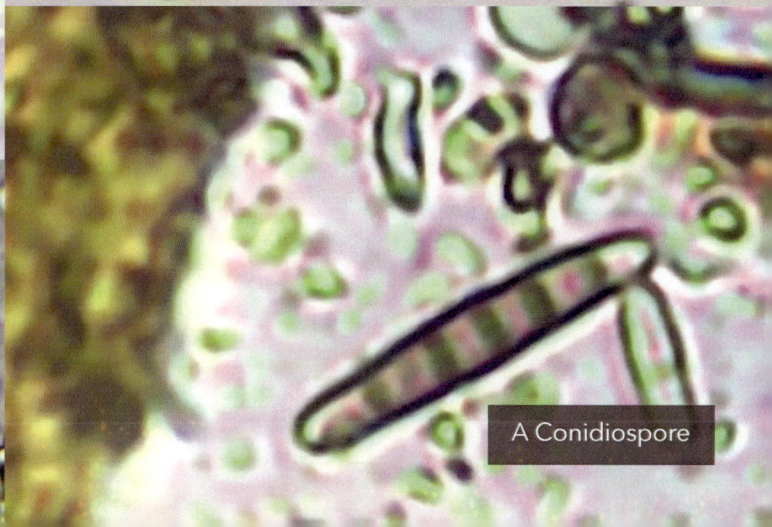

A Conidiospore

Matt Powers © 2023

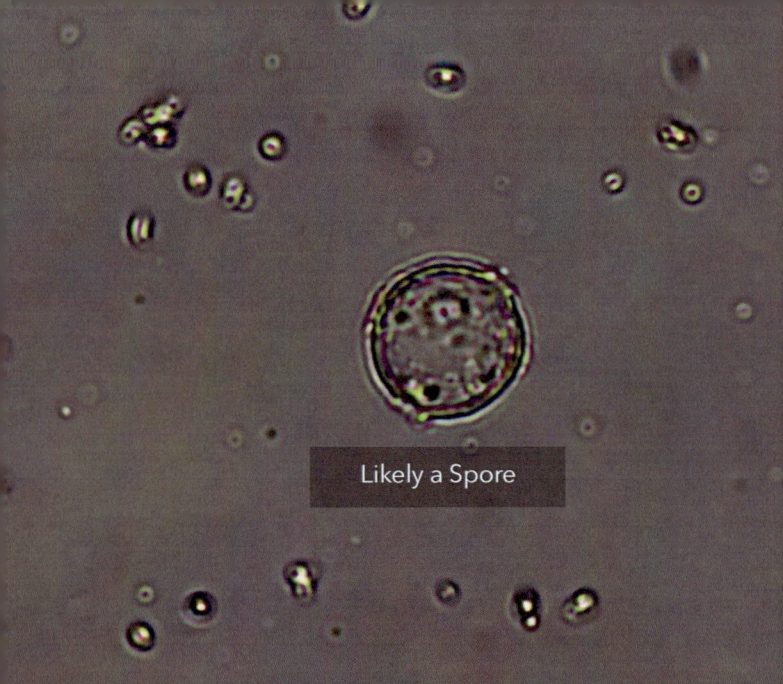

Likely a Spore

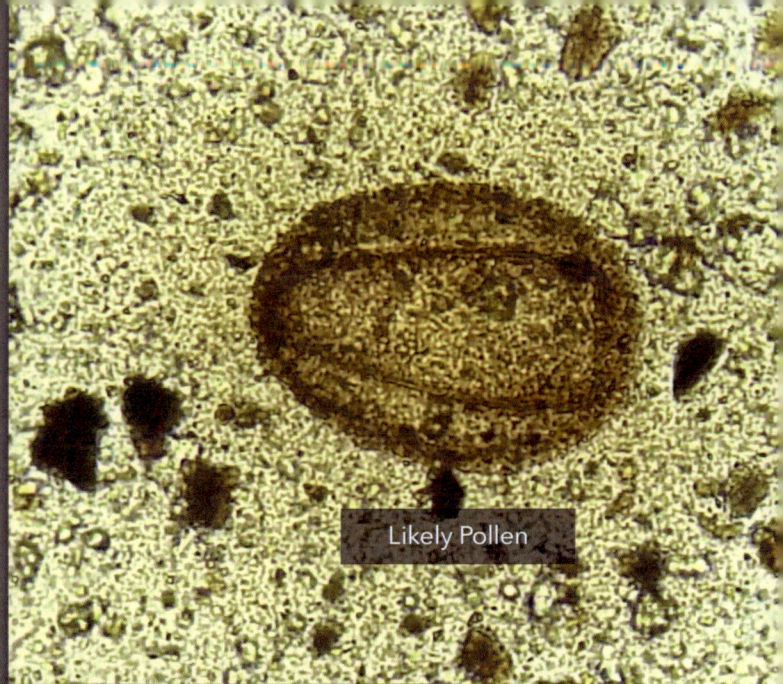

Likely Pollen

The clearer the image, the more certain we can be.

Either a spore or an encysted microbe

Matt Powers © 2023

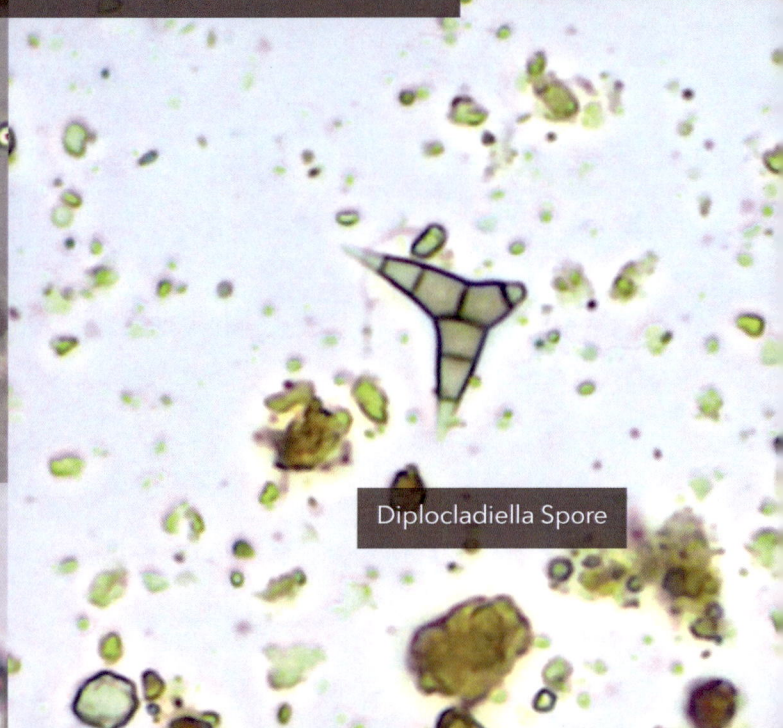

Diplocladiella Spore

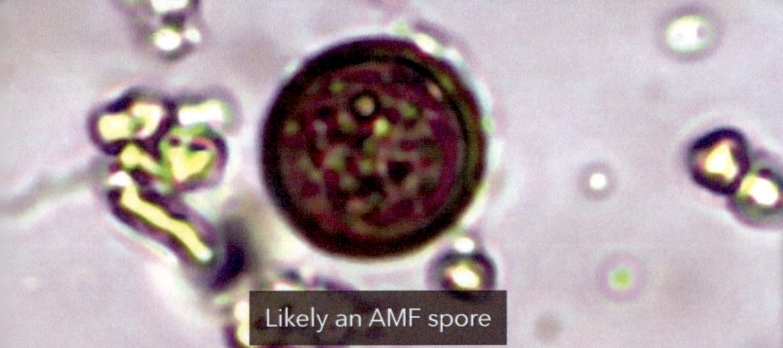

Likely an AMF spore

An Encysted Protozoa

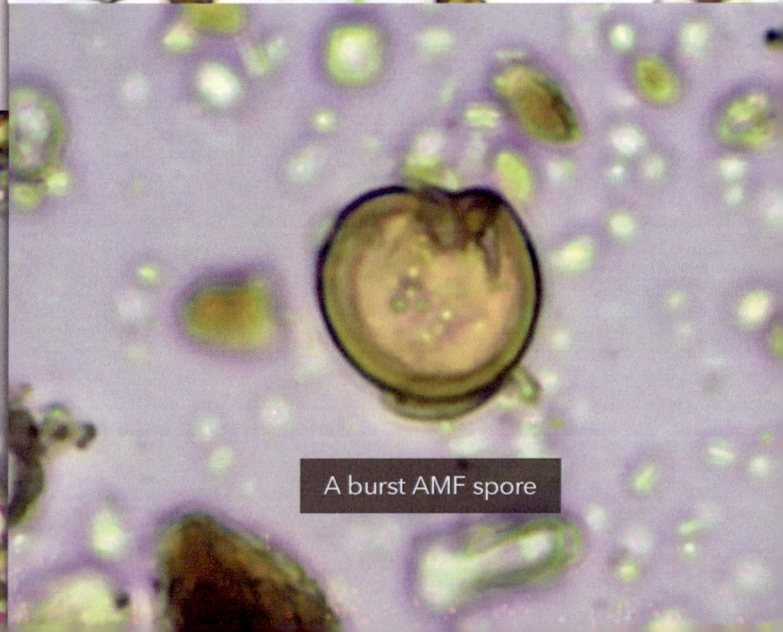

A burst AMF spore

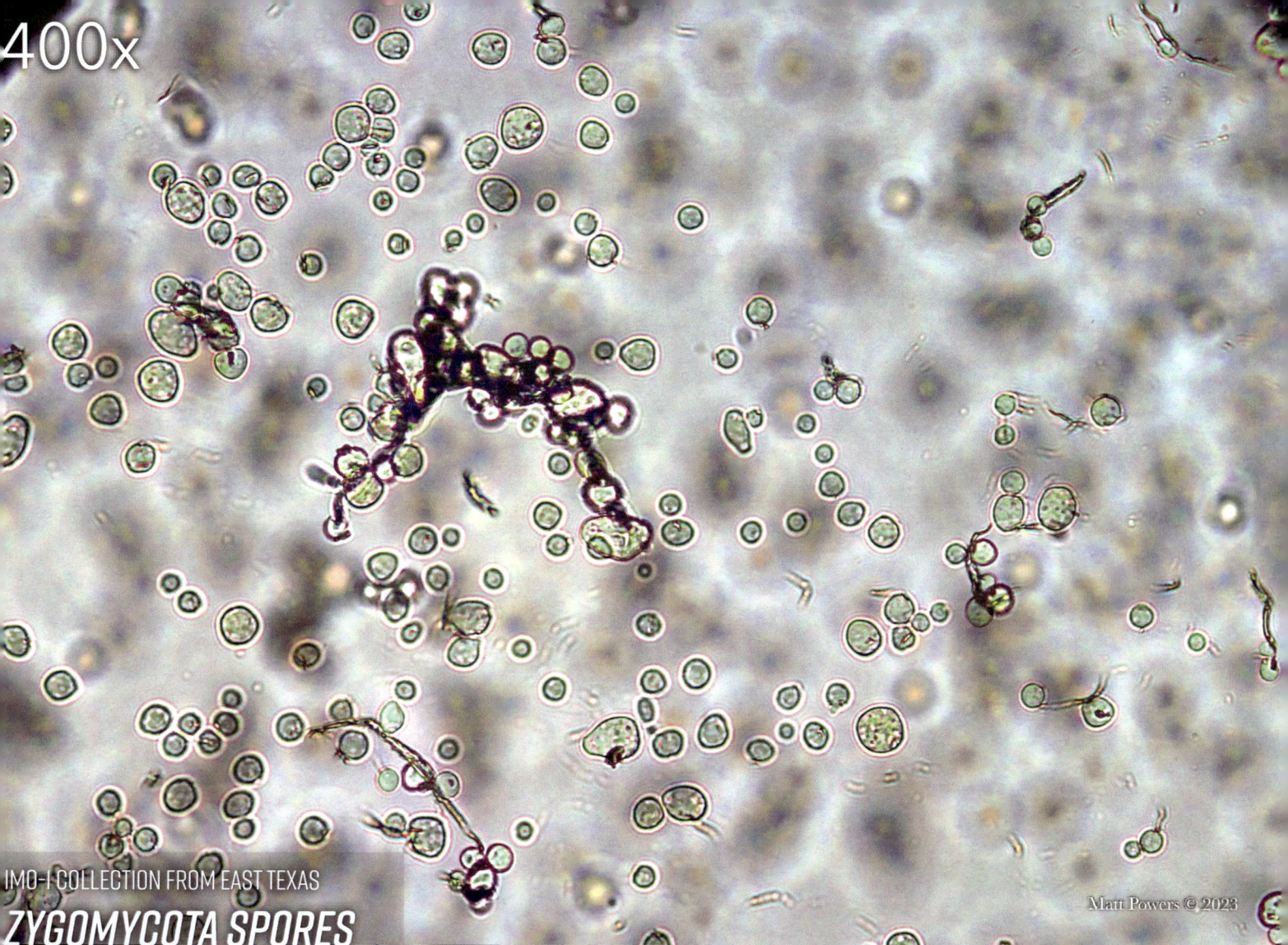

400x

IMO-1 COLLECTION FROM EAST TEXAS

ZYGOMYCOTA SPORES

SAPROPHYTIC MOLD SPORES

Matt Powers © 2023

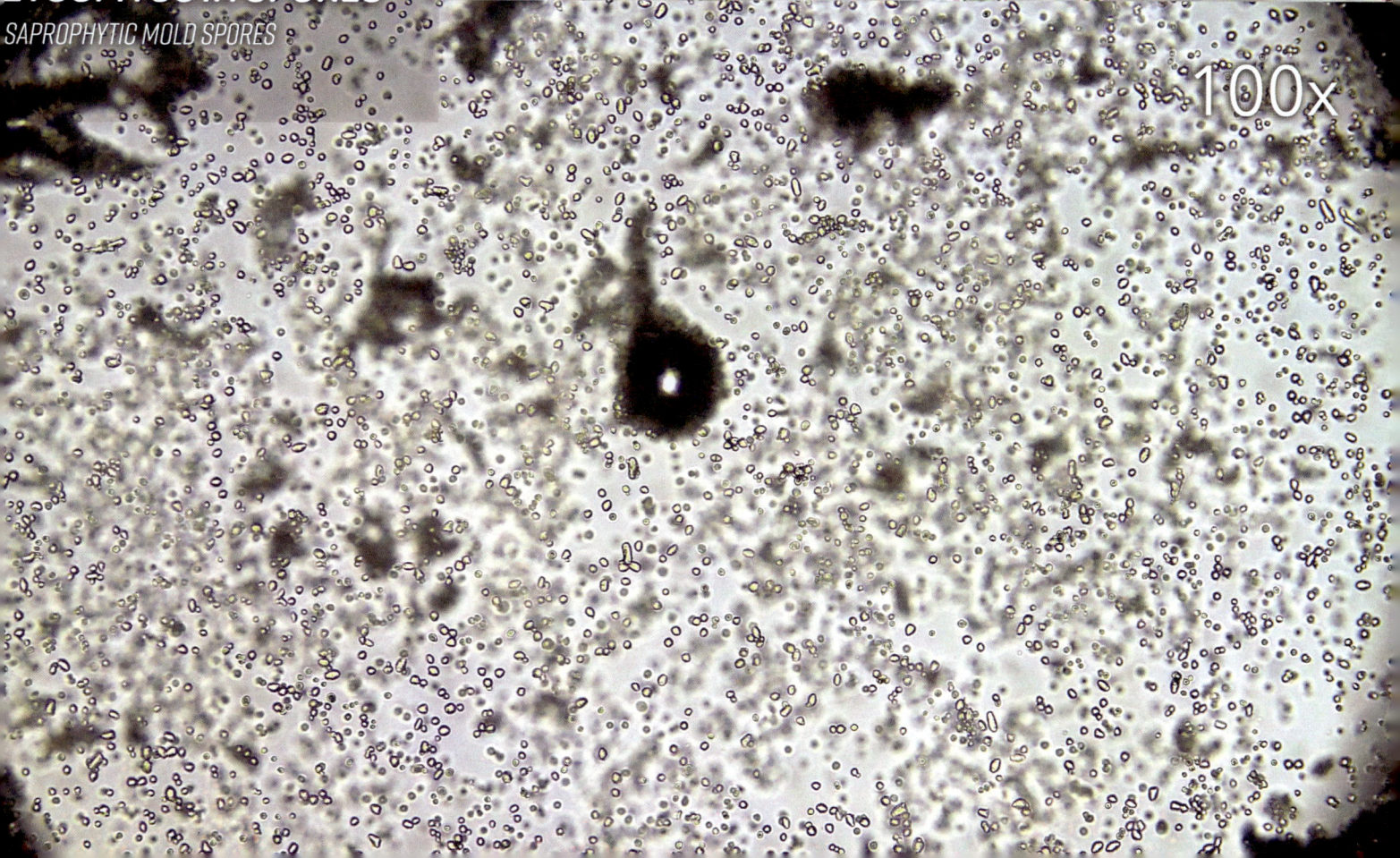

100x

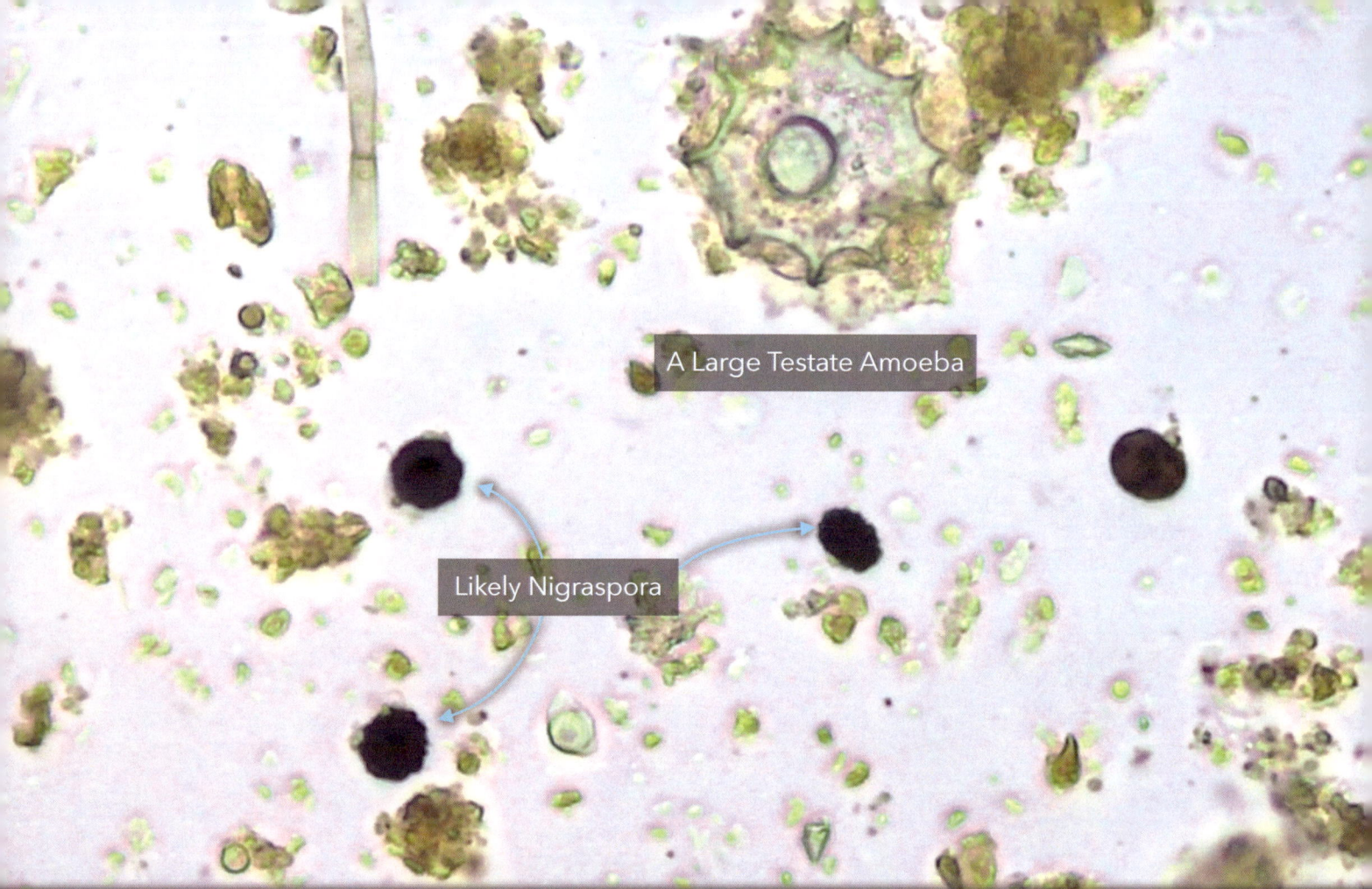

A Large Testate Amoeba

Likely Nigraspora

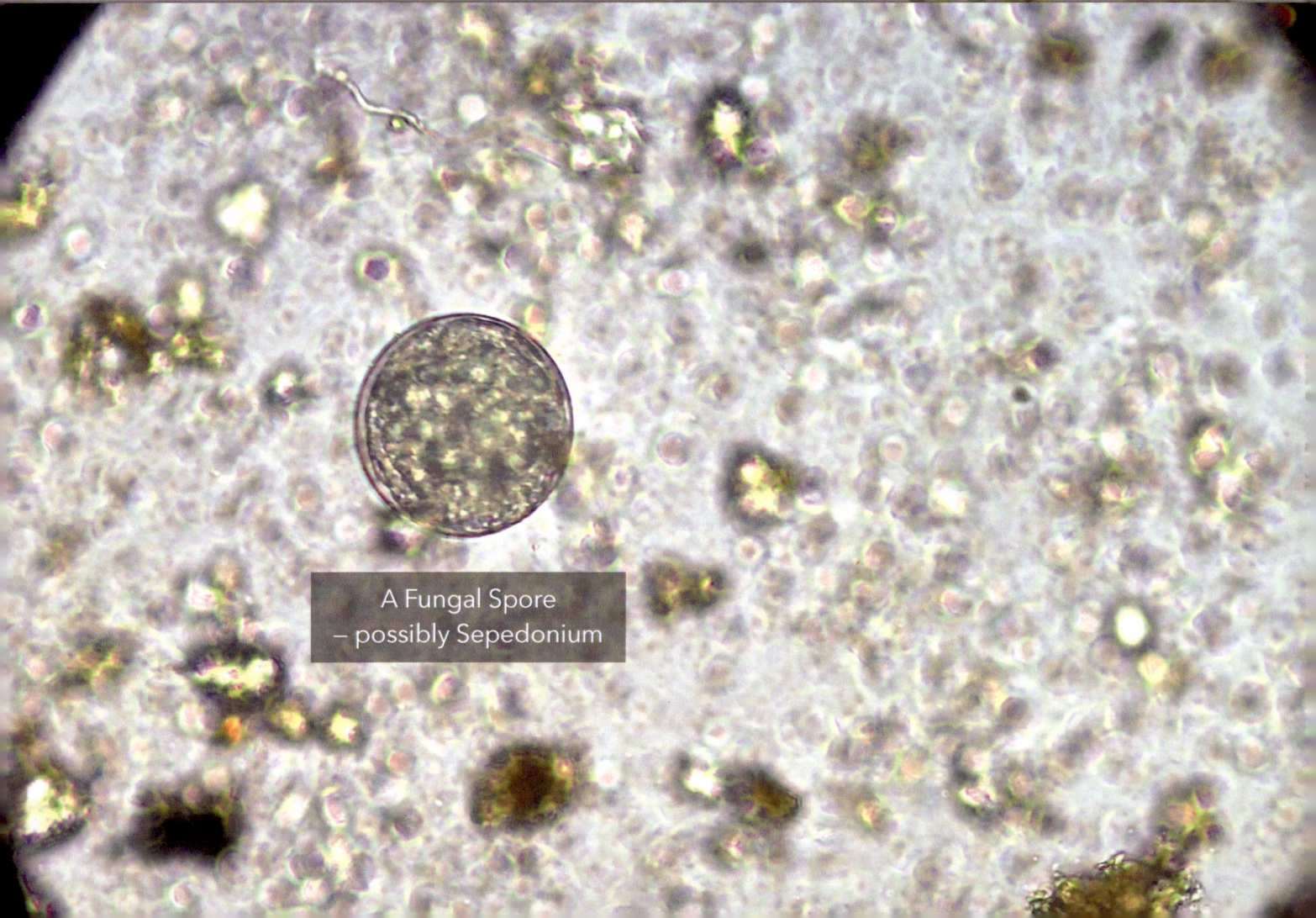

A Fungal Spore
– possibly Sepedonium

soil or compost: it wakes up the spores and cysts quite often, revealing who we couldn't see in a fresh sampling. This shows us how resilient a soil or compost is, how truly complete or imbalanced their nutrient cycling is biologically, and it alerts us to any problems that might emerge later on.

Recommended Morphological Keys for Spores:

Pictorial Atlas of Soil and Seed Fungi: Morphologies of Cultured Fungi and Key to Species. 3rd ed. Watanabe, Tsuneo. 2010.
Fungal Spore Identification and Information Guide. Cialdella, J. & Wonder Makers Environmental. 2022.

Nematodes

Nematodes are round worms that can sometimes be visible to the naked eye but most often we need a microscope to view them. We'll need to be able to magnify our samples by at least 400x (if they hold still go to 600x). Filming them is usually advantageous since they thrash so frantically that they are easy to identify as nematodes even at 40x but hard to identify as to the type of nematode even at 400 – 600x because of these same movements. Dr. Elaine Ingham has suggested using a flame passed under the slide 10 times – or you can blast them with epifluorescence light. If you're filming it, you'll catch them in a moment of frozen shock and what appears to be confusion. It's a very readable physical expression – it may make you feel less inclined to heat the slide. They also hide from the light by tying themselves in knots – this can make it harder to identify them as well, but persistence, patience, and a good camera make it consistently easier.

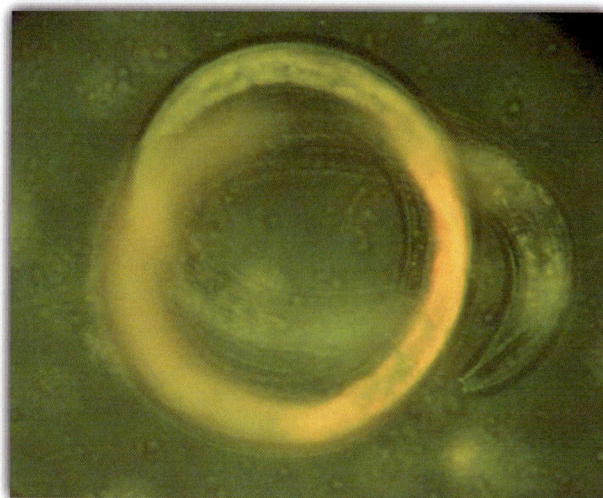

A nematode attempting to hide from the epifluorescence blue lighting.

Nematodes are large which means they consume a large amount of whatever foods they prefer. There are bacterial, root, and fungal feeders as well as omnivores and predators – there are even parasitic leaf nematodes and saprophage nematodes that feed on dead or decaying organic matter. At one point in the agricultural mainstream, it was thought that all nematodes were bad – this is understandable from a medical perspective as round worms are common and dangerous parasites in areas of poor hygiene and a lack of filtered clean water, BUT what we know now is they are vital to the cycling of nutrients in the soil and to making those nutrients available to plant roots.

Do we need to know them down to the species?

No – we only need to understand what they are preying upon to know what they are releasing into the rhizosphere for plant roots to uptake and other microbes to absorb. Fungal feeders are going to release fungal, likely more reduced and acidic, nutrients while bacterial feeders will release bacterial, likely more alkaline and oxidized, nutrients with root feeders releasing even more alkaline and oxidized nutrients, BUT nematodes also show up and reproduce in correlation to the foods present, so if you have a banquet of options and biological rich inputs, you'll usually get a diversity of nematodes to manage it. I say usually because there are folks all over the world in diverse situations making compost – some are on islands, some are within the arctic circles, and some live high up in apartments or on top of mountains, and extreme environments affect microbes in ways we haven't fully mapped yet and nature is known for caveats and special, unique circumstances and expressions. Some folks recommend adding papaya or mango seeds to compost to promote nematodes. Some folks worry about having too many or having bad nematodes (the root feeders) – luckily there are hallmark signs that will guide us in this space.

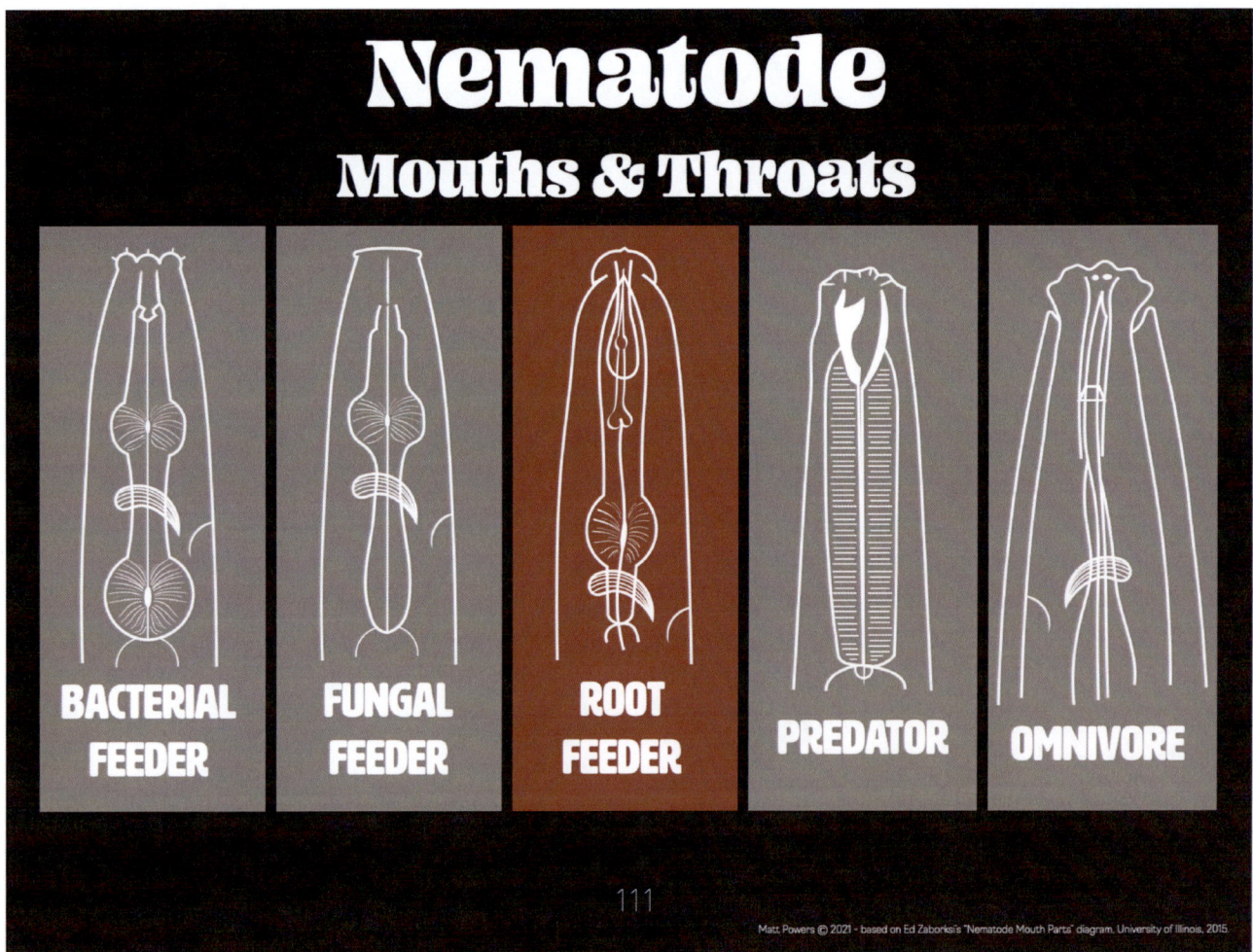

Nematode
Mouths & Throats

BACTERIAL FEEDER · FUNGAL FEEDER · ROOT FEEDER · PREDATOR · OMNIVORE

Matt Powers © 2021 - based on Ed Zaborski's "Nematode Mouth Parts" diagram, University of Illinois, 2015.

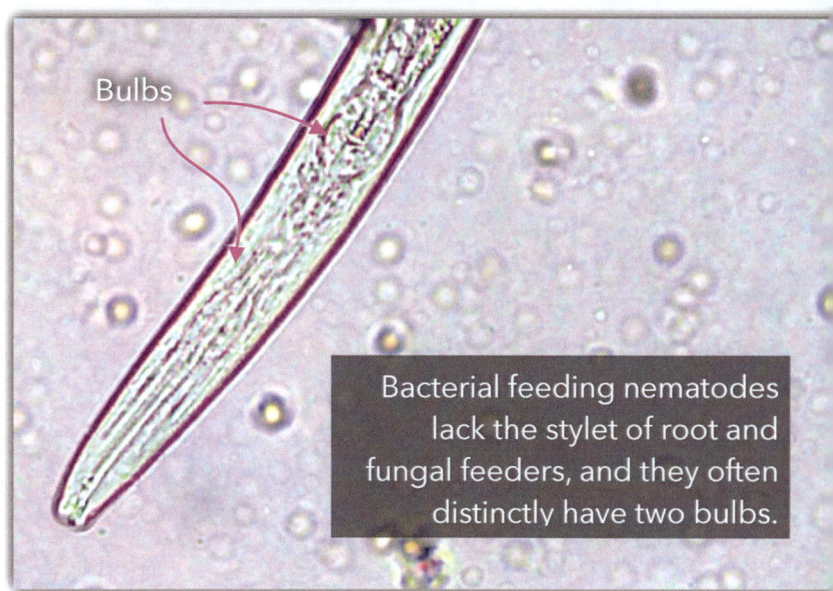

Bulbs

Bacterial feeding nematodes lack the stylet of root and fungal feeders, and they often distinctly have two bulbs.

Morphological identification is not a perfect science – it's been proven that all microscope-based assessment of nematodes is not accurate. DNA testing, again, is required to identify nematodes more definitively. There's several million different nematode species, and they are resourceful: they will eat what they can when they can regardless of their preferences if they are hungry enough and if they can fit it into their mouths. Many nematodes have even been classified as Switchers because they switch their feeding habits according to the foods available. Online keys can help build our fluency, but once we can recognize the mouthparts enough that we can identify who is a root feeder and who is not, things become much easier.

An Online Morphological Keys:

Nematology Lab – University of Nebraska. https://nematode.unl.edu/key/nemakey.htm

Diplogasteridae, a Switcher Nematode that feeds on bacteria, fungi, and even other nematodes

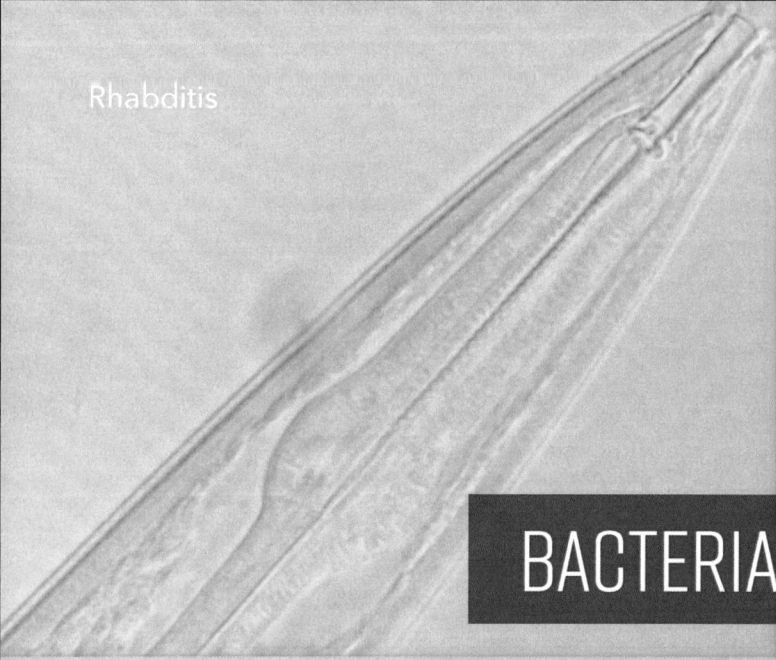

Rhabditis

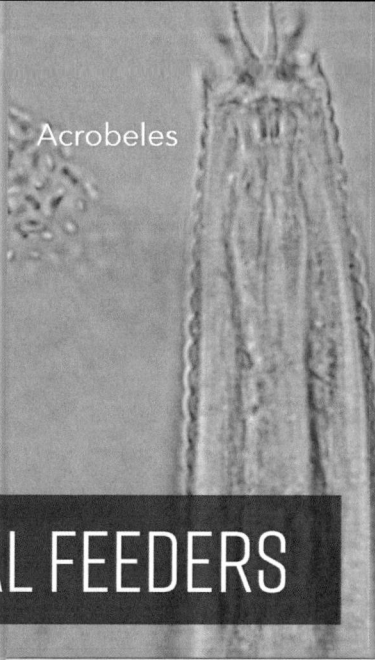

Acrobeles

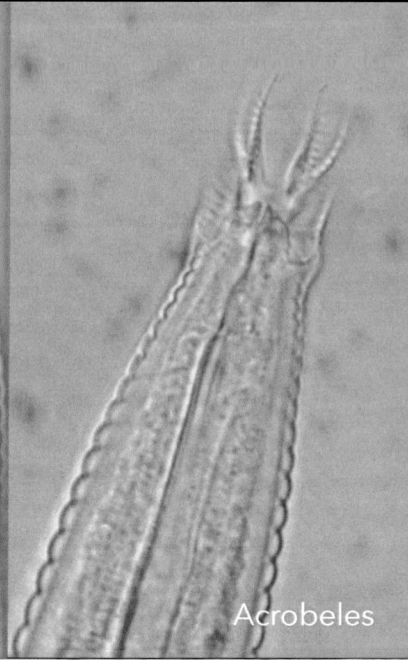

Acrobeles

BACTERIAL FEEDERS

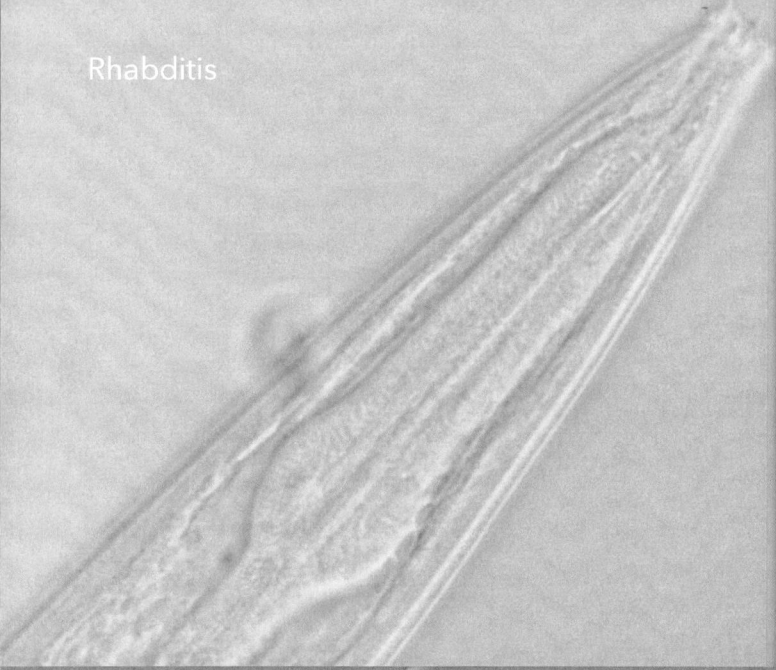

Rhabditis

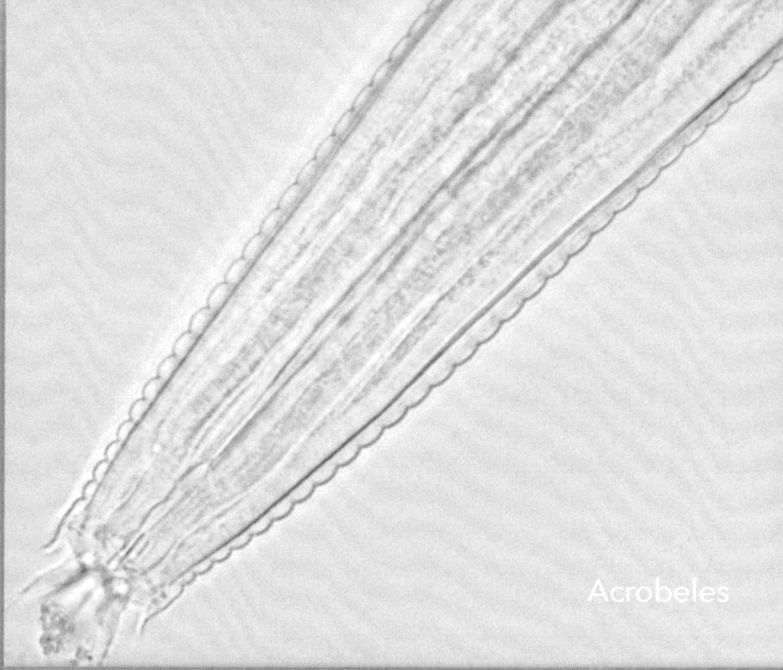

Acrobeles

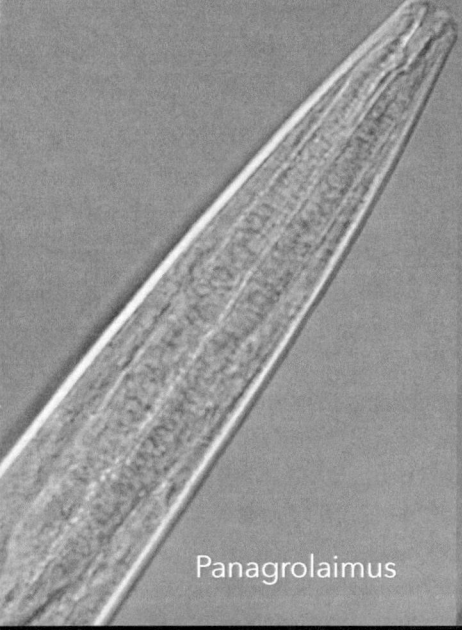

Panagrolaimus

Panagrolaimus

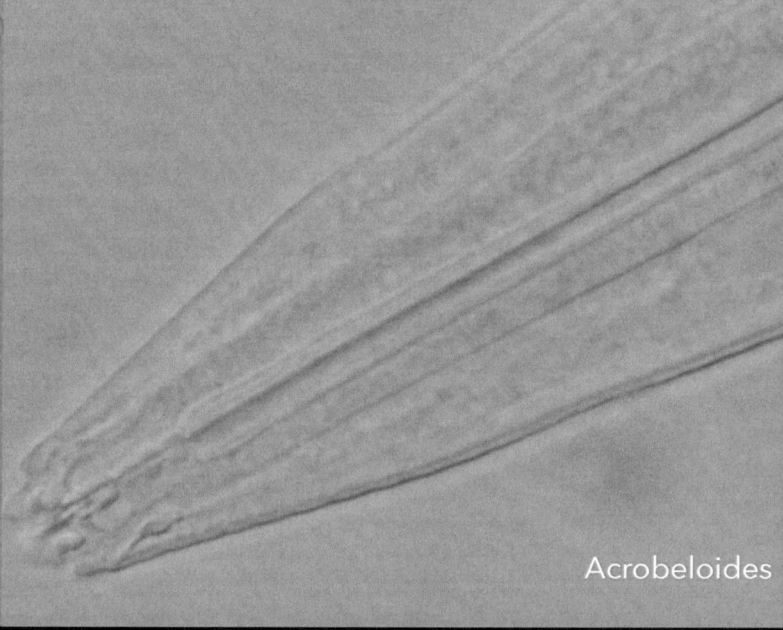

Acrobeloides

Above Image Credit: Xue Qing, Yihao Wang, Xuequan Lu, Haibo Li, Xuan Wang, Hongmei Li, Xiaojun Xie,
NemaRec: A deep learning-based web application for nematode image identification and ecological indices calculation, European Journal of Soil Biology, Volume 110, 2022,
103408, ISSN 1164-5563, https://doi.org/10.1016/j.ejsobi.2022.103408.

BACTERIAL FEEDERS

Behaviorally, these nematodes forage through the organic matter or whip around and catch passing by bacteria with their distinctive and elaborate lips – they even have hairs. They are highly active and mobile. They do not have any teeth, but they often have large cylindrical mouths that allow them to gather in the bacteria and then swallow them whole. They lack the telltale stylet (mouth spear) of fungal and root feeders yet have their own hallmark characteristics: often 2 bulbs (but can be 1 or no bulbs) with their elaborate mouthparts helps them gather the bacteria into their mouths and make them recognizable.

Common Bacterial Feeders: Rhabditis (may also feed on protozoa)., Diplogaster, Acrobeles, Acrobeloides, and Mesorhabditis (can parasitize insects and arthropods) species.

FUNGAL FEEDERS

Having usually a single bulb, sometimes lacking, but always with a stylet that lacks knobs at its base usually means that we are looking at a fungal feeder. If you have fungal hyphae that are missing sections and have punctures (you might not be able to see the punctures), but either way, that's most likely a fungal feeder sidling up to that hyphae and kissing it, then stabbing it with it's spear tongue (the stylet), and then the nematode sucks the hyphae dry. If it's Basidiomycota, their closed septa restrict the attack to just that hyphal compartment, but Ascomycota and others, especially Zygomycota, are

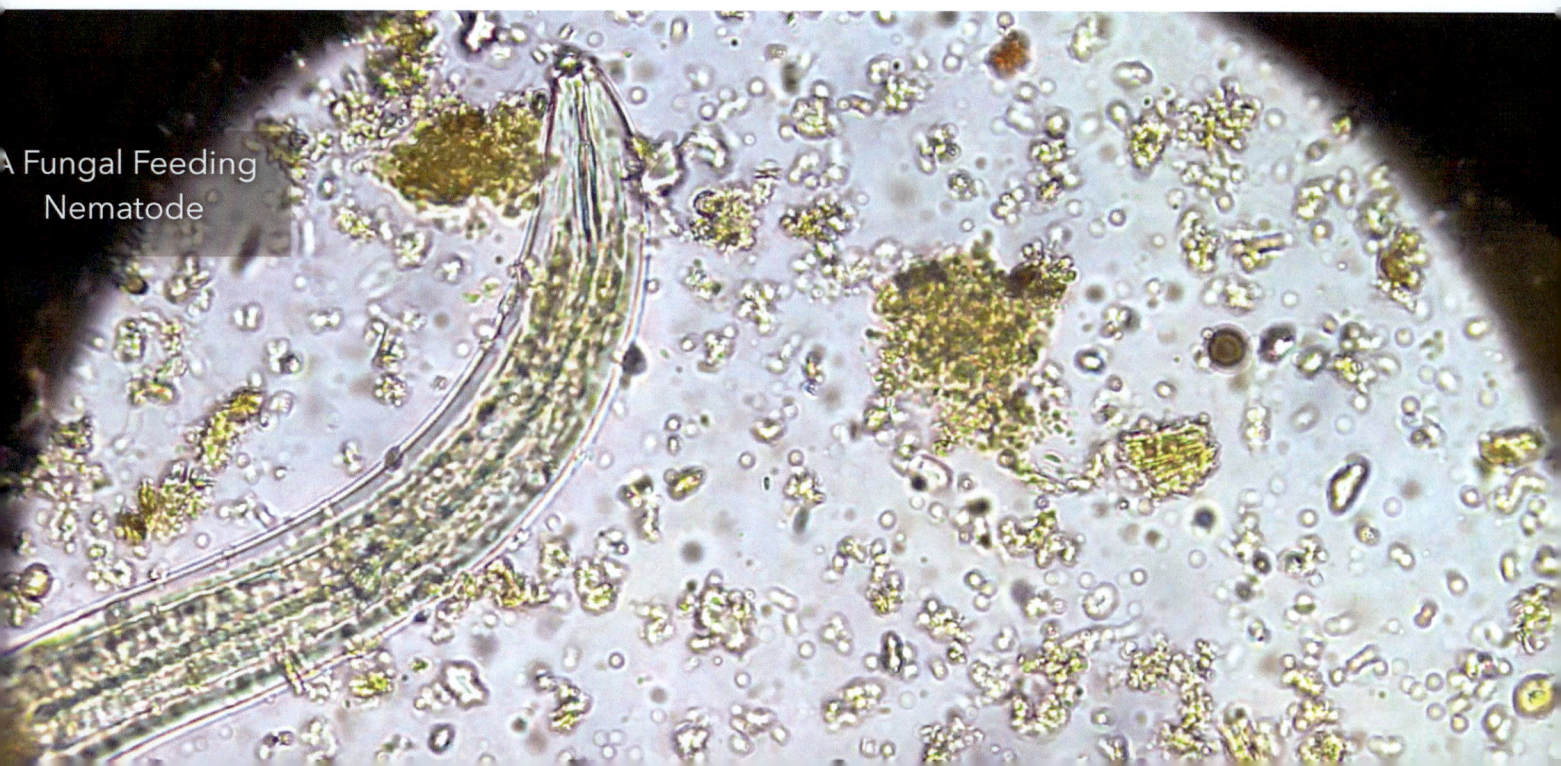

A Fungal Feeding Nematode

more easily consumed in larger sections by fungal feeders because their septa are more open or absent in the case of Zygomycota. This is why in some ways, basidiomycota represents a hardier or more resilient expression of soil fungi because they have fully closed septa, so any predation is always isolated. Juvenile nematodes can be very difficult to identify.

Common Fungal Feeders: Aphelenchoides and Ditylenchus species.

ROOT FEEDERS

Sometimes ringed with round ridges, these root feeding nematodes are agricultural nuisances that have been giving nematodes as a group a bad name for decades. They are often identifiable if you get a clear image – they have stylets most often with knobs at the base: it looks downright phallic, and sometimes they have very large and prominent bulbs, but there are some root feeders with subtle stylets: they lack the bulbs and have a geometric notch at their base or even just lack the knobs seemingly which can mean they are still developing or this species lacks them. The larger bulbs and knobs provide the muscle and leverage needed to stab their larger spears through the root surface – the differences between root surface thickness would like lead itself to local root feeder adaptations like Darwin's Galapagos bird beaks. These are also luckily the target of entire groups of predatory nematodes, so we can control these nematodes with other

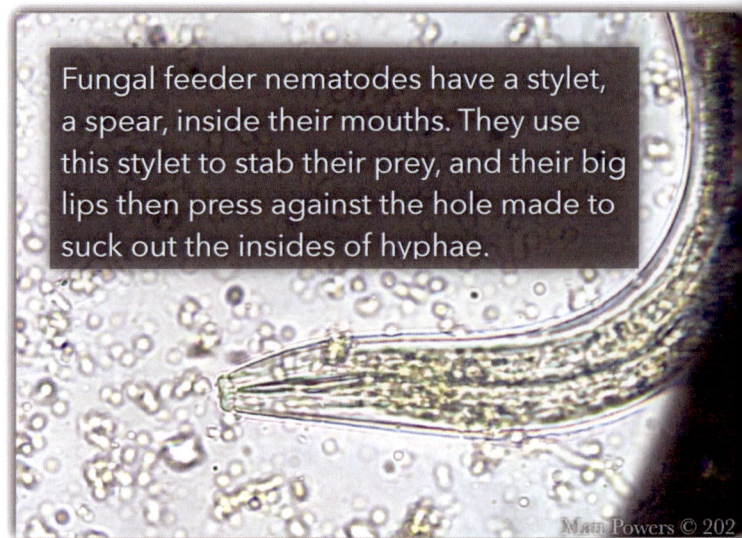

Fungal feeder nematodes have a stylet, a spear, inside their mouths. They use this stylet to stab their prey, and their big lips then press against the hole made to suck out the insides of hyphae.

Matt Powers © 202

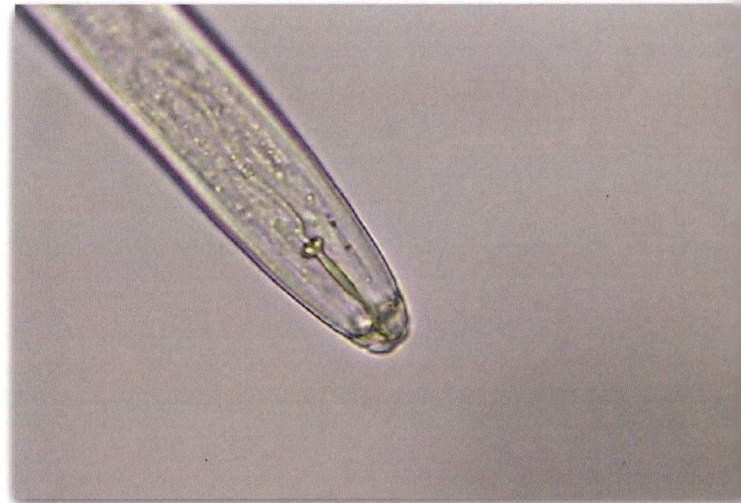

Meloidogyne incognita (root-knot nematode): Adult male, Head: Stylet lips and esophagus @ 400X magnificationScot Nelson (photographer & microscopist). Public Domain. Flickr.com.

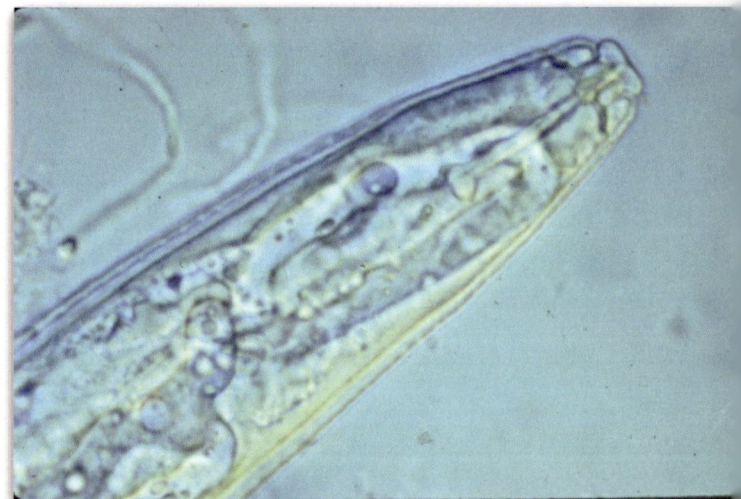

Microscope: Used oil immersion lens (100x objective) | Read: "Awa Root-Knot Disease" www.ctahr.hawaii.edu/oc/freepubs/pdf/PD-20.pdf | Photograph: Mario Serracin, University of Hawaii at Manoa. Scot Nelson (photographer & microscopist). Public Domain. Flickr.com.

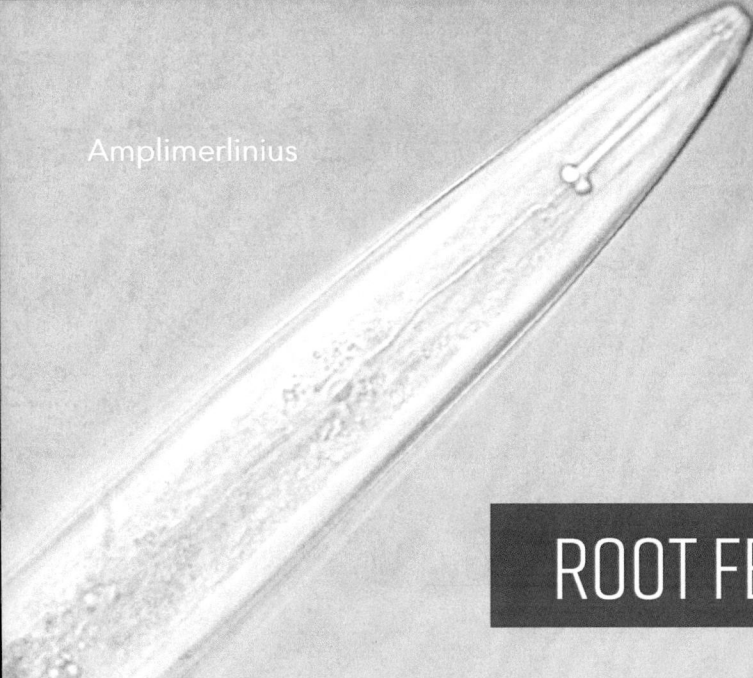

Amplimerlinius

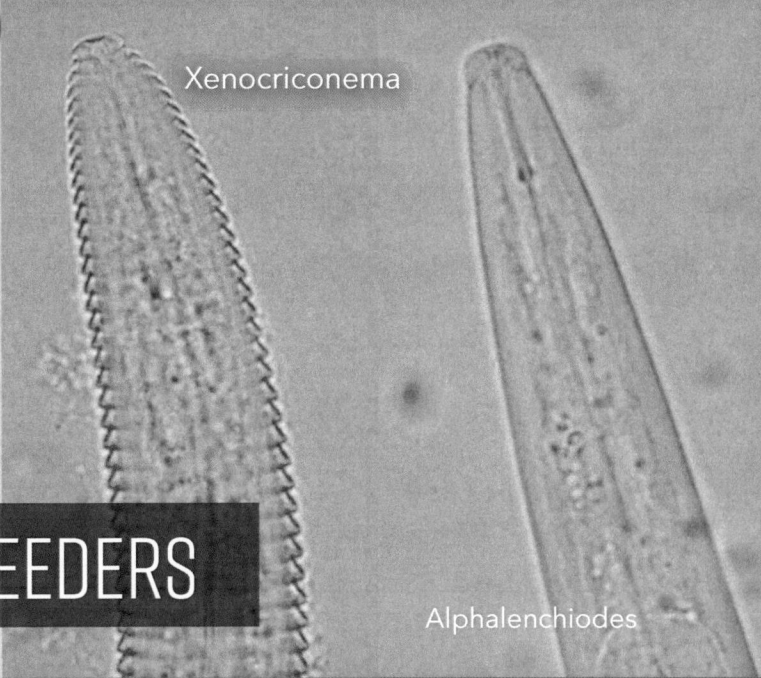

Xenocriconema

Alphalenchiodes

ROOT FEEDERS

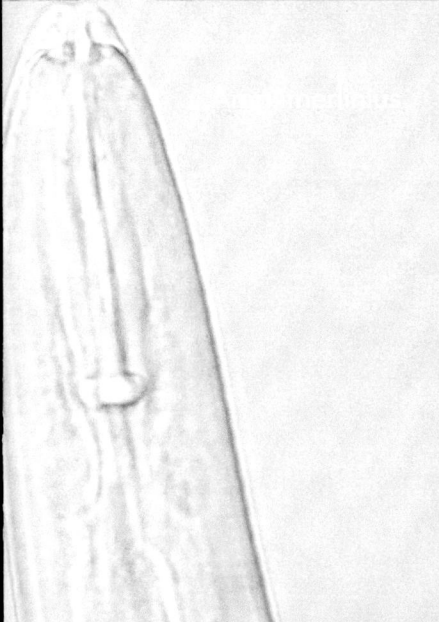

Aporcelaimus

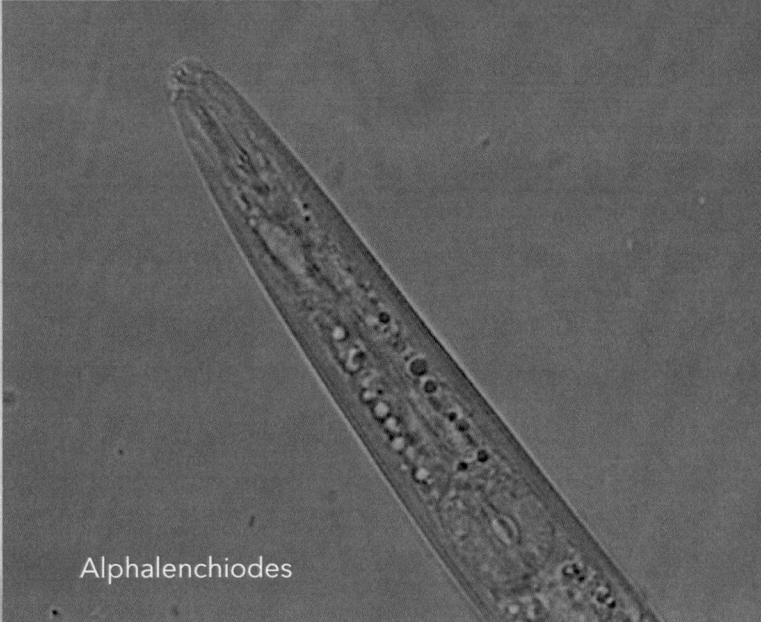

Alphalenchiodes

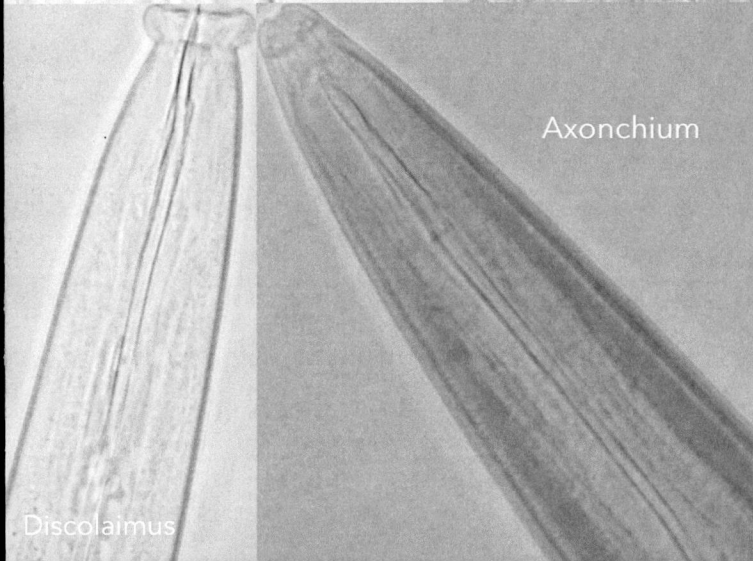

Axonchium

Discolaimus

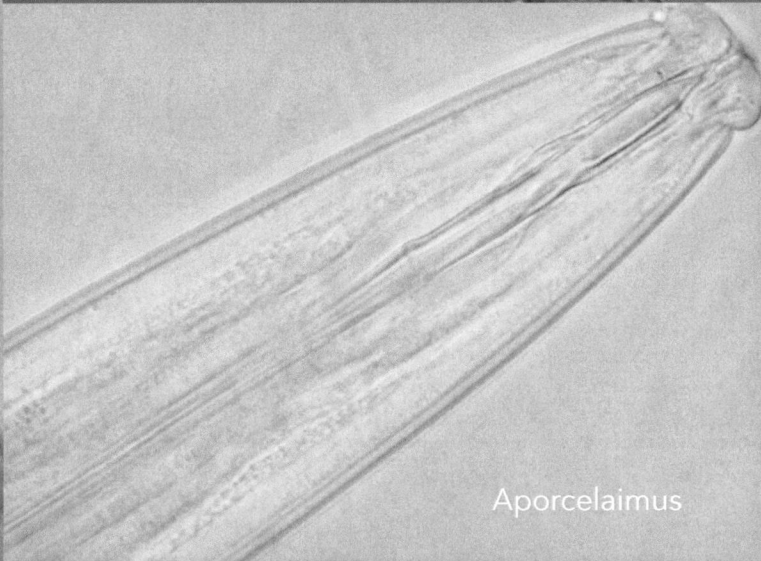

Aporcelaimus

Above Image Credit: Xue Qing, Yihao Wang, Xuequan Lu, Haibo Li, Xuan Wang, Hongmei Li, Xiaojun Xie,
NemaRec: A deep learning-based web application for nematode image identification and ecological indices calculation, European Journal of Soil Biology, Volume 110, 2022, 103408, ISSN
1164-5563, https://doi.org/10.1016/j.ejsobi.2022.103408.

nematodes. I have yet to see one of these in my soil and compost samples.

Common Root Feeders: Tylenchida (some species also feed on protozoa), Xiphinema, Pratylenchus (root lesions), Heterodera glycines (soybean cyst), Alphalenchiodes, Amplimerlinius, Ditylenchus, and Meloidogyne (root knot)

PREDATORY NEMATODES

Predatory nematodes are either insect parasites or feed on other nematodes (and often other microbes like protozoa). The ones sold commercially are primarily insect parasites, so they are effective biocontrol of many pest species including caterpillars, cutworms, and root weevils. Several species that feed on other nematodes specifically seek out root feeders. Predatory nematodes (those that feed on other nematodes) are very large in comparison to their prey; they can be 5x as thick in diameter! They also have a large, wide mouth with a large tooth that can be used to stab into nematodes, amoebae, or ciliates.

Common Predatory Nematodes: the insect parasites: Steinernema and Heterorhabditis, and those that feed on other nematodes: Plectus, Neosteinernema, Dorylaimida (specifically for root feeders) and Mononchoides (some species also feed on protozoa).

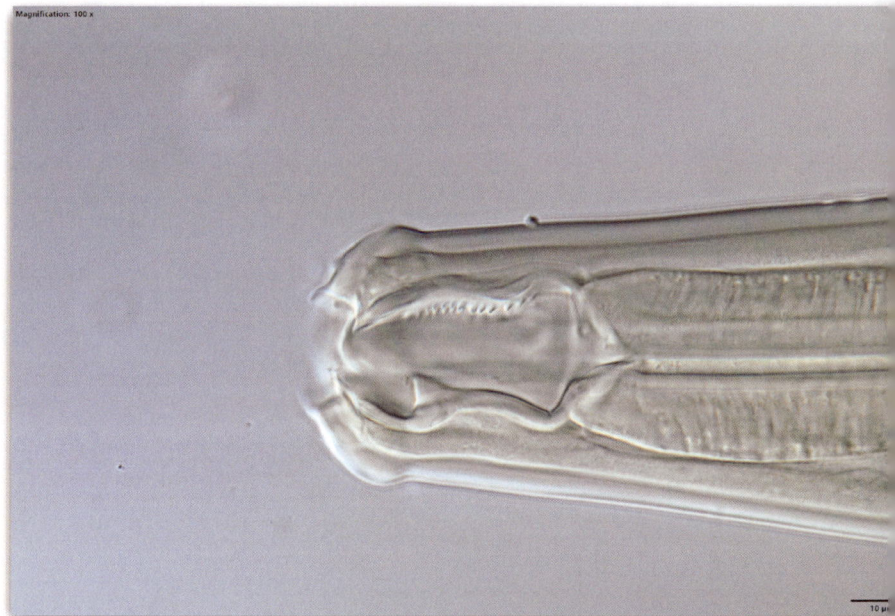

Mononchidae. 1999. Lawrence Perepolkin (photographer & microscopist). Public Domain. Flickr.com.

OMNIVORES & SWITCHERS

With millions of species, it's no surprise that we have some confusion. There are cases of morphological lookalikes, but there's also the fact that there's millions of species and behavior is adaptive across nature. The environment, the foods available, and stressors will cause life to change behavior. This will always remain true, so we have to hold onto our conclusions lightly and be ready to change our assumptions in this space. DNA testing is possible but it's like ciliate DNA testing but next level: cleansing them of bacteria and then flushing them then sequencing their DNA.

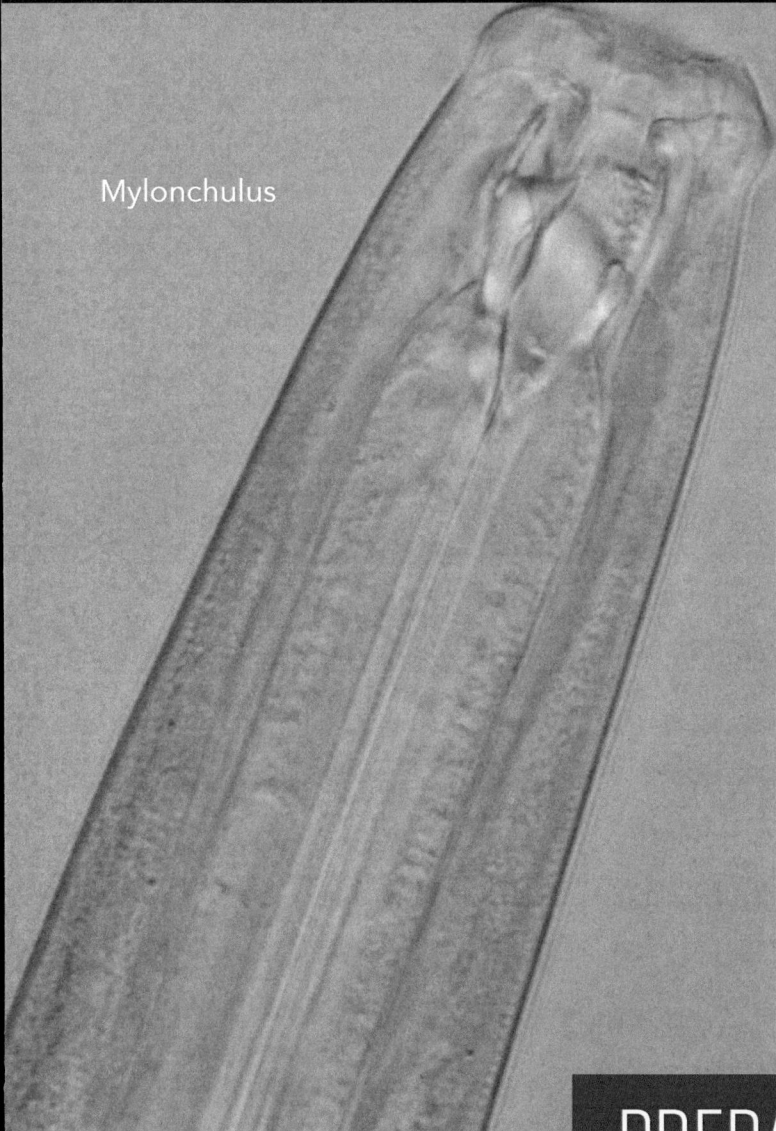

Mylonchulus

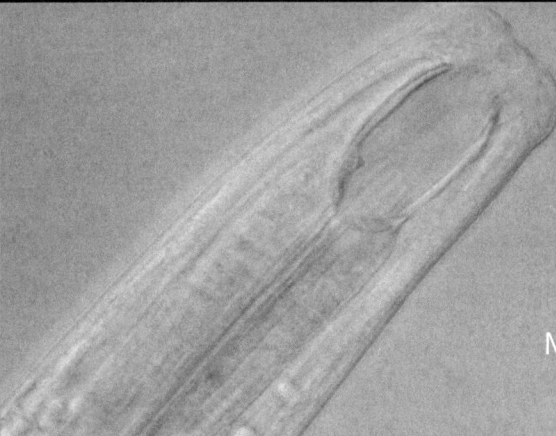

Miconchus

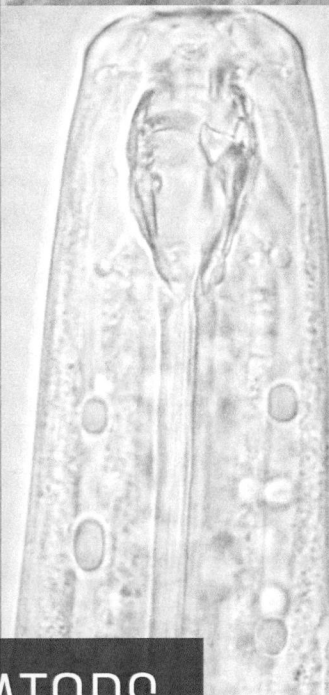

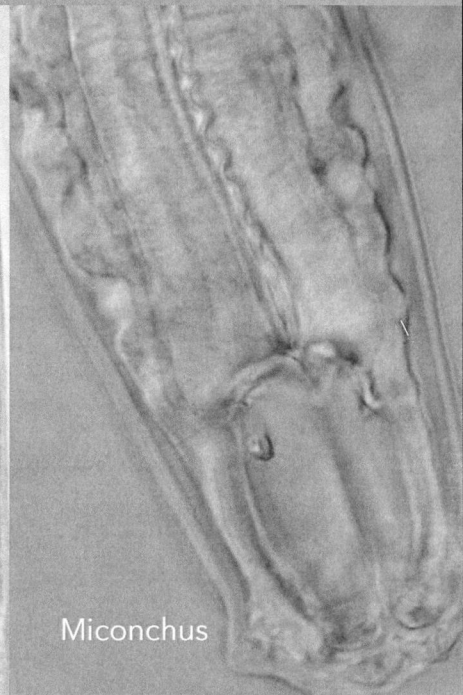

Miconchus

PREDATORS

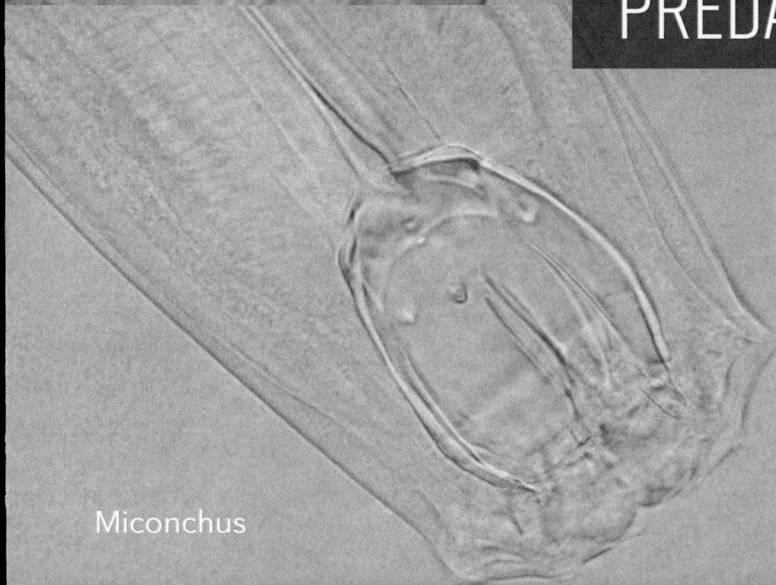

Miconchus

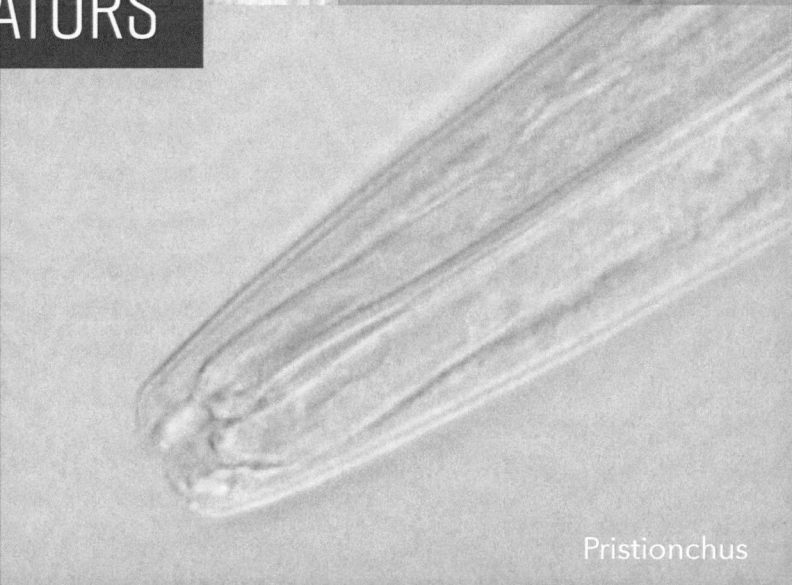

Pristionchus

Above Image Credit: Xue Qing, Yihao Wang, Xuequan Lu, Haibo Li, Xuan Wang, Hongmei Li, Xiaojun Xie,
NemaRec: A deep learning-based web application for nematode image identification and ecological indices calculation,
European Journal of Soil Biology, Volume 110, 2022, 103408, ISSN 1164-5563, https://doi.org/10.1016/j.ejsobi.2022.103408.

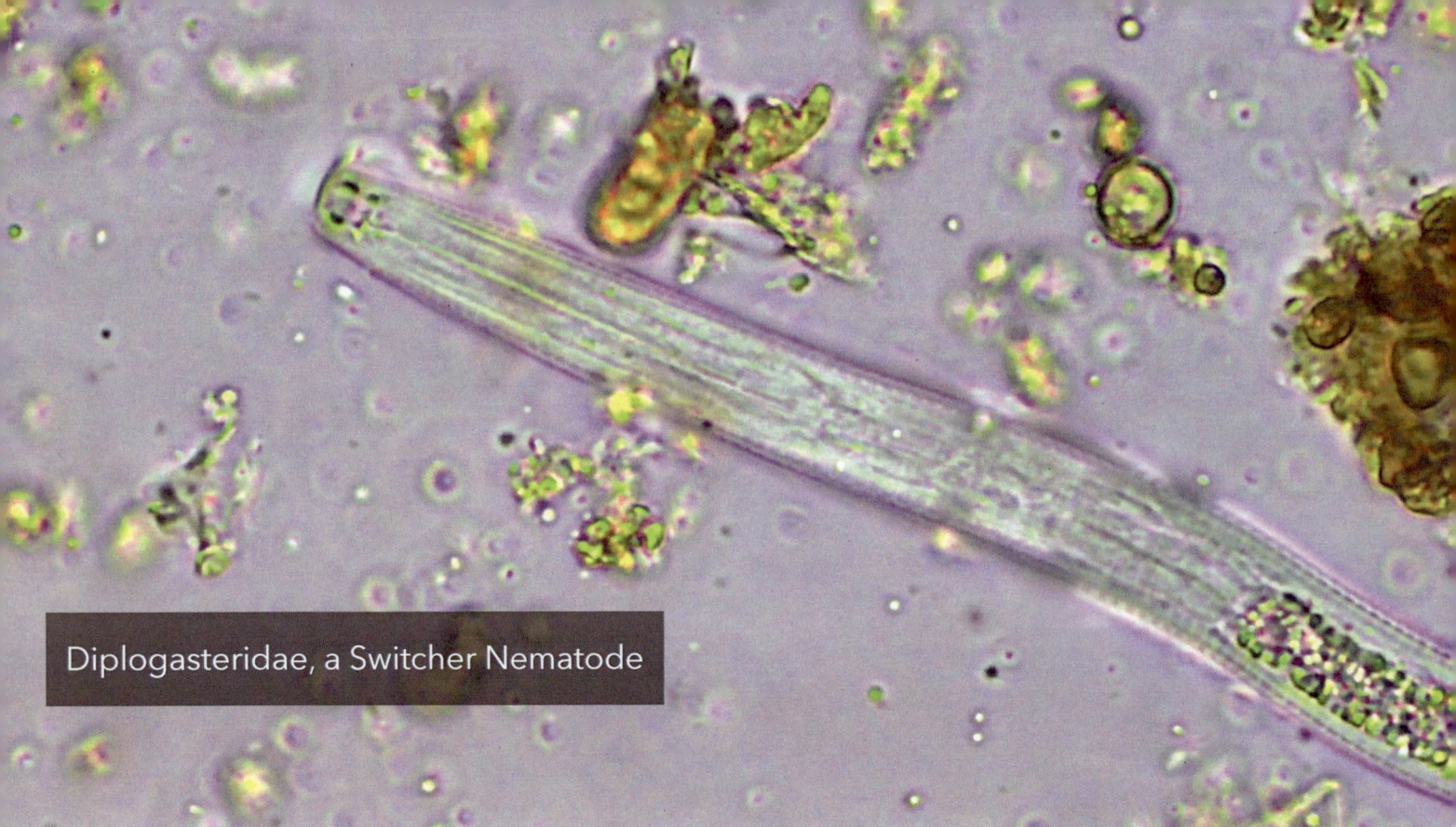

Diplogasteridae, a Switcher Nematode

Some bacterial feeders, like Axonchium, have been found to eat fungi, algae, and even protozoa! This makes sense when you compare Axonchium to Aporcelaimus and they seem nearly identical. These organisms have their preferences but the tools they carry with them can feed them in multiple ways. For instance, fungal feeders have also been known to feed on bacteria and algae. These omnivores can sometimes look like a cross between bacterial and predatory nematodes, but look for the two bulbs! I am of the opinion that nematodes will eat whatever they can get their mouths around – it appears that they like sharks will test or taste things or that they are easily distracted like an attacking predator can sometimes be. The reason to keep an eye out for these types is: they can be fungal feeders in good conditions, but they can become root feeders in times of plant stress.

Common Omnivore Nematodes: Rhabditida (bacteria and protozoa), Axonchium, and Pristionchus

NEMATODE EGGS

These are distinctive, oval eggs that look like snake eggs – you can even see the nematode curled up inside! I have yet to come across one in my samples, but look for these on the R-Soil Database.

An Omnivore Nematode

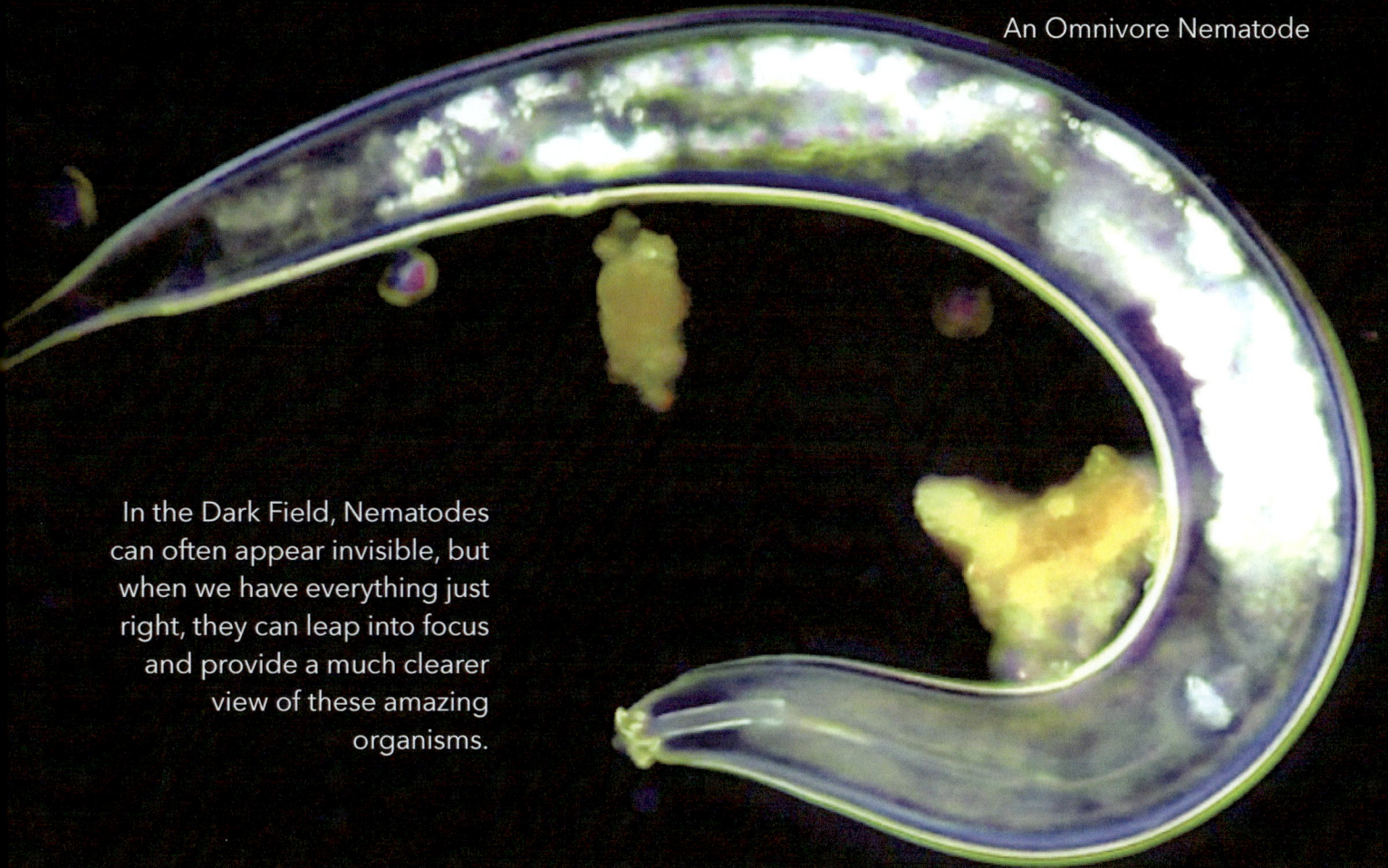

In the Dark Field, Nematodes can often appear invisible, but when we have everything just right, they can leap into focus and provide a much clearer view of these amazing organisms.

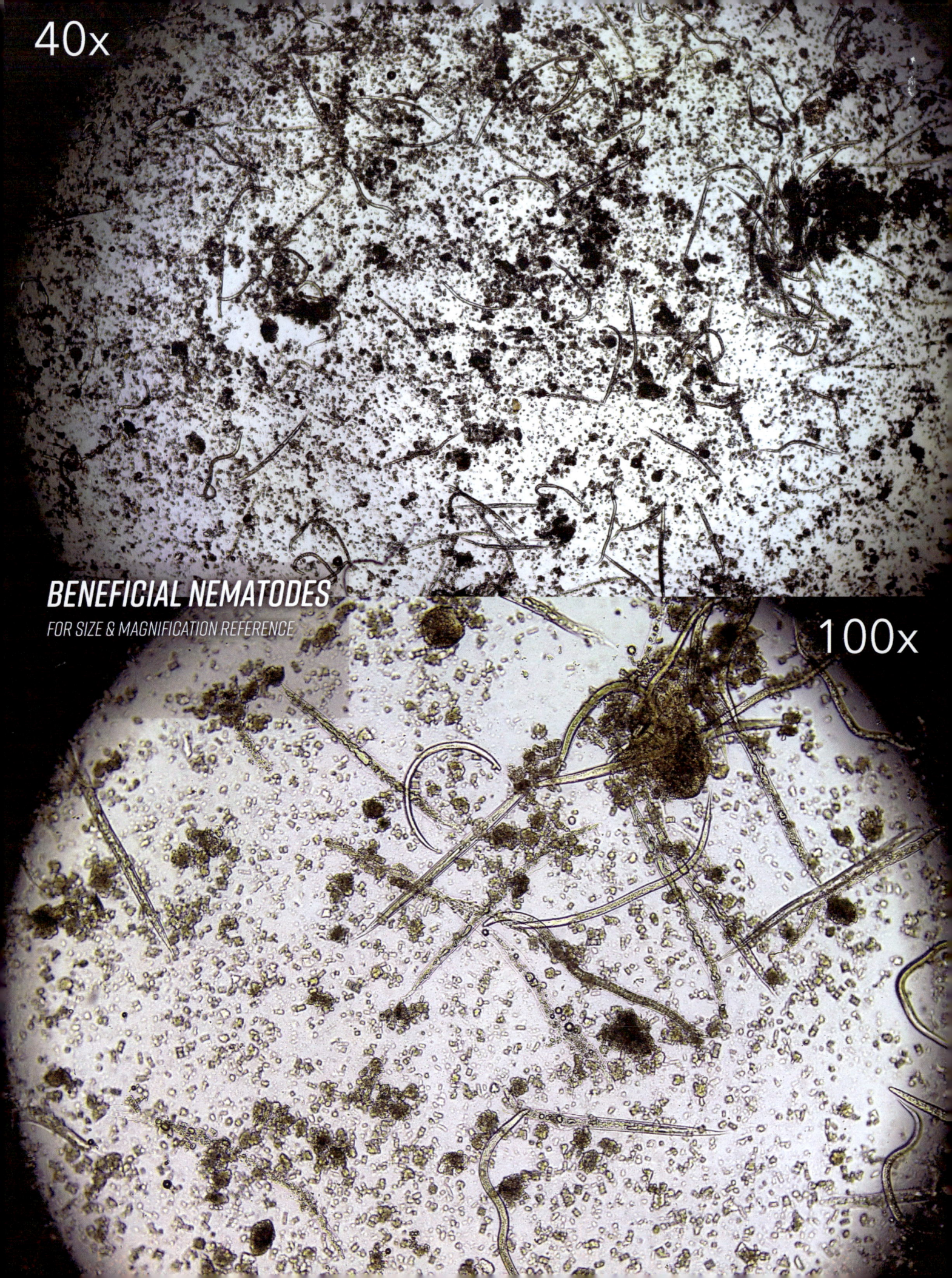

40x

BENEFICIAL NEMATODES
FOR SIZE & MAGNIFICATION REFERENCE

100x

Microarthropods & Worms

Larger than protozoa and nematodes by several orders of magnitude, microarthropods are the shredders of organic

matter. They create new surface area for all the trophic layer of soil food web below them to feast upon, inhabit, and

inoculate. These are springtails (Collembola - good for crops), mites (Acari), sowbug (Isopoda), millipedes (Diplopoda), and

Symphylans (bad for crops). They are visible to the eye but can be tiny still like mites. Potworms (Enchytraeid) are larger

than nematodes but smaller than microarthropods. They are not as common as nematodes, but you may come across

them – if you do, please see if their hairs glow in epifluorescent lighting. I have not seen that particular attribute captured

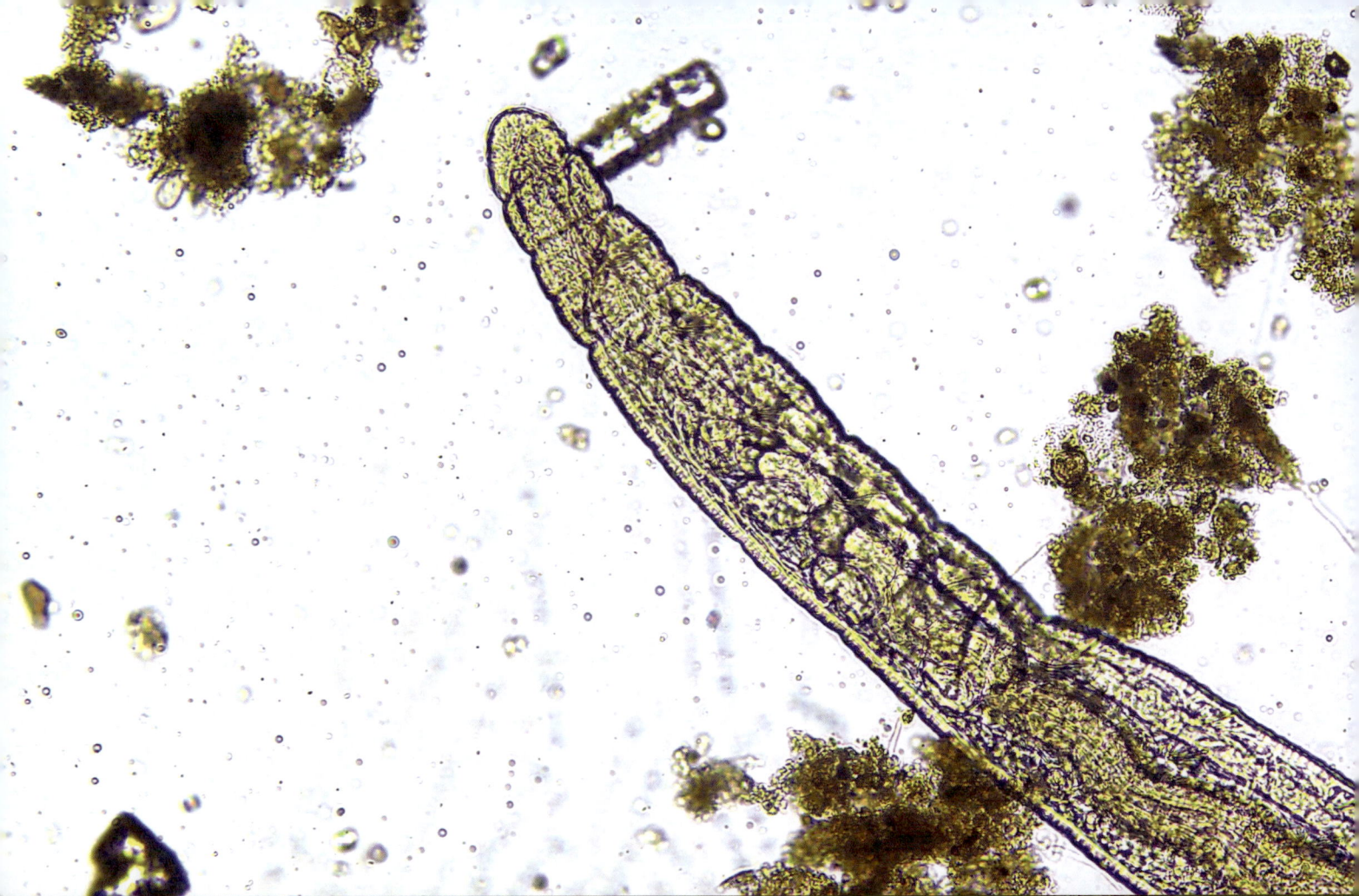

ENCHYTRAEID OR POTWORM

Autofluorescent hairs likely indicate a fungal diet

anywhere else and I'd like to know if it is unique or if it is something that extends across all fungal feeding potworms. Earthworms are bigger than microarthropods, so they are much bigger than potworms.

Identification Key Online:

- Discover Life – Mites:

  https://www.discoverlife.org/mp/20q?guide=Mites

- Mites and Other Microarthropods. University of British Columbia.

  https://zoology.ubc.ca/~srivast/mites/

Minerals, Organic Matter, & Microplastics

Even if you don't do a jar soil test to see the clay/sand/silt composition of your soil, you will notice under the microscope the difference between compacted and structured soil. If it feels cramped in your FOV even at 1:100 dilution, that's compact soil, likely silt or clay that you are looking at. Sand is rather large under the microscope and looks like sea glass more often than not. This makes sense since we make glass out of sand! Silt is smaller than sand but often composed of mostly silica as well – these are tiny glass-like chunks under the microscope, usually clear. Clay are tiny specks are sometimes slightly rod-like – we can easily mistake these for bacteria – this is why doing a viability stain makes our bacterial counting much easier and more precise.

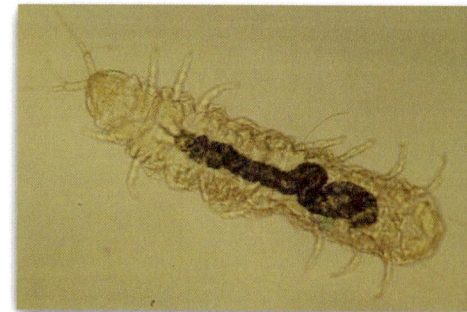

Pauropoda, a microarthropod well adapted to soil. Cristina Menta, CC BY 3.0 <https://creativecommons.org/licenses/by/3.0>, via Wikimedia Commons.

Orchesella cincta, Hairy-back Girdled Springtails. Martin Cooper. 2016. Flickr. CC by 2.0 https://www.flickr.com/photos/m-a-r-t-i-n/24522398794

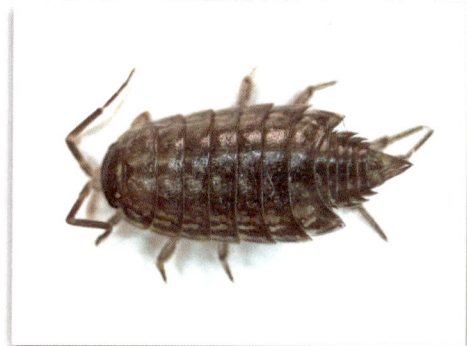

This image is created by user B. Schoenmakers at waarneming.nl, a source of nature observations in the Netherlands., CC BY 3.0 <https://creativecommons.org/licenses/by/3.0>, via Wikimedia Commons

This is most likely a fragment of plastic blue tarp.

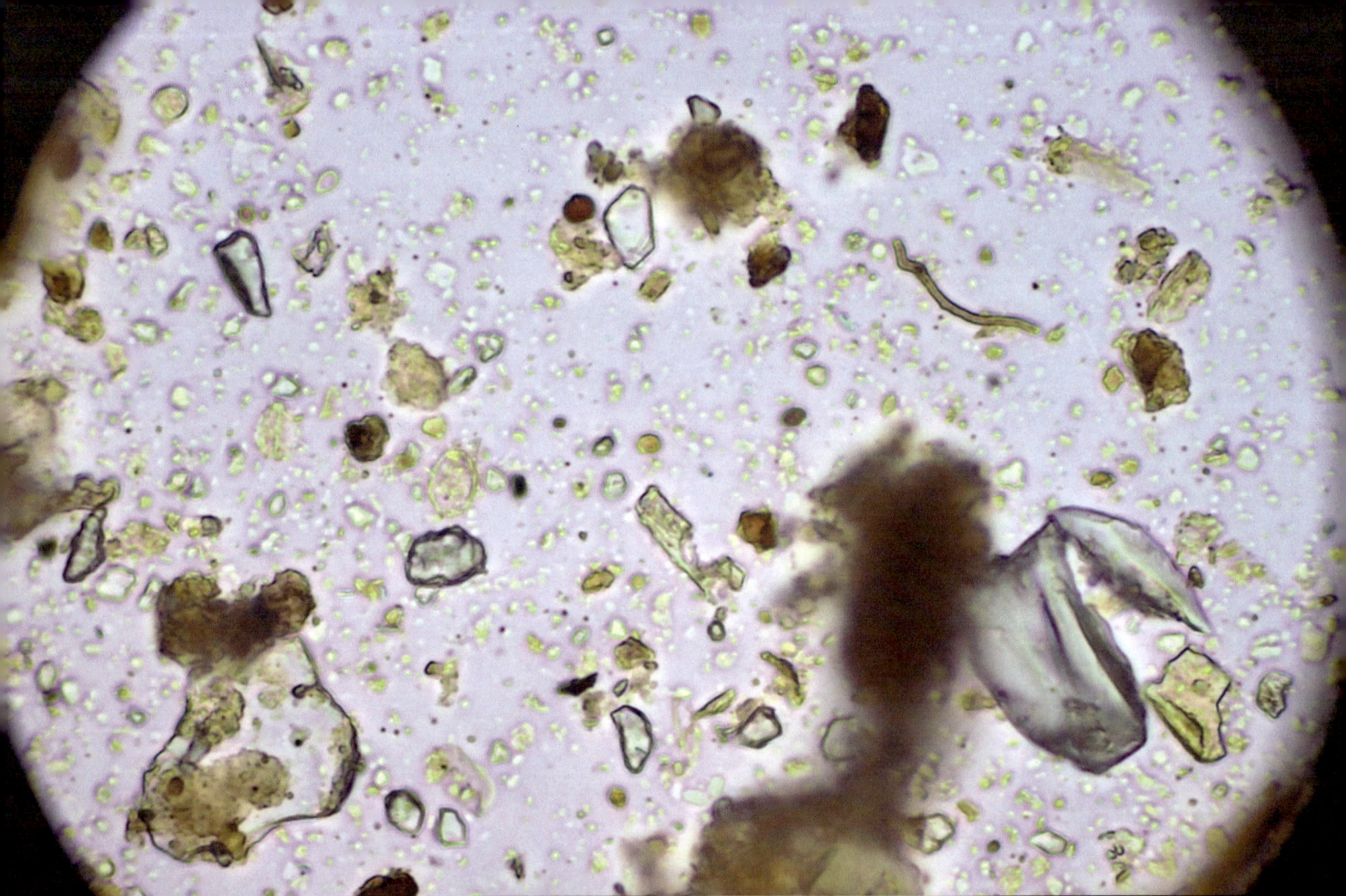
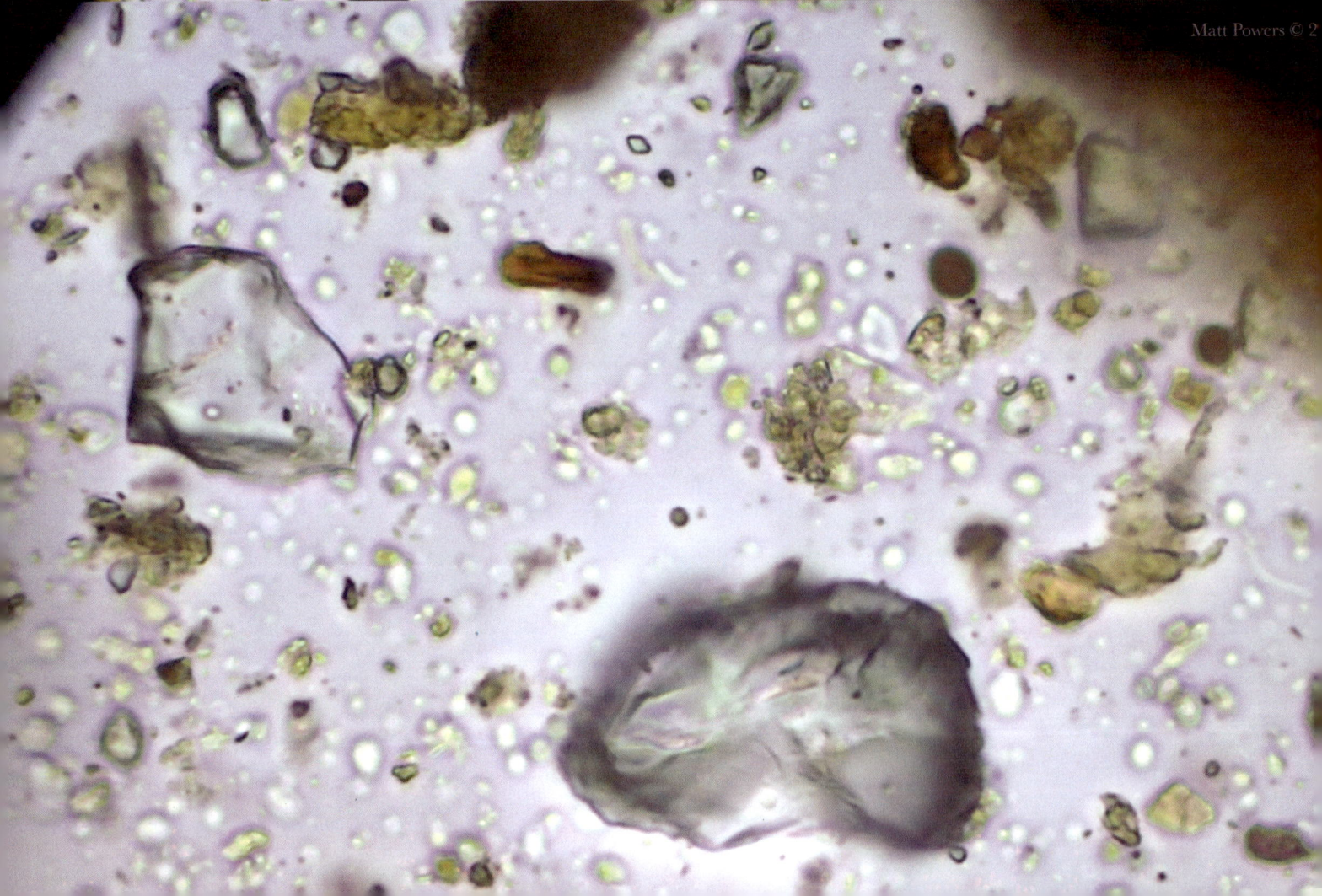

All the fragments that look like pieces of broken glass are silicates. We classify silicates by size: sand, silt, and clay. It is said that sand in relation to the others is like the size of basketball while silt is like the size of a golfball while clay is size of a chalk dot, but you can see here that there is no clear gradient between, only a spectrum of mineral expression in color, size and opacity.

Microplastics shine distinctly, usually orange and yellow, under epifluorescence lighting in a way that is different from fungi and phosphorus bearing minerals.

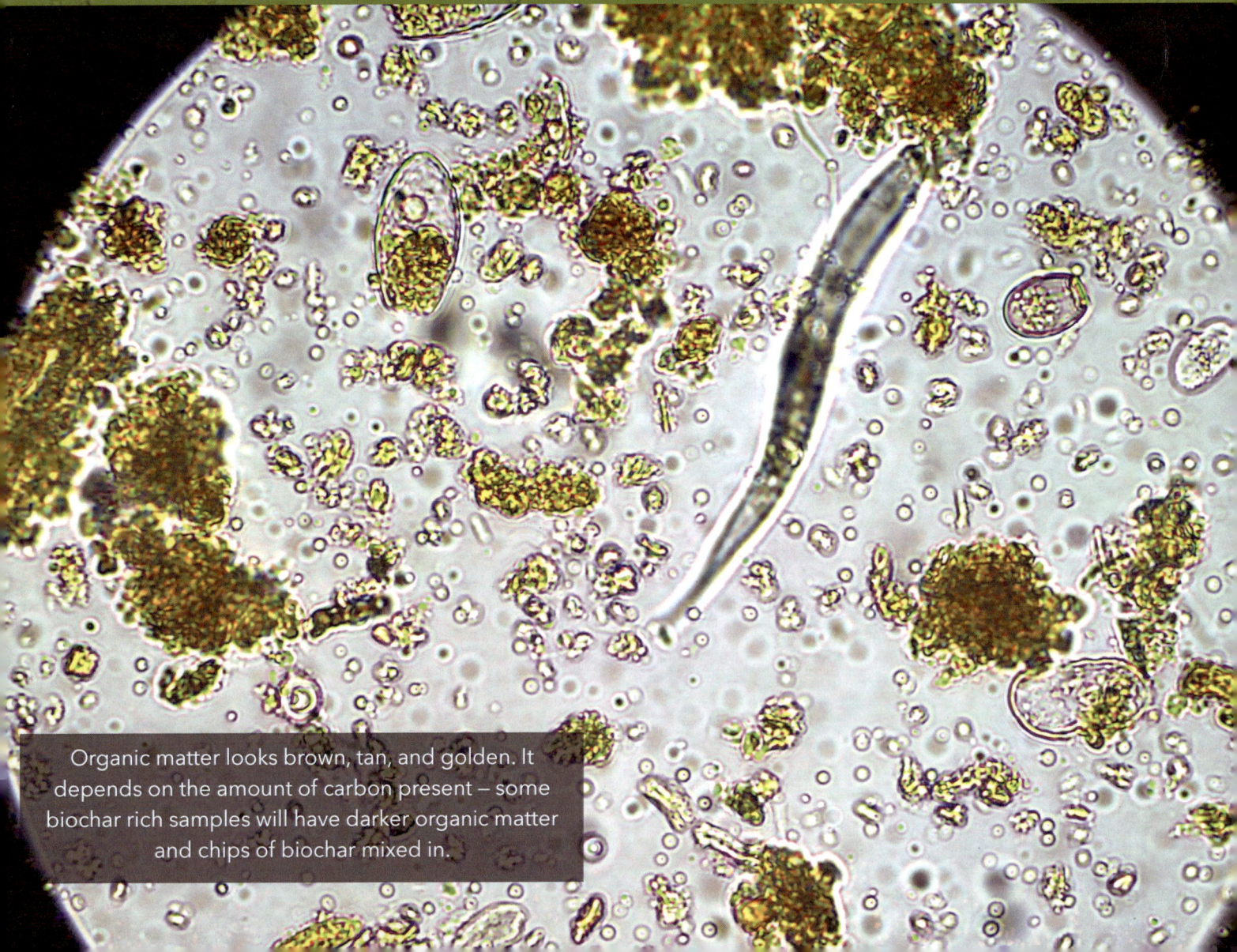

Organic matter looks brown, tan, and golden. It depends on the amount of carbon present – some biochar rich samples will have darker organic matter and chips of biochar mixed in.

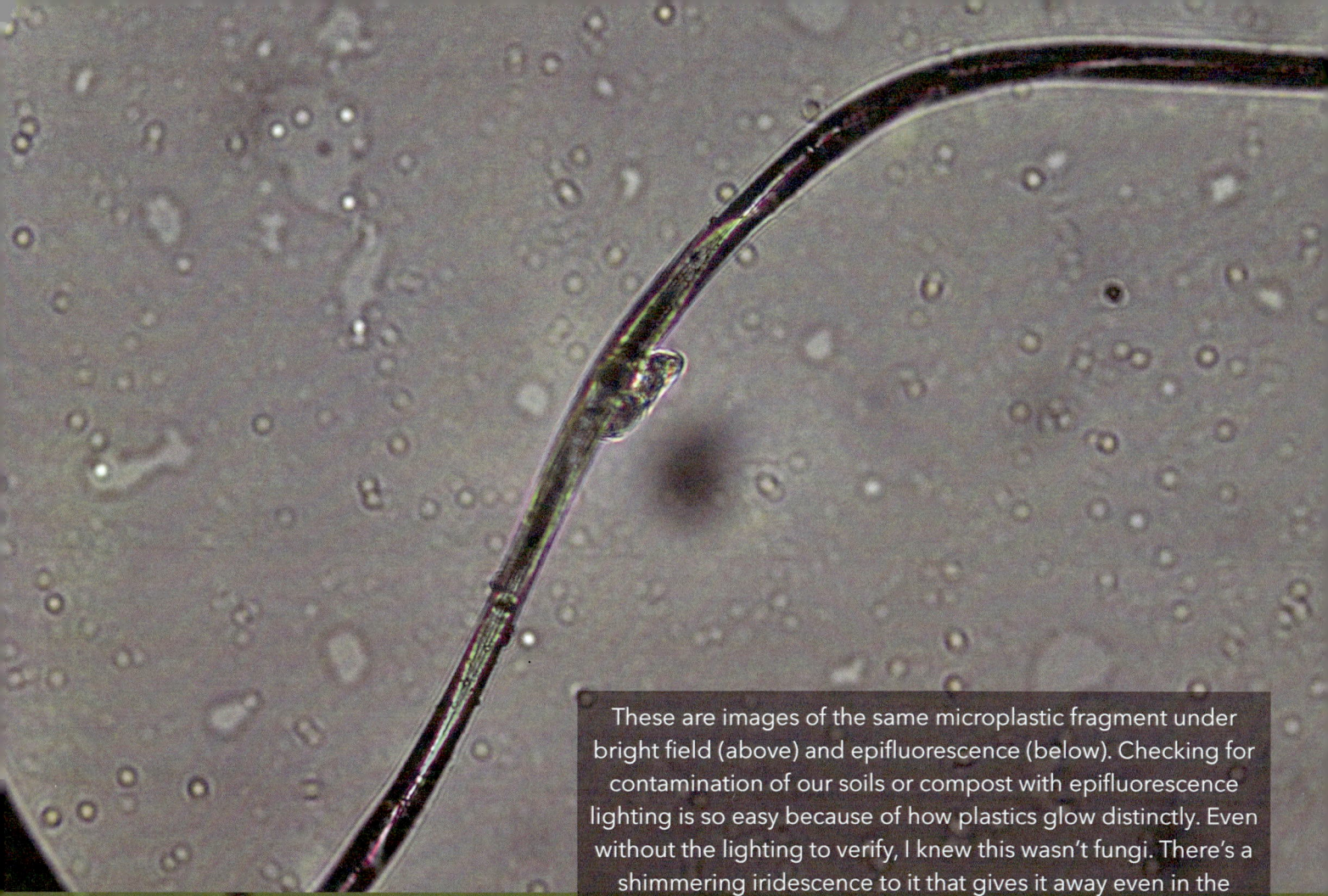

These are images of the same microplastic fragment under bright field (above) and epifluorescence (below). Checking for contamination of our soils or compost with epifluorescence lighting is so easy because of how plastics glow distinctly. Even without the lighting to verify, I knew this wasn't fungi. There's a shimmering iridescence to it that gives it away even in the bright field.

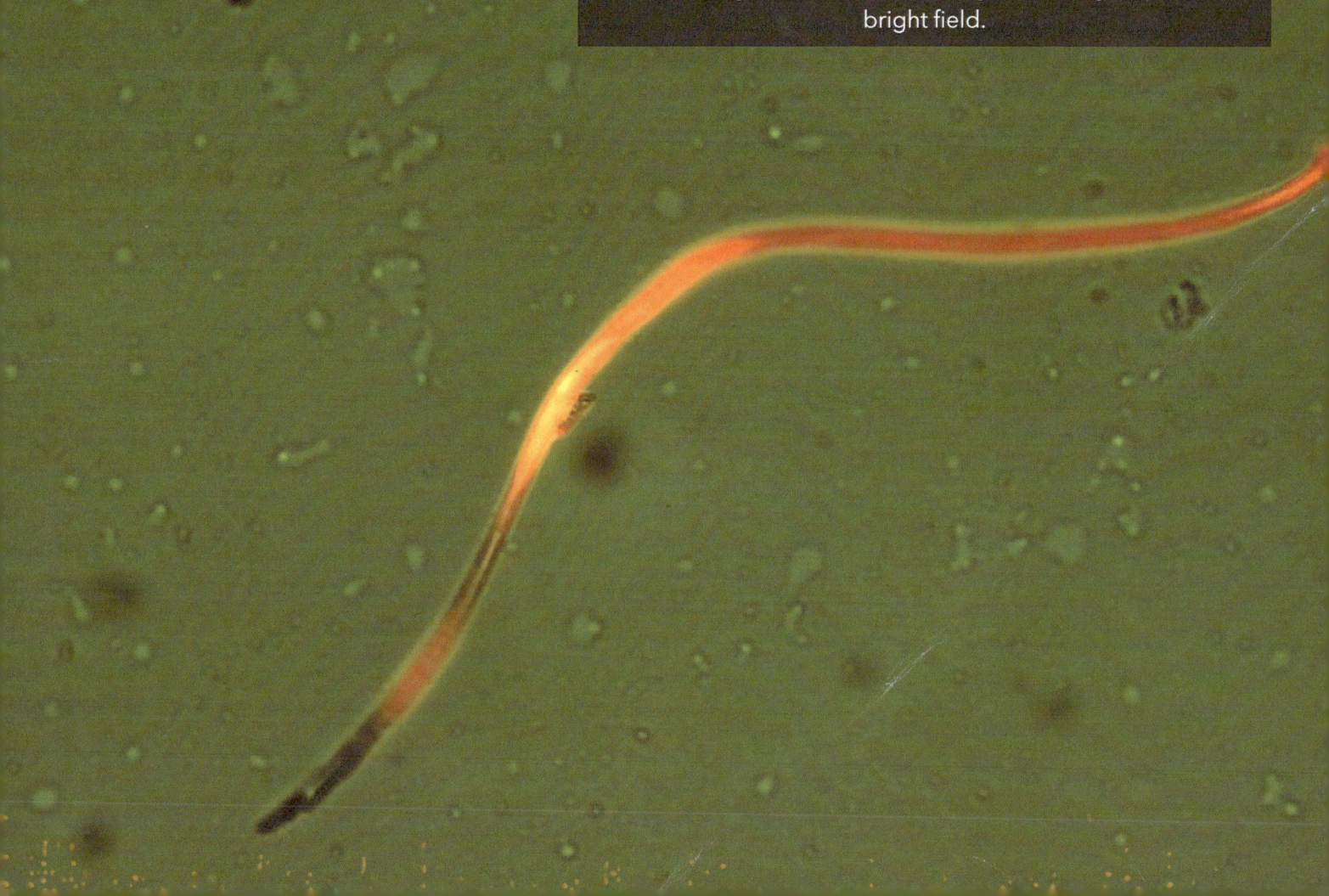

Humic compounds are clusters of organic matter that are brown to golden in color – there are many edges, and it's messy looking, but overall the edges have a rounded feel to them, giving the organic matter a slightly puffy or squishy appearance. The edges are usually lighter in color with the darkest colors being found in the center. Organic matter can allow light to pass through it easily, and we can see inside it, including all the microbes, spores, and hyphae.

Sometimes micro plastics look blue or aqua neon, but it's always somehow off, somehow unnatural in comparison to everything else we normally see.

Matt Powers © 20

In the dark field, the natural colors come through much stronger than through the bright field which either creates a silhouette or shines through the humic compound. Dark field lets us be gentle with the light. Both are useful but can see different aspects of the same space. Using epifluorescent lighting, we can see instantly if our organic matter is or has been inoculated by fungi, and we can see if fungi has been there recently as well. (This does not include basidiomycota which do not always fluoresce under epifluorescent lighting). The advantage of being able to quickly visualize the fungal activity is profound – with the flip of a switch, you can see if

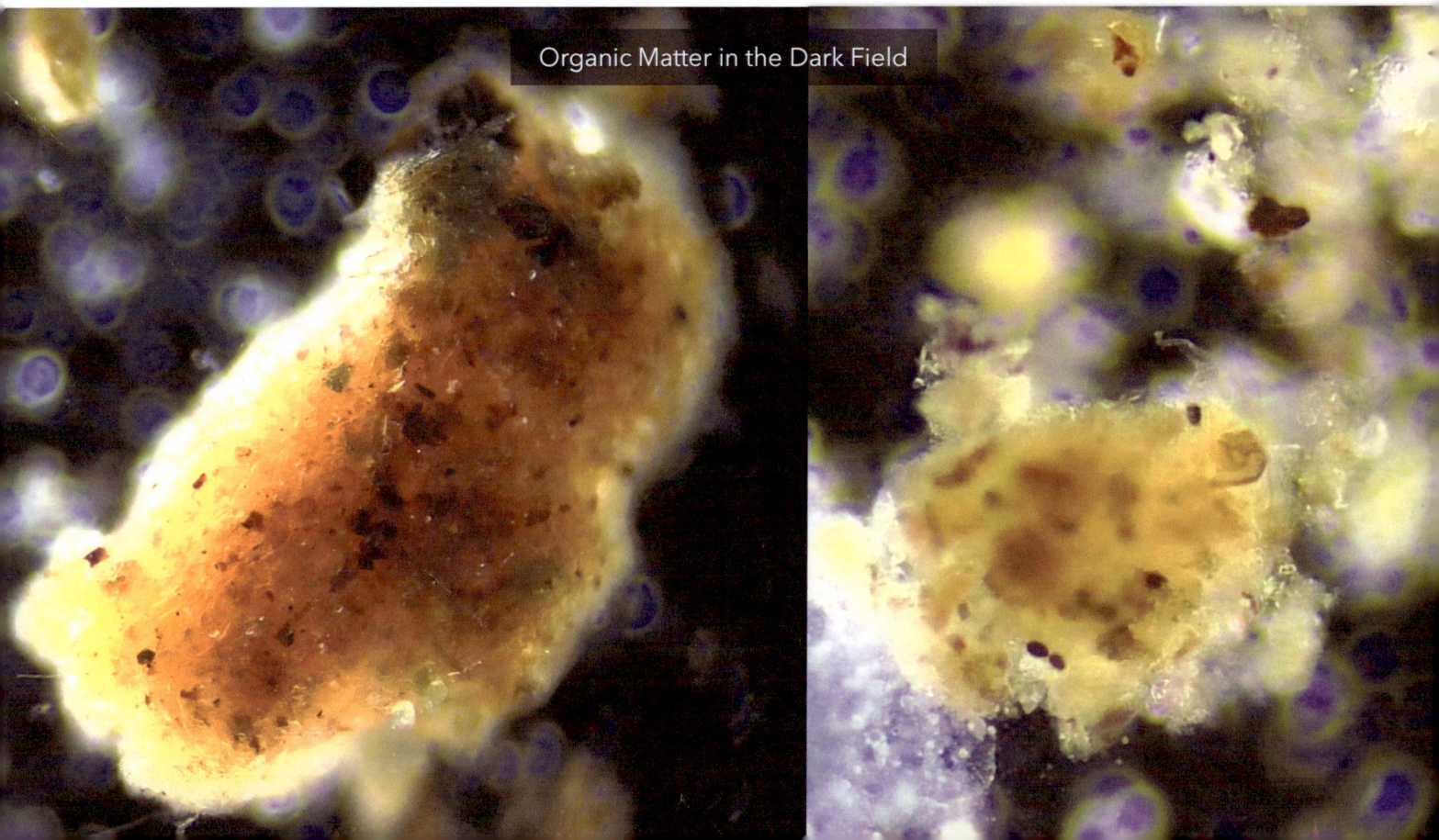

Organic Matter in the Dark Field

your organic matter is being digested by fungi. Even if you don't see any fungal strands of basidiomycota, if your organic matter is glowing like these images (top right and bottom right), you know that your pile is fungal. The reductionist method of using only basidiomycota to determine fungal levels is over. We can fool ourselves too easily if we rely upon only one form of testing.

Sometimes we have fragments of biochar or black minerals that do not allow light through them or into them at all. It's hard to know what they are, but I've come across them a few times. They could be large biochar chips or pieces of rock like obsidian though I think obsidian is likely translucent microscopically.

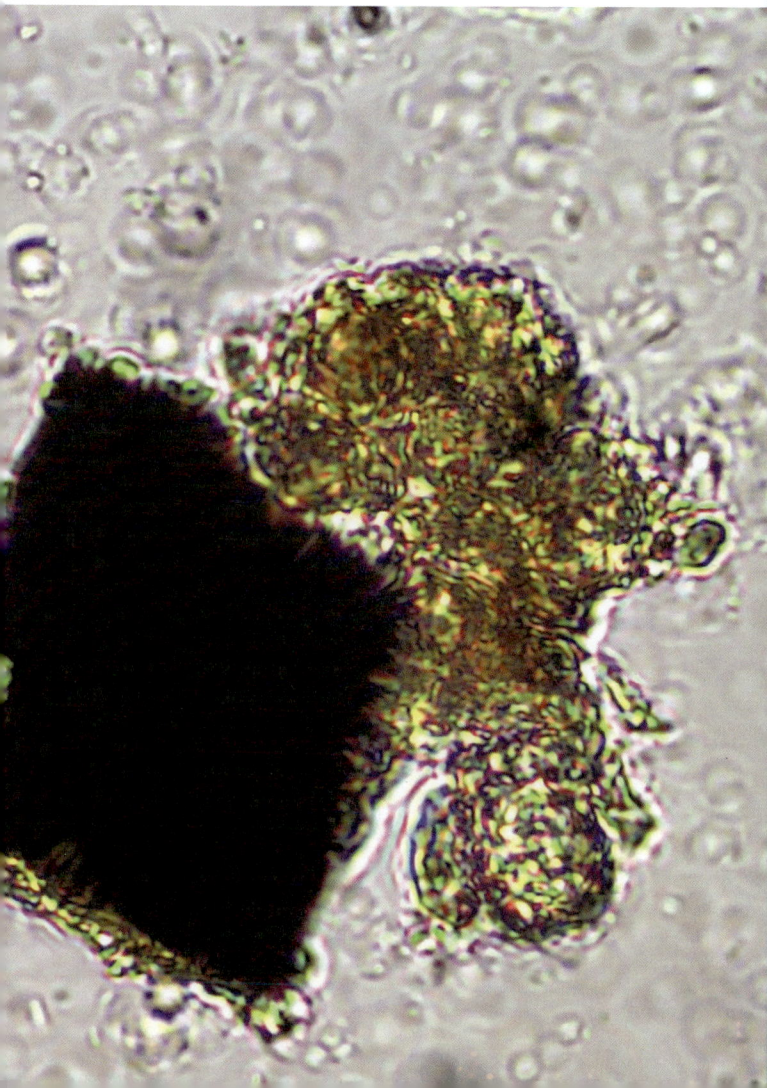

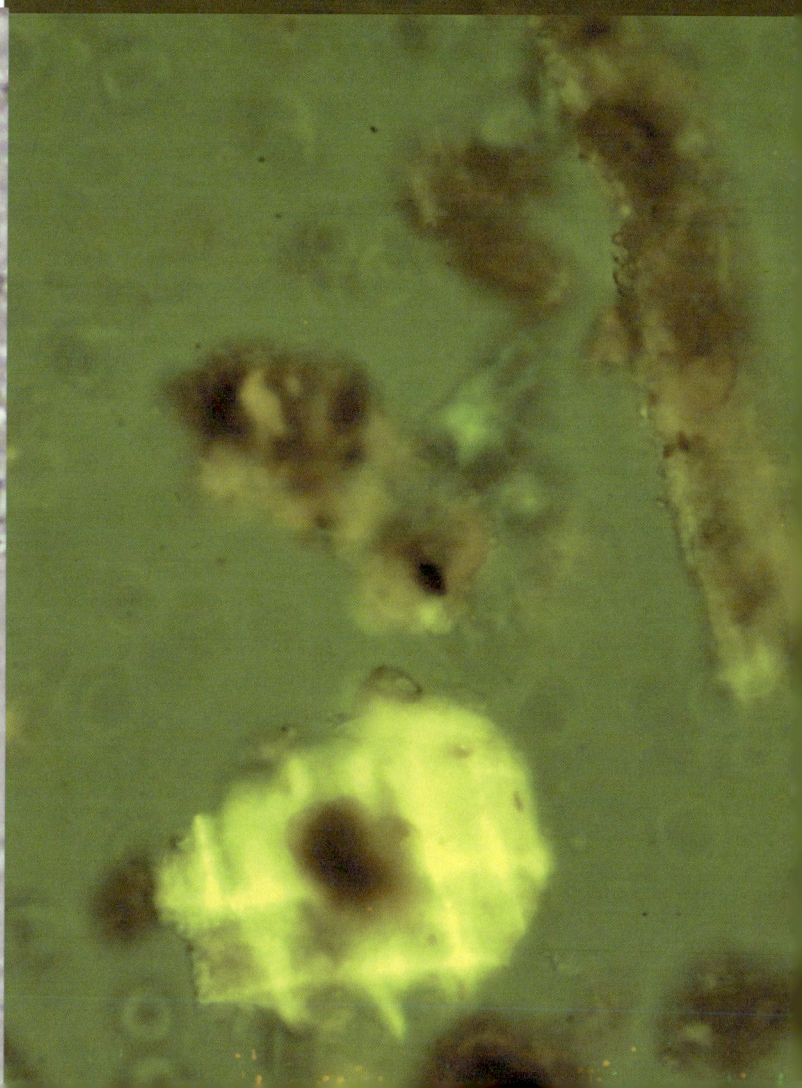

THE TEST PROTOCOLS

WHAT ARE WE TESTING?

What's our context? Are we looking at compost or soil? We can expect our compost to be more rich than our soil ALWAYS but sometimes it doesn't work out that way: that's why we test, and that is also why we have to understand our context. It's not just MORE fungi and bacteria we need either. If we just look at fungi and bacteria in general, we'll fail to differentiate the beneficial microbes from the detrimental ones. This is why I created this book and began this journey.

WHAT ARE THE VARIABLES?

While it may seem obvious that the variables are the water, the soil or compost sample, and the way it was prepared before it was applied to the slide, BUT even if we have the same water, same sampled soil or compost, and prepare it the same way we still have variables at play. A droplet of soil solution differs in its size depending on the force applied to the pipette and the viscosity and surface tension of the liquid – in other words, everyone squeezes out a different size drop from the same pipette and the composition and dilution of your sample will also affect the drop size. When you are counting down to the exact microbe and then multiplying your numbers up to microbes per gram, this matters a lot and will shift numbers dramatically which in turn shifts what we perceive is happening in our soil or compost. We can all use calibrated micropipettes THOUGH I think we'd have to clearly mark that micropipette and dedicate it to that purpose: I wouldn't want to use it for DNA work ever again because of the potential for compost and compost tea to release microbes into the air inside the pipette and for that to get into the upper parts of the pipette (I can zap it in my UV chamber but that's only where the light touches, and there's a limit to the times you can do that, so I want to do it minimally). Even if we calibrated them all, the viscosity would still be a variable as would dilution rates, the water, the ambient temperatures, the humidity, etc. All have an effect, so we have to do our best to work through or around these variables to get a clear picture of what is really happening. If we rely heavily or solely upon numbers generated through microscopic counting and calculating up to microbes per gram of soil or compost, we'll be consistently fooled. Only when we take a step back, introduce other forms of testing to calibrate our microscope numbers can we truly understand them. The great news is – we can do that right now with this book.

WHAT SIZE SLIDE COVER AM I USING?

22mm x 22mm square and 0.16mm thick – depending on the size droplet this slide size can surf, rest on grit, or be just perfect. This is the size I've used for the entire book and all the images. Online it is seemingly the standard, so you shouldn't have any trouble finding them, but even if you can only find 18mm x 18mm, a drop is a drop for our purposes.

What size pipette am I using?

3 ml disposable pipettes usually – because of the variance in size, it's much more important to measure the soil microbe ratios and look to the dilution rates than it is to try and perfectly count each bacteria and then scale up that number as if were a perfect representation of truth: it is fundamentally misleading to do that because of the variance in droplet size and fields of vision vary in their count depending on the size. If things are balanced in their ratios then they can be diluted in the soil or concentrated in the compost, but each will have a variable rate of expression in terms of numeracy in the soil or compost profile always.

How many fields of view per slide?

It differs depending on your camera type/style, the magnification setting on your camera, the width of your objectives, and if you have a 0.5x reducing lens for a better field of view (which I have, and I picked a spot for it and left it so it doesn't change). This means that fields of view per your microscope may slightly differ from my own, so again, the ratios of soil microbes are the best and most trustworthy guide for us to base our conclusions upon because even if your microscope is

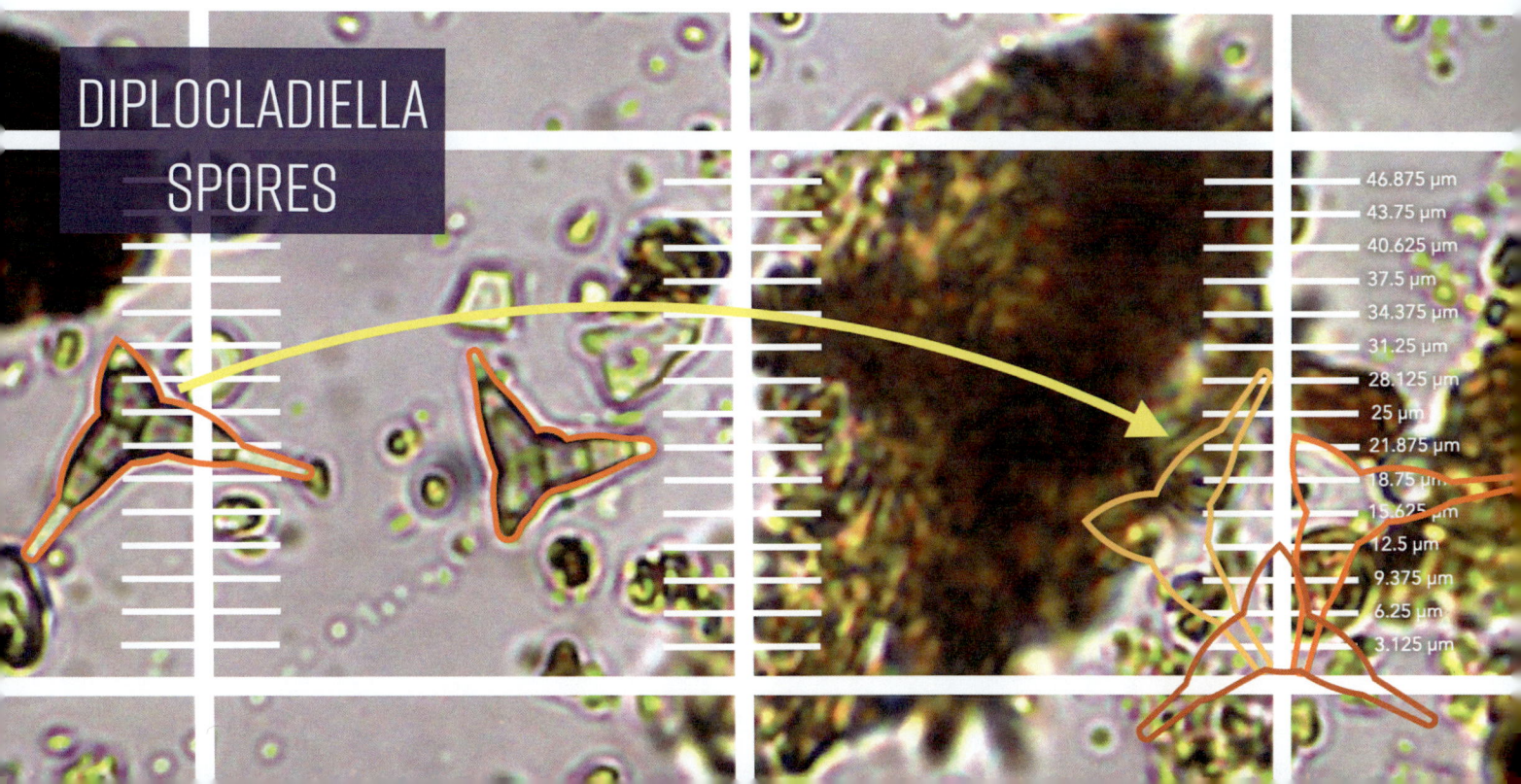

DIPLOCLADIELLA SPORES

46.875 µm
43.75 µm
40.625 µm
37.5 µm
34.375 µm
31.25 µm
28.125 µm
25 µm
21.875 µm
18.75 µm
15.625 µm
12.5 µm
9.375 µm
6.25 µm
3.125 µm

different from mine or anyone else's, the ratios can be scaled up or down and still remain useful benchmarks. This also means that we need to use rulers or grids like a hemocytometer or micrometer ruler for quick counting and measuring especially as we are developing our eye – we can develop digital versions of these rulers and grids to use for our own setup by taking images of them at each magnification then tracing them along with the field of view in a graphic editing program like Keynote (Mac) or other graphic editing or presenting software: you can then export your drawing only as a transparent .png file and then super impose that over any microscope image, match the field of view for that magnification, and then measure or count anything you like accurately – we just have to develop a pattern for documenting our images that keeps track of our magnification levels (objectives), and sometimes, we have to verify the legitimacy of our rulers by lining up several rulers against each other and seeing if they match. Usually, the more expensive guides are better, but you can also find cheaper guides that are nearly as accurate. In those same programs, we can trace our spores, and measure the traced shapes. We can isolate parts of images so we can more easily bring attention

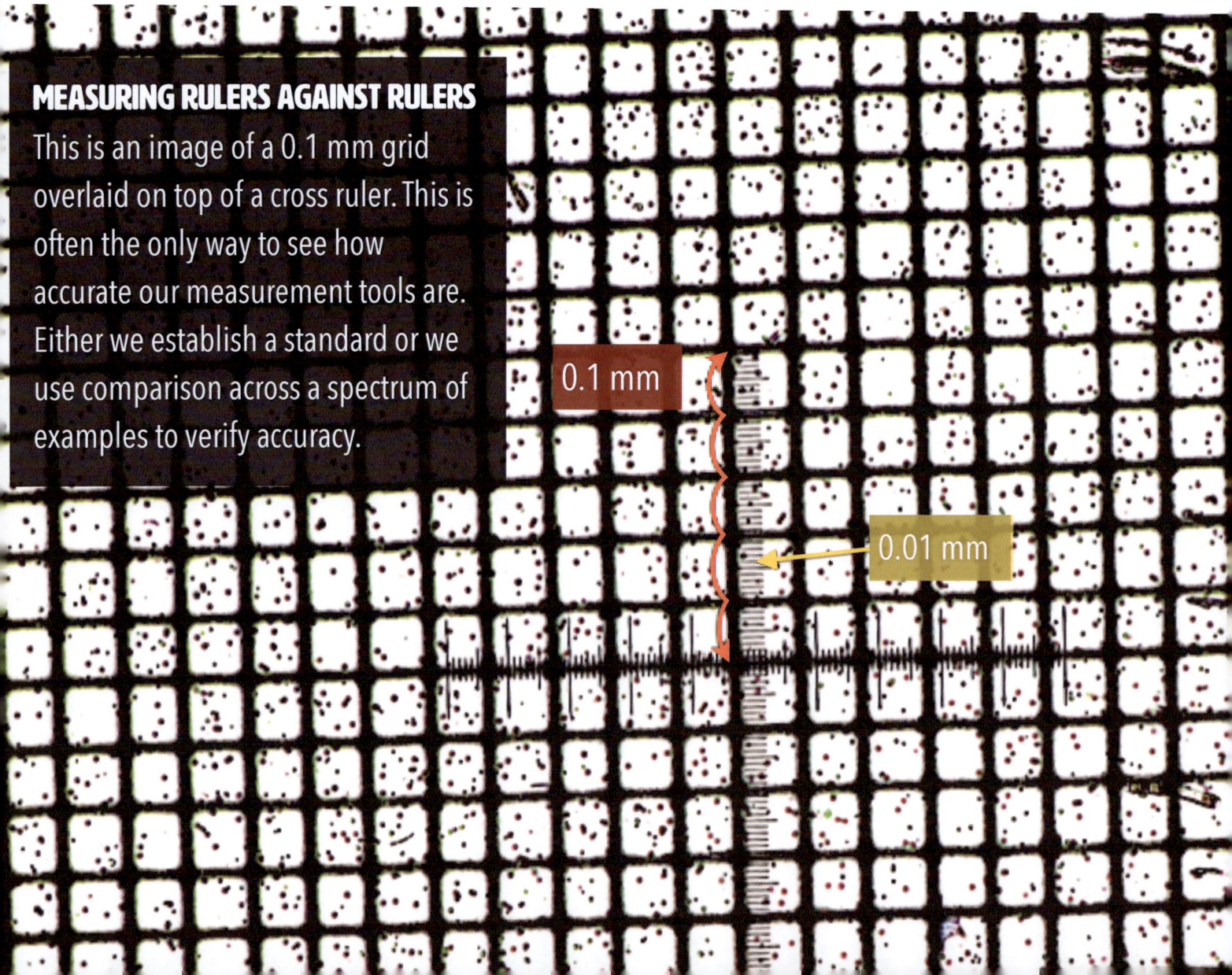

MEASURING RULERS AGAINST RULERS
This is an image of a 0.1 mm grid overlaid on top of a cross ruler. This is often the only way to see how accurate our measurement tools are. Either we establish a standard or we use comparison across a spectrum of examples to verify accuracy.

0.1 mm

0.01 mm

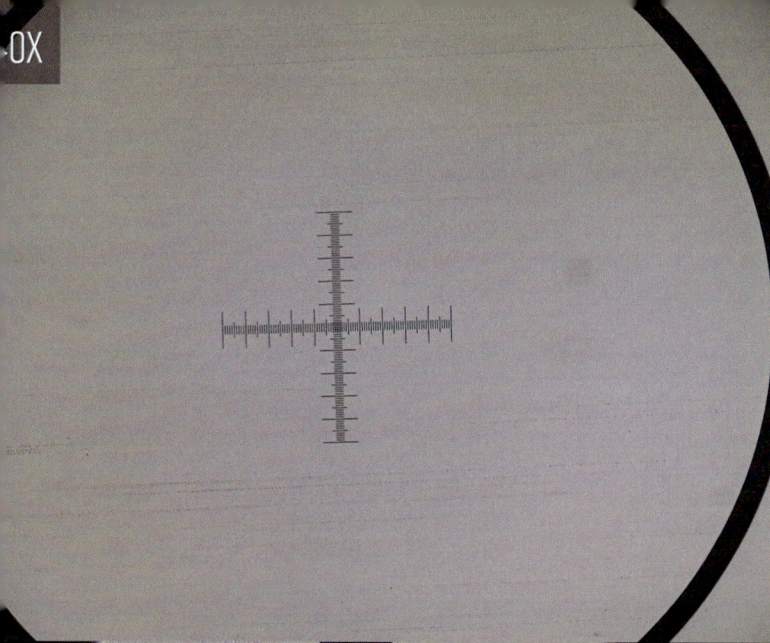

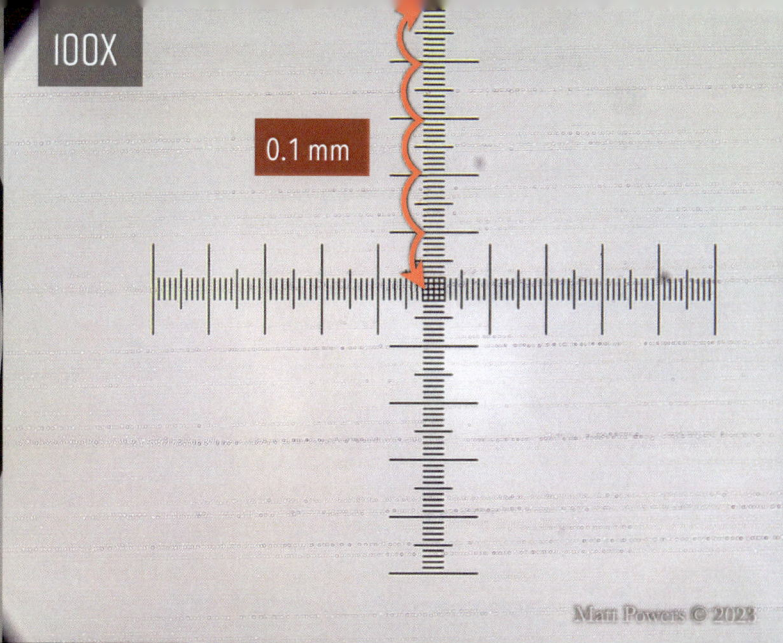

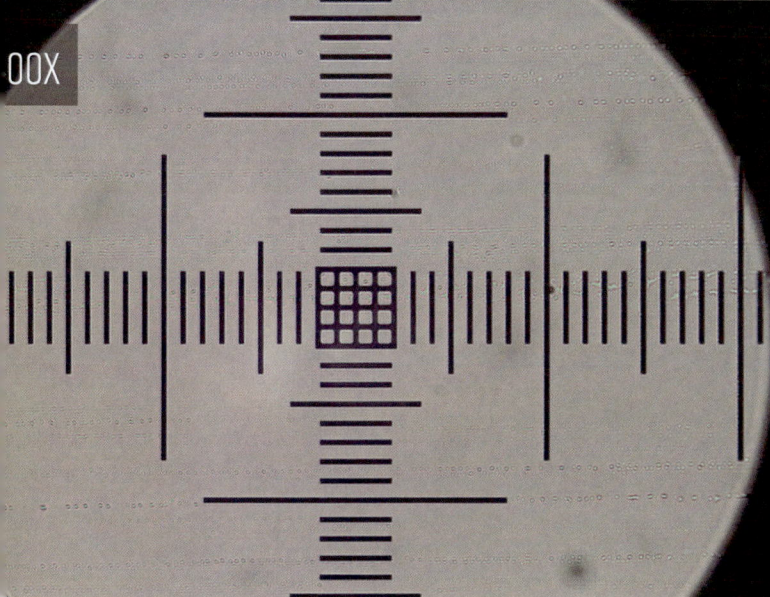

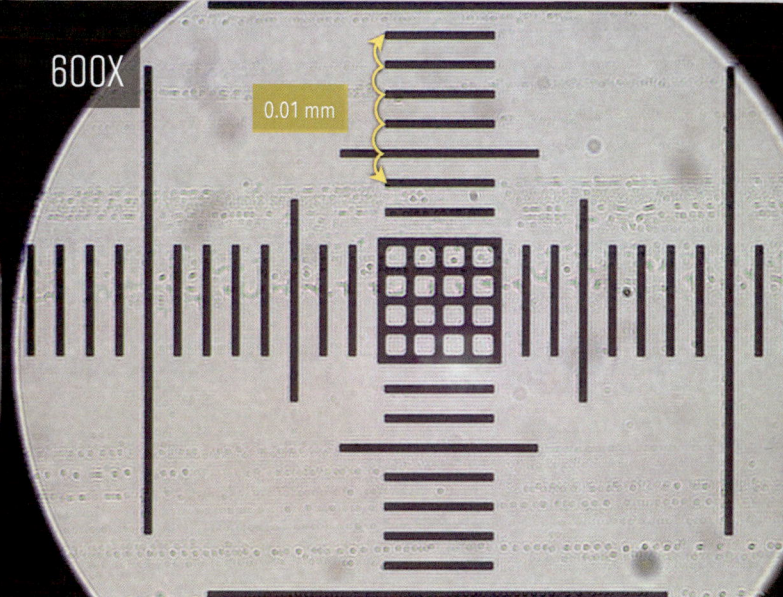

to them or discuss them. Helicon imaging software allows us to make composites of the same image by combining the in-focus portions into a whole image. This is incredibly useful software to help us better visualize this space. You may develop your own patterns for working through and around the variables, for measuring, for double checking, and that is excellent – as citizen scientists, we should be always testing, comparing, and seeking to know more.

TROPHIC RATIOS = THE THREAD COMMON TO EVERY CONTEXT

Keeping the soil food web in mind, all the trophic layers, all the ways that plants are getting nutrients, and the nutrient cycles themselves, a dynamic picture emerges of our soil and compost. We can't approach things linearly and we have to be careful how we generalize though we can still do some of that. We have to consider all these pathways of nutrition and exchange when we are examining soils, plants, or compost (check out the images on the following pages and check out the first book in this trilogy for more details). We can examine indirectly and sometimes even directly witness some of these processes using our microscopes with the methods in this book. If we don't run our thinking through these cycles and pathways, we may miss how our plants are feeding themselves or we might miss how they'd prefer to feed!

Just Remember: Your droplet, slide size, objective width, and camera setup may be different, but your trophic microbial ratios allow us to work around those hard realities and limitations.

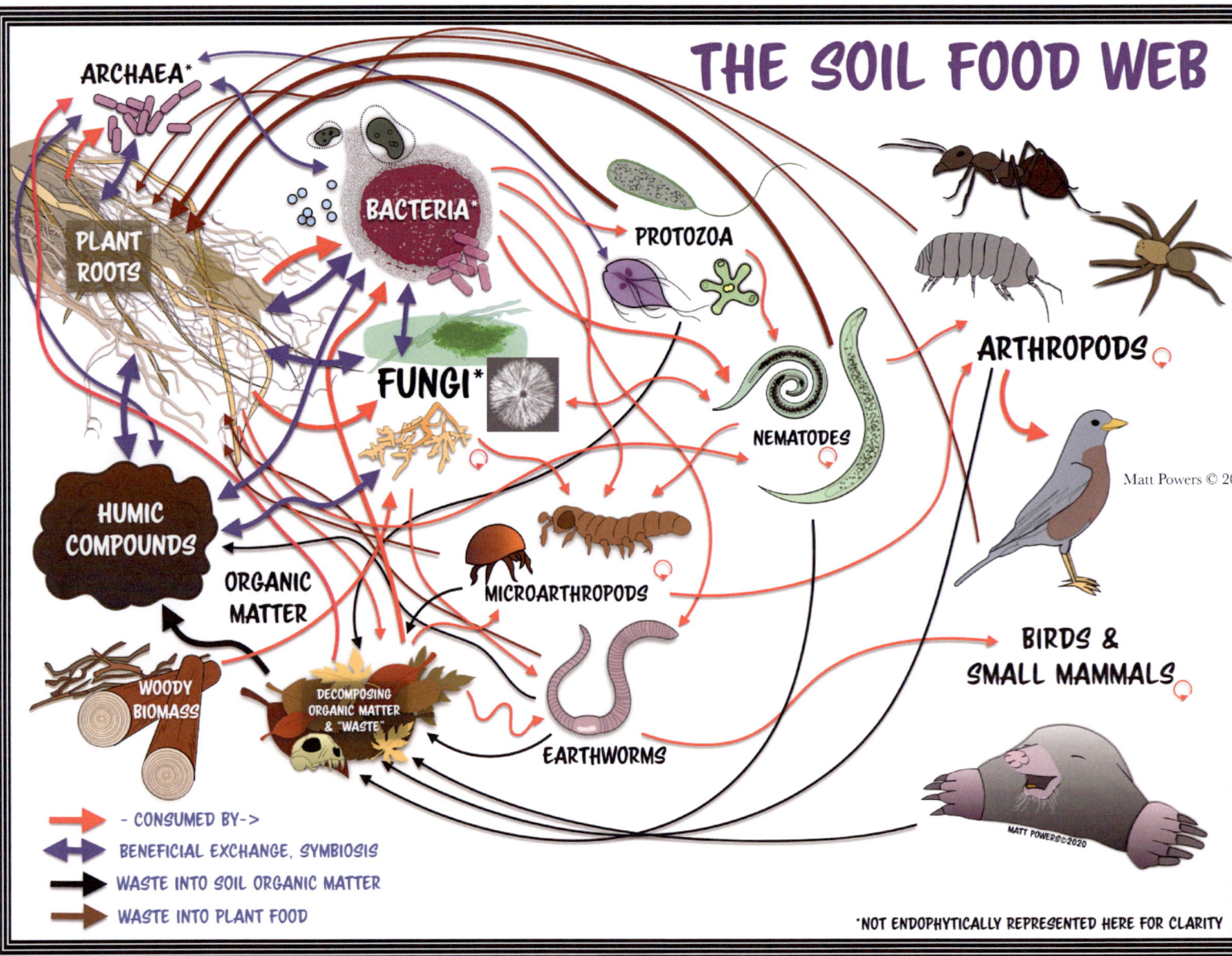

THE SOIL FOOD WEB

ARCHAEA*

BACTERIA*

PLANT ROOTS

FUNGI*

PROTOZOA

NEMATODES

ARTHROPODS

HUMIC COMPOUNDS

ORGANIC MATTER

MICROARTHROPODS

BIRDS & SMALL MAMMALS

WOODY BIOMASS

DECOMPOSING ORGANIC MATTER & "WASTE"

EARTHWORMS

Matt Powers © 20

MATT POWERS©2020

- CONSUMED BY->
BENEFICIAL EXCHANGE, SYMBIOSIS
WASTE INTO SOIL ORGANIC MATTER
WASTE INTO PLANT FOOD

*NOT ENDOPHYTICALLY REPRESENTED HERE FOR CLARITY

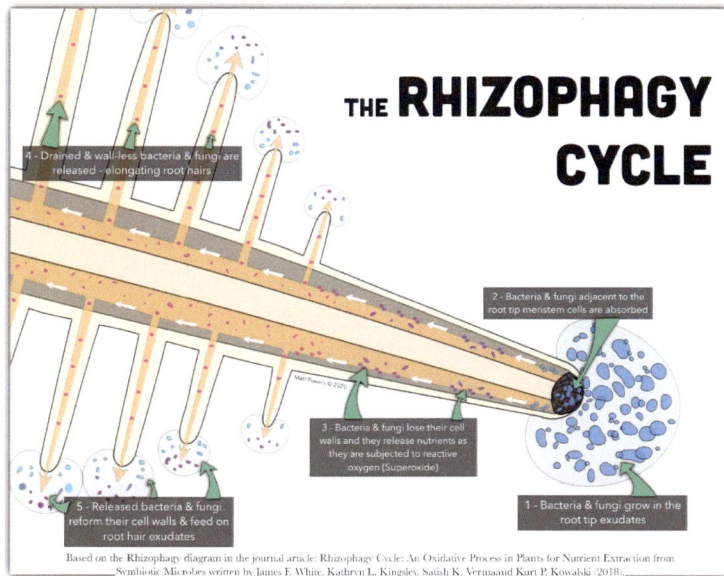

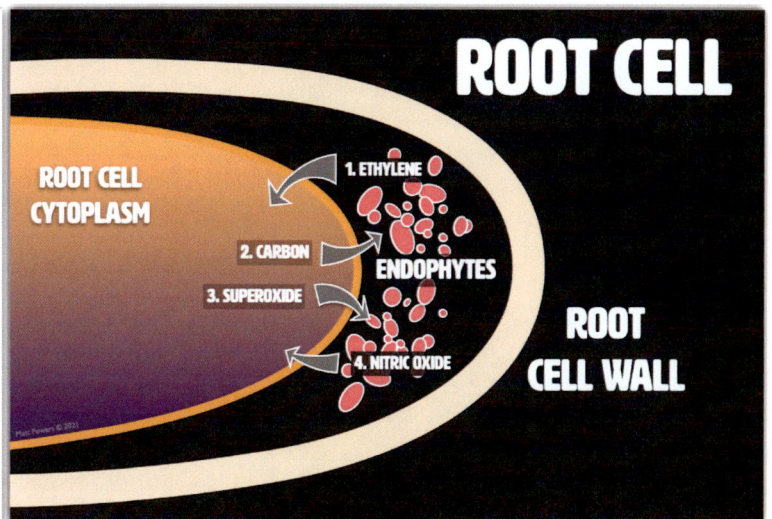

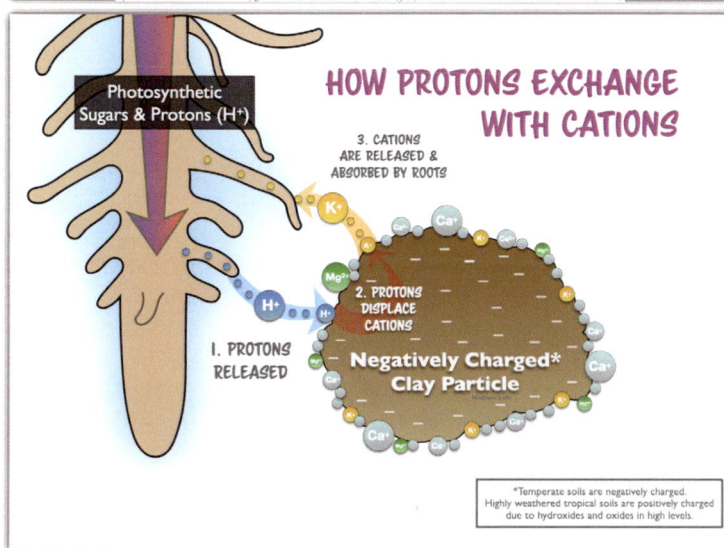

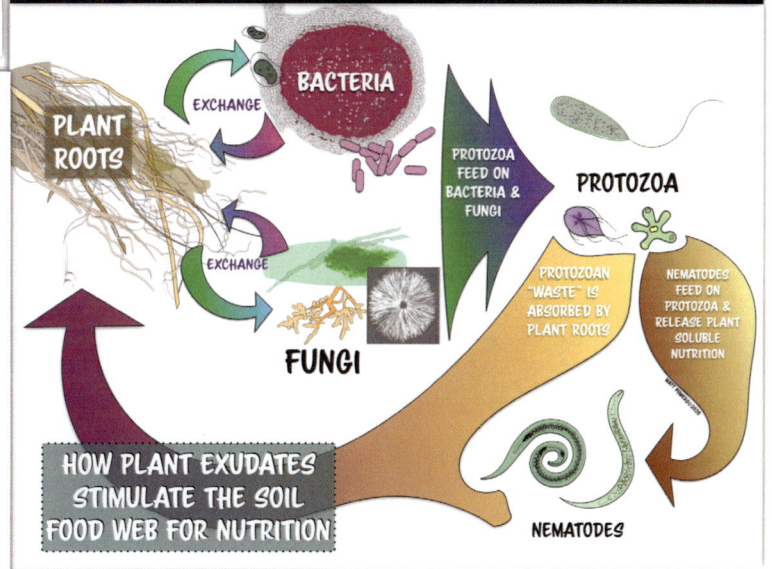

These images are from **Regenerative Soil** the 1st book in this trilogy.

The Soil Food Web Trophic Levels: Organic Matter & Plant Roots [the Trophic Foundation] – Bacteria [Holding Nutrients] – Fungal [Holding Nutrients] – Protozoa [Releasing] – Nematodes [Releasing] – Microarthropods & Worms [Releasing] – Larger Animals & Earthworms [Releasing]

KEY QUESTIONS:

- Do you have plant roots & organic matter at sufficient levels? Can you improve these? How?

- Does your soil or compost contain bacteria at sufficient levels? Too much? Too little?

- Do you have a diversity and sufficient abundance of fungi and fungal spores in your soil or compost?

- Do you have protozoa at high enough numbers? Nematodes? Microarthropods? Worms?

- Do you have potential pathogens? Do you have an imbalance in the nutrient cycles?

FOUNDATIONAL BEST PRACTICES

- Always Try To Have a Control for Comparison (another sample, a reference book, an online key or database, etc.)

- Always (When Possible) Allow Microbes To CYCLE through their LIFE STAGES (give things time)

- Always Film or Capture Images of your Work for later reference (reflect back on your work with new eyes)

- Always Do MORE TESTING Than Just Microscopy (combine as many tests as possible for higher levels of accuracy and understanding)

- Always Look To The Context To Guide Your Interpretation (look to the holistic context)

RULES OF THUMB:

- Avoid Definitive Conclusions – keeping an open mind allows us to account for lookalikes, potentially new exceptions, and the limitations of morphological identification as a practice. Bringing in other forms of testing, like DNA, allows us to develop more definitive conclusions.

- Control as many variables as possible – even a drop is variable in its contents.

- Dilute to 1:100 to count bacteria using an accurate hemocytometer .

- If you lack protozoa and nematodes, you are missing the main nutrient cyclers.

- Beneficial compost is always going to be more diverse than soil because of the organic matter levels: expect your good soil samples to be on average 2 – 4 times less diverse and abundant microbially than your good compost samples.

- Open septa hyphae are considered non-pathogenic when teeming with bacteria.

- If your lighting is off as you move up in magnification through your objectives, you will need to manually manipulate the condenser in its slot to align the objective with the light – especially with dark field. If your light is not aligned initially you won't see anything clearly.

- If your sample is too dense with life or minerals, it will create layers in the field of view, upper and lower or even upper, lower, and middle. Make sure to view each of them – different microbes prefer different layers in the slide as well as in the test tube.

- Turn the lights down in the room for better images – start always with no or very little light and slowly increase until you reach the ideal levels.

- Don't be afraid to move the condenser up and down for fine tuning – wiggling everything a bit to tighten things up is always how I tend to do things.
- Don't be afraid to dim the microscope LIGHT source and BOOST the BRIGHTNESS on your Camera!
- Don't be afraid to put things under the microscope WITHOUT A COVER SLIP!

RSM Soil & Compost Testing Protocols

GATHER YOUR SAMPLES

Now that we've spent the time to learn how to identify and recognize the microbes and minerals we will see under the microscope's magnification, we need to know how to measure, calculate, evaluate, and catalog what we find. We need to collect our samples without damaging the microbes as much as possible and without contaminating them. Avoid taking just one sample as soil diversity is so great that we can hit a sampling that only has one piece of the large picture: we could find that particular sample to be a poor representative of the actual soil as whole field or garden. Take samples from across an evenly distributed matrix or grid across the area to be tested. Soil samples are typically from the top 6 – 8" of soil. While some research will say they've examined the first foot of soil, that will skew our readings a bit as the top 6 – 8" is where almost all the bioavailable nutrients are, so adding in lower horizons of soil will only make them seem less diverse and confuse things a bit – they are different areas entirely and plants treat them differently sending tough tap roots down into the subsoils and more sensitive and tender roots out laterally. Tap roots are for water primarily while lateral roots are for nutrition. This plays into how we sample roots – we need to understand the context there as well: is it young or old? Is it the actual tip or an offshoot from the main root body? Is this your garden soil? Is this the fill soil from a construction event on that site? Even the time of year will have an effect on what we will be seeing as we dilute and take things down to one drop to examine them.

A soil coring tool is often recommended and used to take soil samples of the top 6-8 inches of soil. Often samples are taken from a grid for mineral testing, then mixed thoroughly in a large container or bucket, and then a new sample is made out of the combined samples which is tested to give a generalized picture of the entire area. This avoids outlier samples and even physical distribution of differentiation: you can't see how things play out on the land – you might have a dead spot or a fertile patch, but by mixing like that, you'll always miss things. It's more expensive if you are paying someone else to do the work, but if we are doing it for ourselves, we can get a home NPK kit from a reputable source like

LaMotte, and we can do these tests on our own without having to mix everything up into an average that can confuse or trick us. The university or commercial lab can tell you the ppm of this and that, but doing a high to low gradient test we can do ourselves more regularly could actually give us more accurate information in terms of what actions to take where and when. The same principles are at work with testing biology: if we mix it up, we'll destroy things and muddle the picture we are trying to form. We can take cores, but the sampling is rather narrow, i.e. it doesn't contain a large enough intact volume of soil. I much prefer the trowel or shovel method where we excavate a section of a larger section of soil that has a greater diameter than a coring tool. 2 – 3 cups of soil in a loosely closed baggie can preserve it's soil life populations for a surprisingly long time: months if not a year, depending on the microbiology and the organic matter in the sample (all other variables accounted for). I've seen it – this is a miracle for soil biology-based businesses: your products maintain their value! As long as they have that undisturbed core environment maintained, they can go on for longer than expected. We can't store these samples in a hot room – try to keep them at ground level (that seems to also really help).

SAMPLE PREPARATION

I gently use the test tube itself to sample from the baggies. You can also use a spoon or another clean tool. The idea is to get a sample that has been handled as little as possible. I take samples from the top, the middle, and the near the bottom of a sample. Each has its own test tube. The 4th test tube is for making our 1:100 dilution, and the 5th test tube is for our Acridine Orange (AO) viability stain prep.

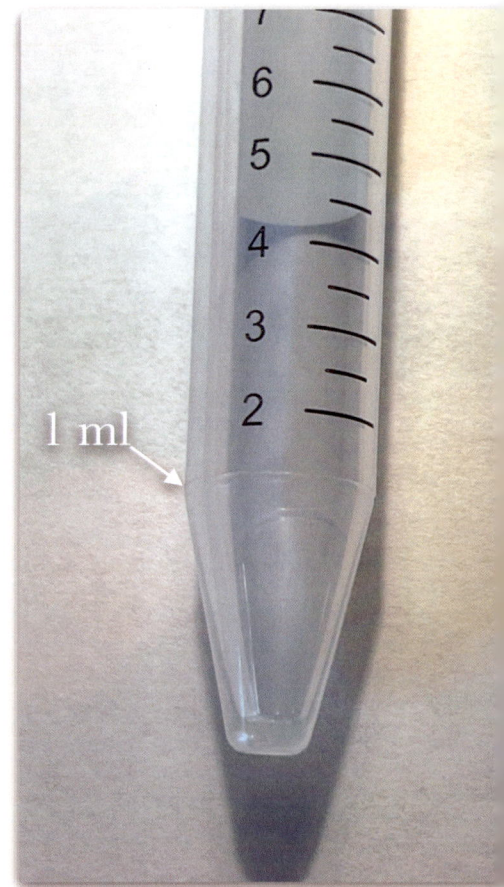

1 ml

- A Compost or Soil Sample
- 5 x 15ml test tubes
- Distilled/Filtered Water
- Diluted Acridine Orange (still fresh) – you'll need 2 – 4 drops *(epifluorescent lighting required)*

With 1 ml of soil in each of the first 3 test tubes, dilute them to 1:10 with filtered or distilled water. That is, fill the test tubes with soil or compost to the line where the test tube narrows (the implied 1 ml line), and then tamp it down gently until the air pockets are gone. Fill the test tubes to the

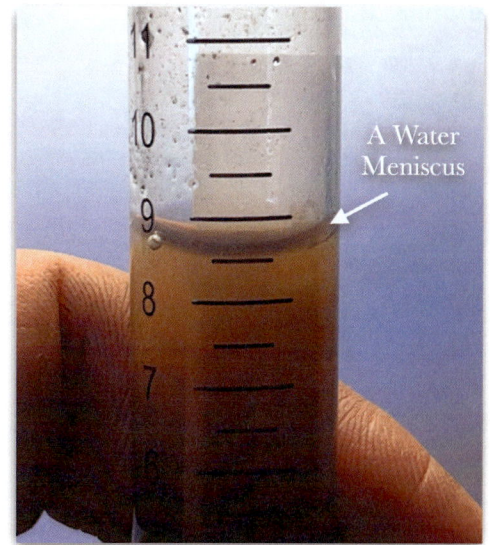
A Water
Meniscus

10 ml line with distilled or filtered water – the water will form a classic meniscus convex on the inside of the tube: use the bottom of the line of the convex curve to determine the ml.

1 ml is equal to a cubic cm and 1 gram when it comes to WATER, so it's not quite the same for soil but it's close enough for this work which is all approximations, characterization, and samplings rather than definitive testing – we cannot test all the soil in a field and count all the microbes, and we cannot know for certain who is really in there without DNA testing, so that's what I mean here. The only way to be truly definitive is to test everything from all possible angles, and that's not possible, so we make approximations, generalizations, and use multiple tests to ground our tentative conclusions in greater certainty.

Next we invert the test tubes, grab all three and do them all at the same time to save time. Keeping our upper arm stiff, we slowly extend our forearm with the samples and then retract it so the samples go from upright to sideways, sloshing the contents about but not shaking them or using a machine to agitate the sample – some have suggested timing the motion to 1 second and doing the motion for 30 seconds total, and then to let the samples rest 10 – 30 seconds, and after that, to take your sample with your pipette. More often than not I watch the heavy particles that the cover slide would rest on fall to the bottom and then I take my sampling when it is clear of those particular particles regardless of how long that takes – sometimes it's only a few seconds. The test tube mixing process is another variable – everyone's arm is different from the length and the speed at which we invert the tubes, so the amount of force will always differ and does differ for each of us as we do this throughout a sampling. Not only that, but the mineral content of the sample can also play a part. If you have mostly coarse river sand or just sand for that matter in a soil sample and you invert a 1:10 dilution for 30 seconds versus a compost sample with no sand diluted 1:10 that you invert for 30 seconds: the sand will grind up and cut everything to pieces – biochar is capable of the same thing. Depending on our sample's mineral content, we may need to be more careful and gentle with how we mix things that 30 seconds at 1 second arm inversions. An extremely gentle inversion event may yield more secrets than a thorough shaking – it depends on what you are measuring and looking for. Adding a surfactant or even a bit of soap can make it so spores are dislodged form organic matter, making it easier to count them, and likely combining this with a vortex mixer will yield even higher reads, but we also have to have systems we can all participate in so we can compare our results in a clear and consistent manner. I think using the above method as a point of departure is wise – by being observant, we can see samples that settle out quickly need to be sampled earlier than 30 seconds. We can also observe the distribution of the organic matter and particles in a test tube as we initially and gently move the water back and forth in the tube. You may have soil high in silt and sand, so you will have it distribute evenly quickly once the hydrophobic bonds are broken.

These 1:10 dilutions are for examining nematodes, protozoa, microarthropods, spores, and fungi. I usually do these examinations first and then move on to the 1:100 dilution, bacterial examination, and hemocytometer work.

Sample with a clean pipette from a variety of depths in the test tube to see the variety that the sample contains. It's important to create movement to prevent microbes sticking to the inside of the pipette: we do this by intaking and releasing liquid from the sample from our target area 2-3 times while the pipette is submerged in the sample. You don't want any air, so you don't want to overdraw and you don't want large pieces of organic matter or minerals in there, so you want to be careful not suck them up. You are essentially mixing the solution one final time before sampling a small amount to put one drop from that amount on a slide. The slide floats on that drop. If you have organic matter that is large and upsetting the slide floating on top, you can draw the drop back and forth with the cover slide and remove the debris.

Some folks put two drops and two coverslips on one slide (on opposite ends) to save time, but you can run out of room and sometimes your slide can slip into the opening for the light path. It doesn't matter how you do it as long as you are comfortable with it – taking a drop from the middle, top, and lower sections of each test tube and looking at them at 1:10 dilution means 9 drops of testing at that dilution. Nematodes tend to sink to the bottom because they are heavier while flagellates tend to be near the top with testate amoebae throughout. The different depths when combined will give us the average and the range of expression for protozoa, nematodes, spores, fungi, and microarthropods. At first counting these will help us develop our eye, but over time, we'll be able to look at a soil sample and know if it's bacterial or fungal dominant without counting. Most samples are starkly missing members of a particular group, and the few that have everything are amazing to witness and hard to not recognize.

SAMPLE EXAMINATION

Document your work as you go to keep track of your samples, the magnifications, and the numbers. You may not be adding this to a database or preparing a report for a client, but things can easily get mixed up if we don't keep things clear.

The Hemocytometer

Tiefe · Depth
Profondeur
0,100mm

Neubauer
improved

0,0025 mm²

LW
Scientific

Simple yet Powerful

It's just a 3mmx3mm grid, but under the microscope and in the chaos of the soil solution, there is often no static frame of reference. Even digital grid overlays and eyepiece grid overlays do not offer the same instant verification and concrete reliability of the hemocytometer.

Trusted and reliable enough for medical applications all over the world, this is a tool that every serious soil lab already has on hand and uses daily.

What is a Hemocytometer?

A hemocytometer is a special microscope slide with a very precise grid etched into it — it's how we count and sometimes measure microbes. Used in a medical laboratory setting primarily, hemocytometers are an essential tool for anyone doing work in this field.

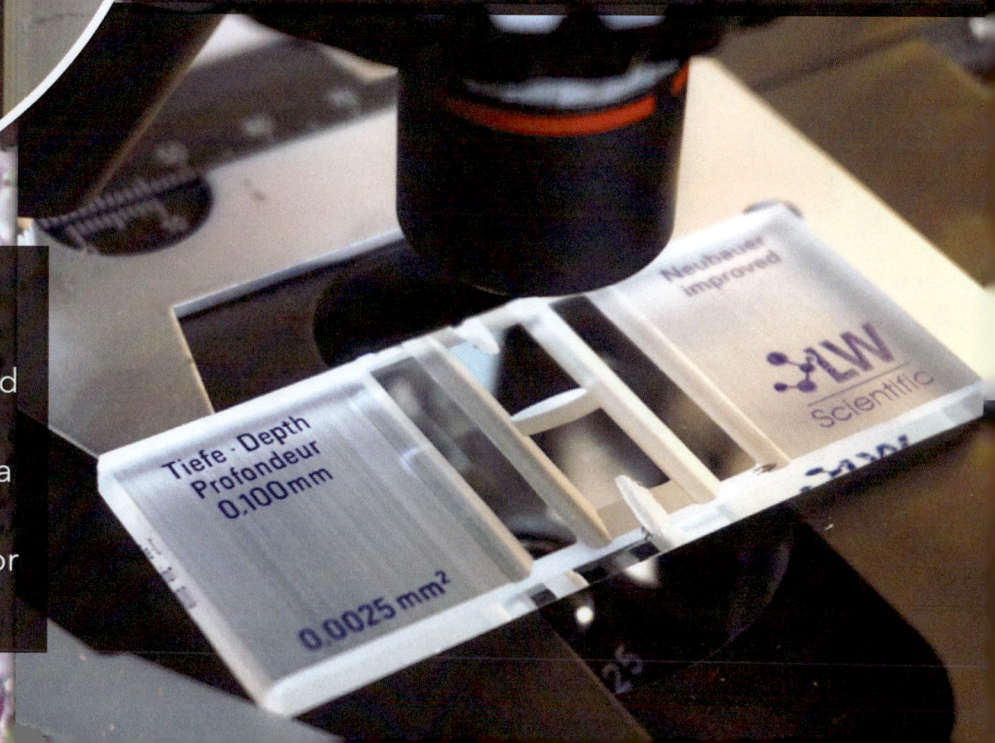

Tiefe · Depth
Profondeur
0,100mm

0,0025 mm²

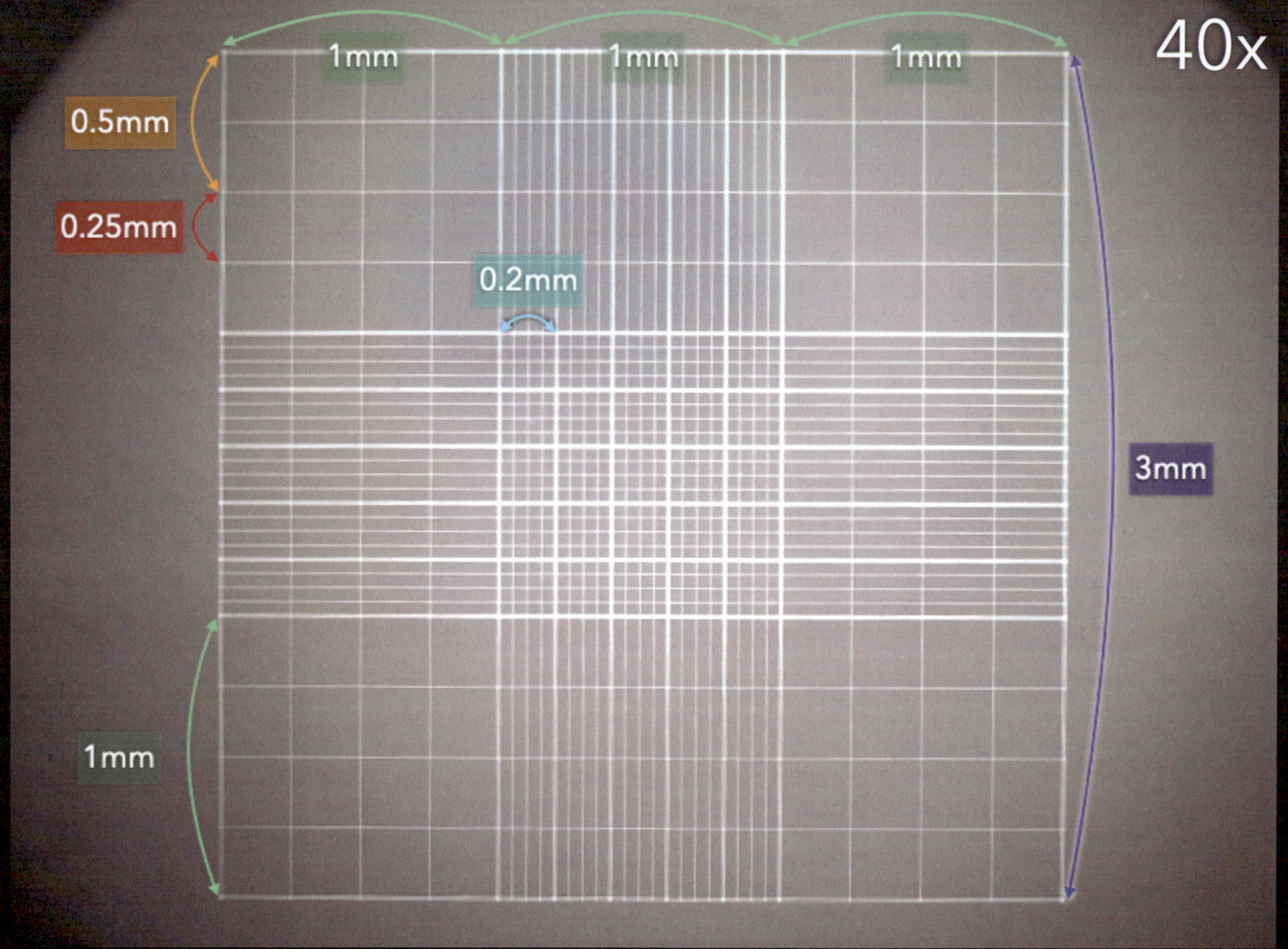

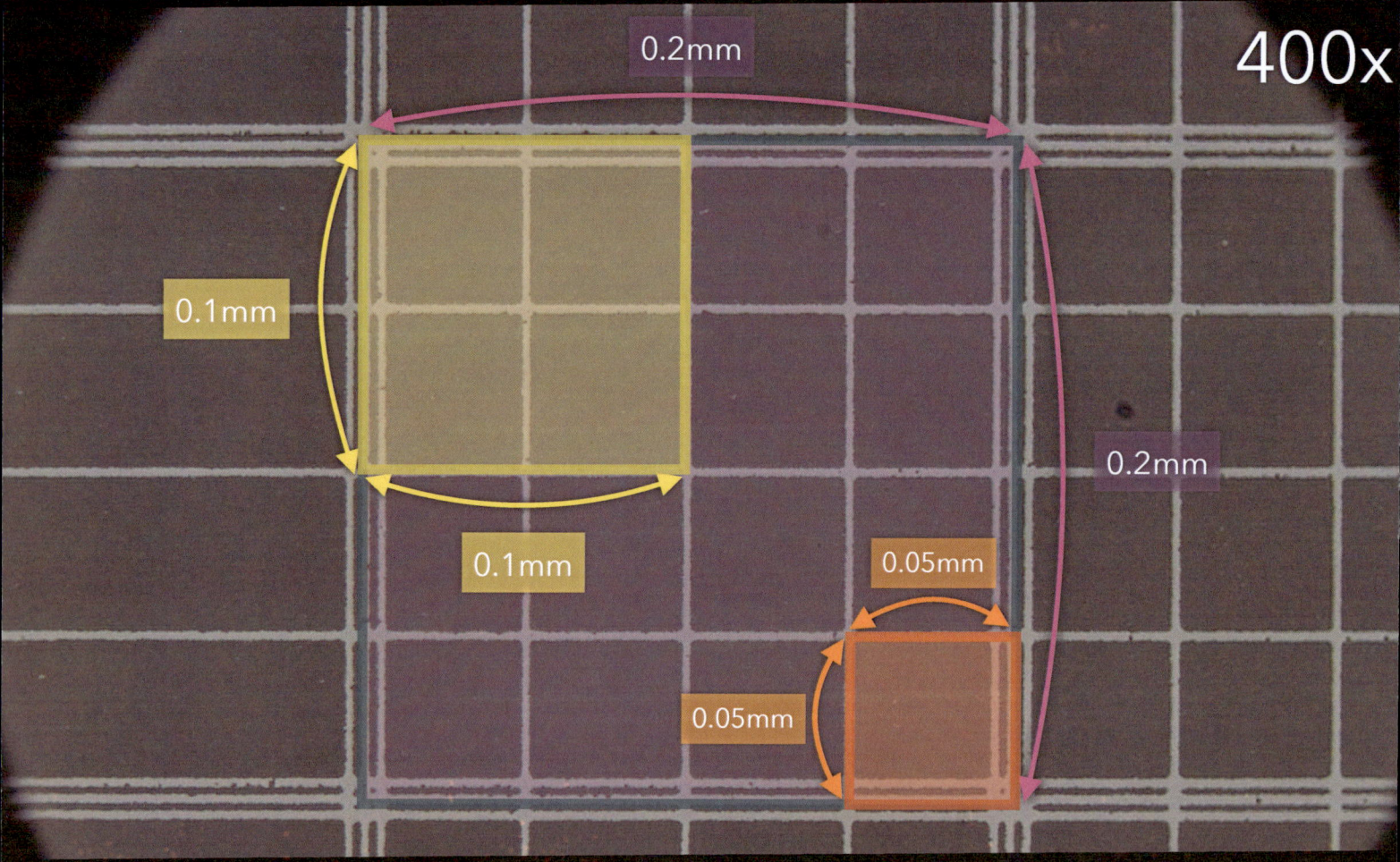

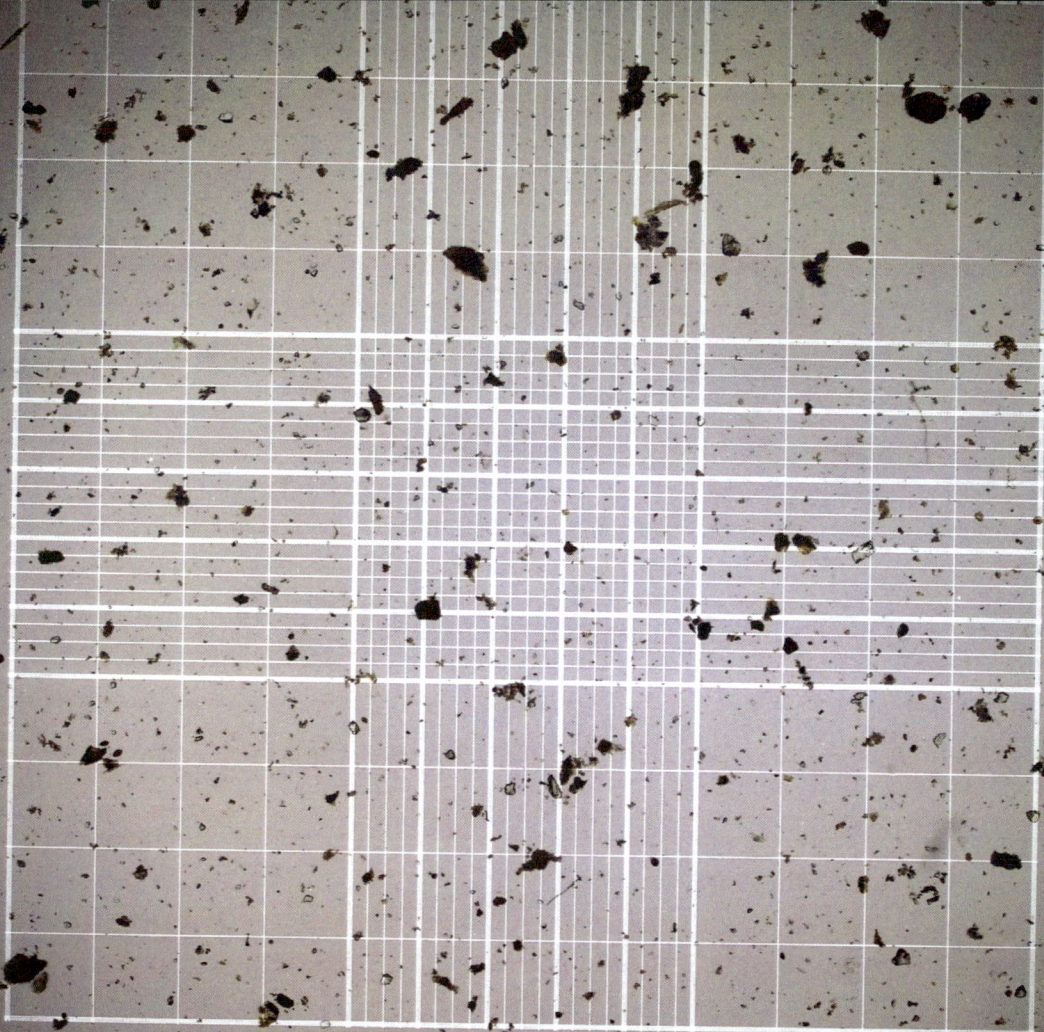

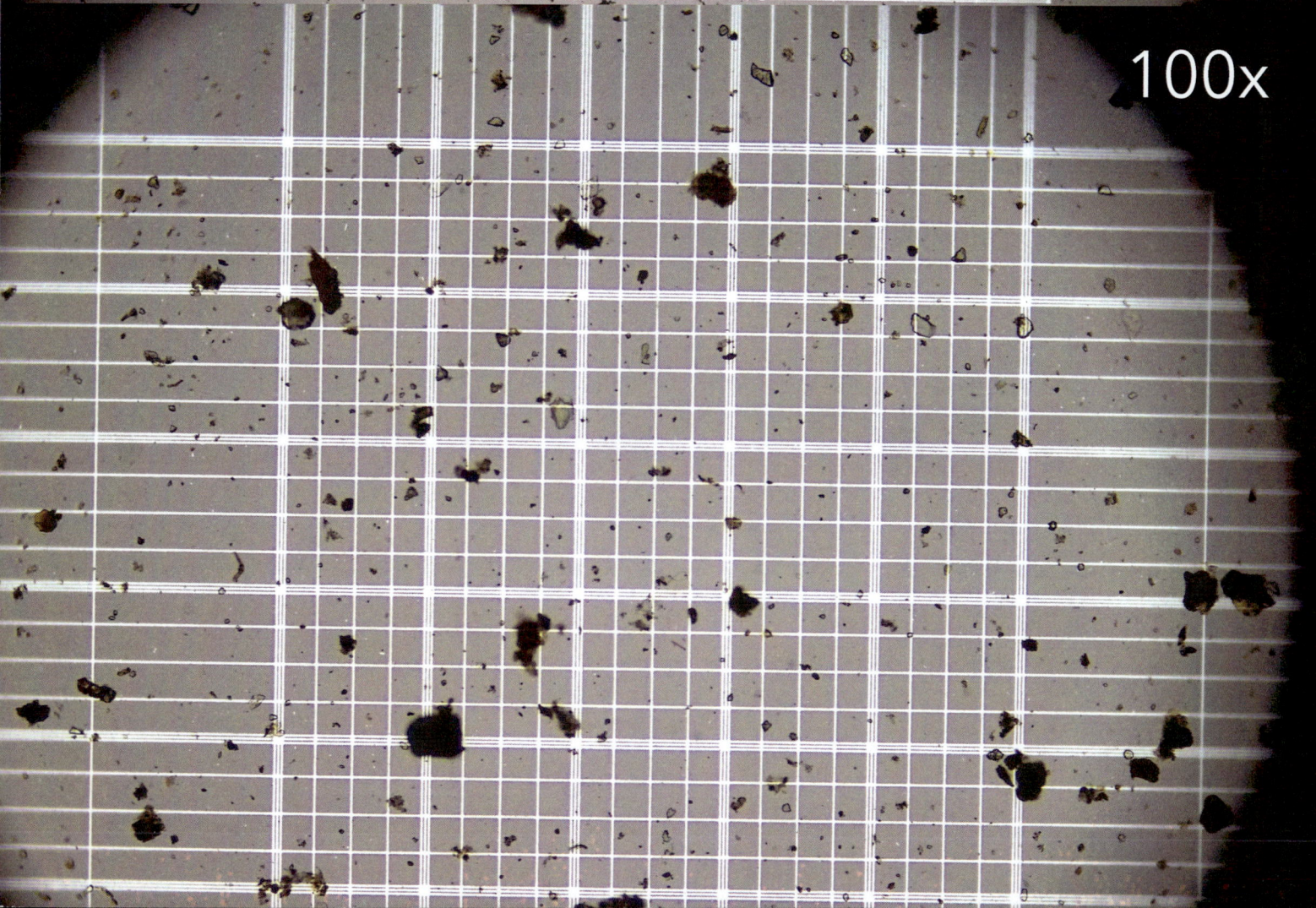

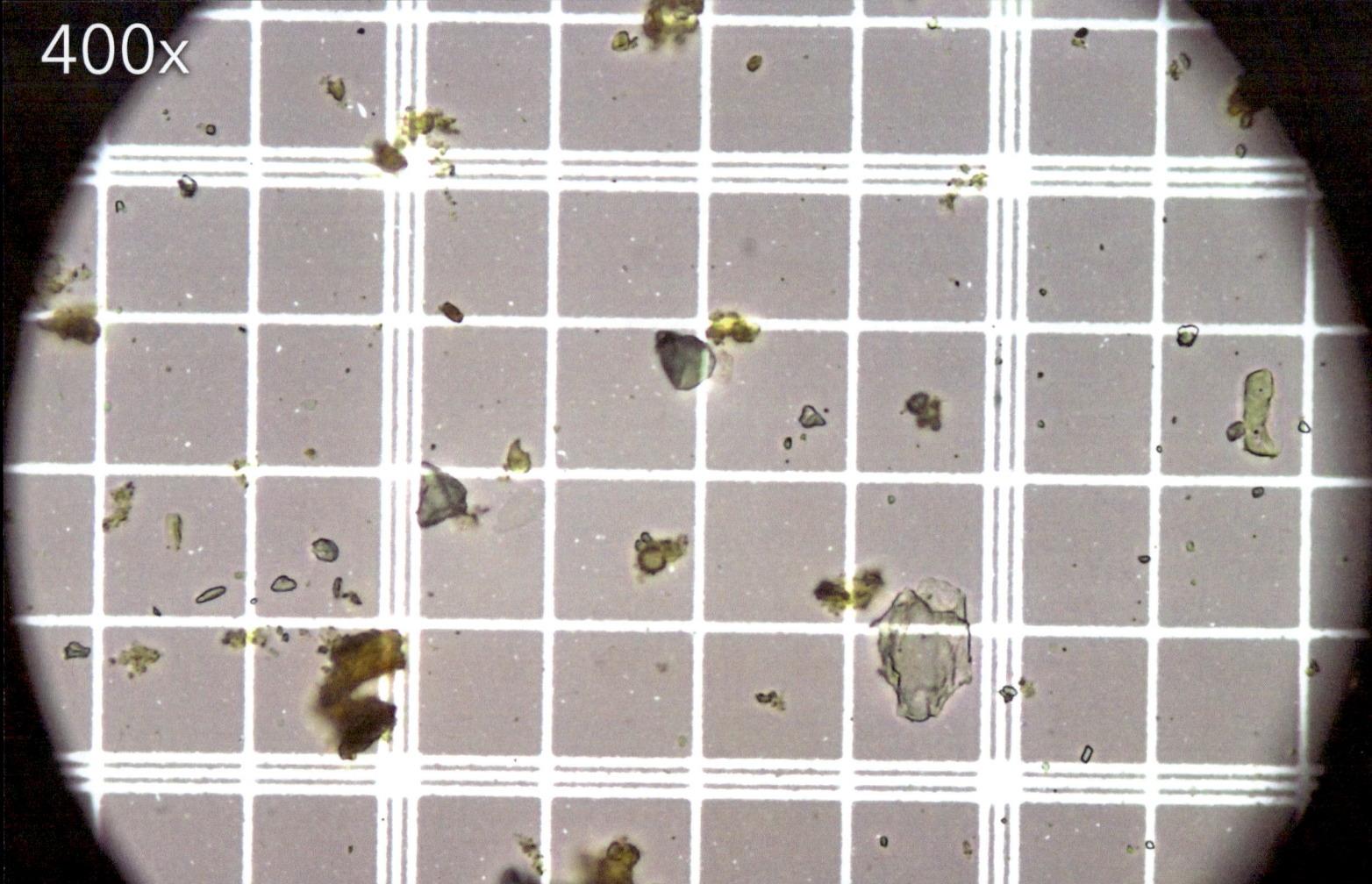

400x

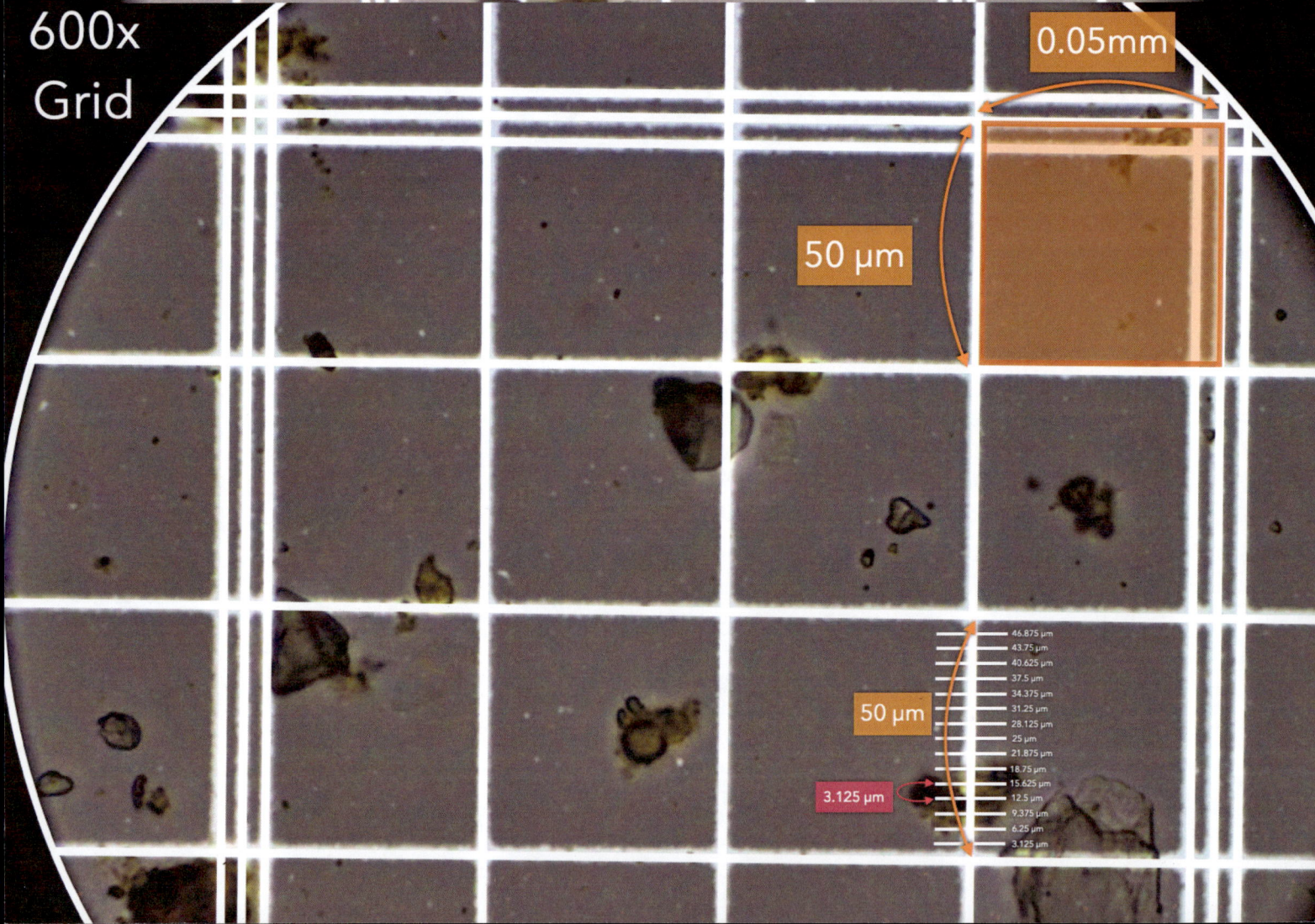

600x
Grid

0.05mm

50 µm

50 µm

3.125 µm

46.875 µm
43.75 µm
40.625 µm
37.5 µm
34.375 µm
31.25 µm
28.125 µm
25 µm
21.875 µm
18.75 µm
15.625 µm
12.5 µm
9.375 µm
6.25 µm
3.125 µm

0.05mm

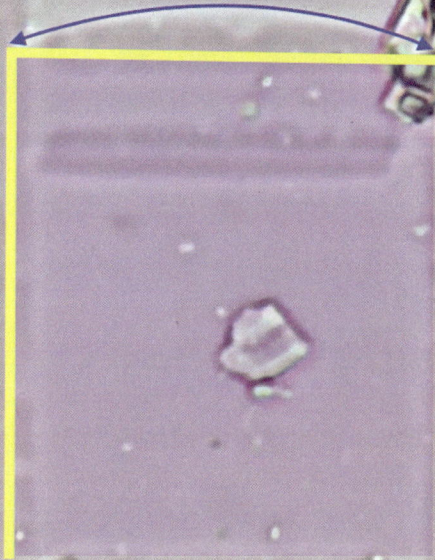

Count Everything in the 0.05mm² Section + the Upper & Left Boundaries

0.05mm

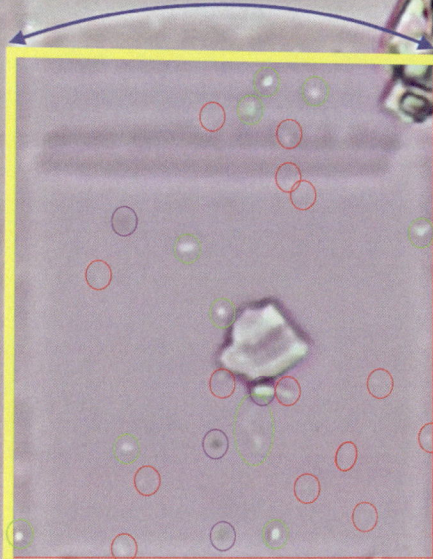

Method 1 - Bright Field

Likely Not a Microbe - Examine Further

Likely Bacteria - Examine Further

Bacteria - Note & Characterize

Method 2 - Acridine Orange Stain

| 10 | 14 | 4 |

Acridine Orange Examination

| 18 | 10 |

Evaluate & Count Everything
Focus In & Out to Examine Everything

If acridine orange (AO) is used in combination with propidium iodide (PI), a LIVE/DEAD stain that shines GREEN/RED is possible. I don't recommend PI as it is not as safe as AO and I don't think we need it to do the work, but it is possible and easy to do.

These are different vermicompost samples - this is actually a still from video segments of each sample. The far left is a natural worm casting from our dirt road here in Texas (the same dirt road featured later in this course), and we compare it to 3 different store-bought bagged worm castings.

THE BRIGHTER THE AO DYE, THE LESS ABSORPTION BY LIFE = LESS LIFE.

ROAD   WIGGLE   BRUT   HARRIS

Have a journal or use the RSM workbook. Count all the total nematodes per drop you see, magnify, film, and photograph them as best you can to determine what they feed on to know how they participate in nutrient cycling in your soil or compost. Note the likely types they are. Note the ones that remain unknown. Determine the average protozoa per FOV by tallying their numbers for 10 randomly selected fields of view then dividing by 10 for the average. Magnify, film, and photograph them as best you can. It is easy to note a passing flagellate or a zooming ciliate, but identifying them is often impossible, but that doesn't negate our ability to understand our soil or compost more deeply and to make good decisions based on that deeper understanding. Microarthropods will most often be crushed by viewing them under a cover slide on a slide – many are easier to view in a well slide with multiple drops and more freedom for their movement. Some are too large even for that treatment and we will use a burlese funnel trap to examine them. These microbes are much more mobile than any of the other organisms we are examining – no one is adding them in to their systems. When we bring in the trophic levels below them (bacteria, fungi, protozoa, and nematodes) and provide a thriving ecosystem of plants with vigorous root exudation, they arrive on their own in time, so documenting them is less important as they are shredders of organic matter and do not directly release the nutrients within bacteria and fungi in the way that protozoa and nematodes do. You can read more about those tests in **Regenerative Soil**, the first book in this trilogy.

Once the work at 1:10 dilution is documented and complete, I move on by mixing the test tubes briefly again, letting them sit 10 seconds, and then taking a generous 0.3ml from each to add to a 4th test tube, usually this ends up being a full 1ml of combined solution but if you need more, add drops from the middle test tube until it reaches 1 ml. Dilute with distilled or filtered water to the 10ml line – you now have 1:100 diluted soil:water. I usually vigorously shake this test tube because we are wanting it to be distributed and for the bacteria to dislodge from the organic matter as much as possible. I can see how a vortex mixer would be useful but it's not necessary. We can just shake it vigorously.

Having diluted acridine orange (AO) ready is important so you can just take 1 ml from the 1:100 solution, add 2 – 4 drops of AO, let that sit for 2–3 minutes while you do your hemocytometer prep. KEEP IN MIND: the more organic matter you have, the more drops. The less you have, the less drops (or you get a shiny haze). To make it even more complicated, some biology digests the stain faster than others, so always taper on and off the plain light source to verify if everyone moving is glowing – if it is off, perhaps prep a fresh batch of AO stain (it can lose efficacy in diluted form over time), or increase the drops per 1 ml. It is recommended that one wets the slide before applying the cover slide and then to move the cover slide in a back and forth manner pattern until Newton's Refraction Rings form, BUT this introduces MORE WATER! I instead put the slide on dry, and then add 1–2 drops to either side so that the drops wick in-between the slide cover and the hemocytometer slide using capillary action until the grid is fully immersed in solution. View at 400x – 600x down to 0.05mm^2 – the smallest section of the hemocytometer grid. Choose 10 random 0.05mm^2 sections from the grid and tally the bacteria in them and divide by 10 to get your average number of bacteria per 0.05mm^2. To size up our average from 0.05mm^2 to 1mm^2, we need to times that average by 400 (there are 400 0.05mm^2 squares in the 1mm^2 center of the grid). We then times the average per 1mm^2 by 100 (the dilution factor). Now we have the number it is on average when it's not diluted in a test tube. Then we'd divide that number by the multiplication product of the depth of the slide, 0.1mm, and the area, 1mm^2, which is 0.1 after it has been converted to ml by dividing by 1000. The final product is our # of bacteria per ml of soil which is a rough approximation for per gram of soil. This is scaled up number so it can easily be wrong as an average. It is a generalized snapshot of a particular area of your soil or compost at a particular moment in time.

$$\text{Total Bacteria/ml} = \frac{[\text{the Average Bacteria Count per 0.05mm2 section}] \times 400 \times [\text{the Dilution factor}]}{[\text{the Slide Depth (0.1mm) x the Area (1mm}^2)]/1000}$$

Bacteria

HOW USEFUL IS COUNTING?

If you need to count, if it's not obvious that it's bacterial or fungal dominant because both are there in significant numbers, you'll need to dilute your sample 1:100 soil:water, BUT our accuracy dips with every dilution on top of human counting being a method characterized by error and prone to over counting. AI counting and flow cytometers are the industry standards but even they are not considered definitive representations of the soil F:B ratios as everything is a collection of pin pricks we analyze taken from a diverse universe of microcosmos: there are pockets of all kinds of expressions. Animals die in the soil. Plants rot. Mineral deposits are found. Microscopic animals and microbes rearrange it all and transform it constantly. That's not saying we can't learn a ton by observing carefully but we have to have respect for the space to see it properly. There are several examples, some documented as published journal studies, showing that counting can be misleading. It is only a truly useful metric when it's used as a generalization and in the right context.

When is counting bacteria useful?

- When analyzing a sample with noticeably high or low diversity

- When presenting data to a client for emphasis

- When analyzing a biofertilizer or inoculant for establishing efficacy and checking label claims and #'s

- When needing a comparison to the Microbiome Meter's readouts

Knowing if it's fungal, bacterial, or balanced is important, but seeing which is dominant can be recognizable with practice and time without counting. Because no fixed associations with particular #'s have ever been proven, only generalizations around what too few, imbalance, and pathogenicity look like have been established, we can rest assured that it's the ratios, the balance, the relationships, the channels for nutrient cycling, and the identity of the bacteria themselves that matter most. That and the minerals, nutrients, organic matter, other microbes, water, air, and light available – those all influence the expression genetically. And since bacteria are so fast at reproducing and so changeable via HGT, we know that expression crosses species and phylum boundaries, so it can change who's in there rather quickly when it rains, when it dries, when we disturb it, when we amend, and so on. It is a system always in flux – thus, our #'s will always be in flux as well. We cannot hold tightly to a number that is never constant. It's better for us to map out the nutrient cycling pathways and to develop a more consistent observation practice of the soil over time: over the day cycle, over the season, and over the years, so we can better recognize the swing and change. In time, we will have automated readers we can use on the R-Soil database, so all your research can be compared to the community entries, and that will give us nuanced feedback that, while we know we can have wide margins of error in counting and sampling, thousands of readings will give us a clear correlation that we'll very easily be able to see on a graph regardless of our own local and personal variables. This is how we'll overcome the margins for error in sampling and counting that we as individuals will, without question, incur – it'll be automated, easy, and the information will be far more accurate and helpful for the overall understanding of the soil microbe ratios, their cycles, and the associated soil attributes and plant reactions are.

What are the Ideal #'s for Bacteria?

It's not possible to be definitively precise with counting, but for compost, if we are seeing 10–30 microbes per $0.05mm^2$ on average in a 1:100 soil:water diluted sample, we have a strong population of bacteria to build off of. Soil is generally 2 – 4 times less diverse and abundant – otherwise we wouldn't use compost as biological inoculants for our soils. This doesn't mean you can't have vibrant soils already – there are soils that don't require nitrogen to grow acres of corn planted every 12 inches, but it took decades to figure out why (it was microbial biomass from vigorous turnover in the soil). It's also such an easy thing to fix if your numbers are low.

What If My #'s Are Low?

Feed the bacteria simple sugars, and you will raise your bacterial count quickly. If you have bacteria present at all, you can grow them with something like a biofertilizer brew mixed with a fresh dose of sugar (or molasses for more minerals) and kelp (for more fungal balance). Biofertilizers like yeast, rhodopseudomonas palustris (purple non-sulfur bacteria), lactobacillus spp, and streptomyces (actinobacteria) will stimulate the indigenous microbes by providing a fresh food source – it's like a jolt of energy for the soil and can restart the trophic nutrient cycling and wake up encysted predatory microbes that restart the process. Adding compost extract and amendments of compost bring in more organic matter so microbes have staying power, but we can also make a brew and get things going or amend what the compost is doing for an even greater effect.

What if My #'s Are High?

Compaction and a lack of oxygen can often drive this as can decomposition. What your soil needs may be aeration, but in addition to that, more directly, you can add in a fungal compost with sufficient numbers of predators like protozoa and nematodes to cycle the bacteria through the system and establish balance and cycling again.

WHAT IS ACRIDINE ORANGE?

Acridine orange is a coal tar derivative that has been in use for a variety of applications for over a 100 years. It has been shown to kill cancer cells and prevent cancer cells from developing as well as preventing the spread of malaria. It is common, cheap, and well studied. I still wouldn't get it on my skin because it penetrates cells and causes nucleic acids to glow but AO, according to the research, is safer than fluorescein diacetate (FDA). AO is a fluorescent dye so you don't have to use much for it to be effective – it just sticks to living tissue – while FDA is a non-fluorescent dye that has to work its way into the microbe and react to create a fluorescent effect. FDA is not entirely effective and some microbes simply digest and break down the dye.

How did I stumble across AO as an option since it is not within the recommended nm range for my epifluorescence lamp? I figured since AO emits green light when it is excited by 520nm lighting, I would try it out since my LWscientific machine

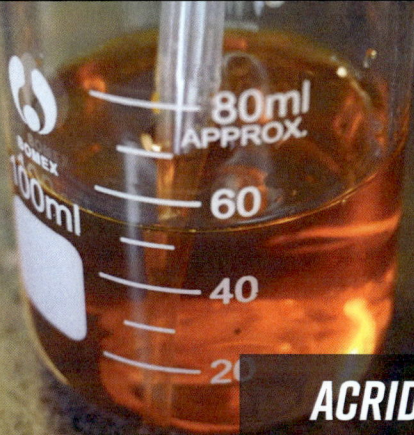

ACRIDINE ORANGE HYDRATION & DILUTION

Matt Powers © 201

Step 1 - **Hydration of AO:**
AO + Distilled water
Carefully mix 50mg AO into 10ml of DW
Store in the Refrigerator

Step 2:
Combine 1ml of hydrated AO with either:
0.5 ml of glacial acetic acid
with 50 ml Distilled Water (DW)
or
50.5 ml of diluted Distilled Vinegar (DV).

Diluted Distilled Vinegar (DV) How To:
Dilute 5% vinegar by 5 to get 1%, so for us:
10ml of DV + 40 ml of DW = 50ml of Diluted DV @ 1%

Step 3:
Sample 1ml of a 1:100 dilution (10ml) of soil:water sample
After it has been inverted for 30 seconds & allowed to sit for 10 seconds
Take 1ml of 1:100 soil solution & add it to a new test tube
Add 2–4 drops of 1% AO to the 1ml of soil solution
Use the pipette to mix thoroughly
Let sit 2–3 minutes

Step 4:
With a hemocytometer, count the glowing bacteria – those are the viable ones
Use the formula provided earlier in the book to calculate viable bacteria per ml. Flash the plain light to see if you have any microbes moving that aren't glowing: add any you find to your counts and adjust your method (fresher AO, higher concentration of AO, more or less time staining, etc. – all are variables that may need to be adjusted).

EXTRA STEP:
Using a standard slide, add a drop of the stained soil solution
and examine the full diversity of the soil solution. Examine behavior, organization, arrangement, overall layout
of color, and the organic matter - what color is it? etc.

has a filter cube releasing at 510nm (and up) and the 490nm of the initial beam. I reasoned it could be close enough – what I didn't expect was how BRIGHT it would be. It washes out all the details because it's SO BRIGHT, so I had to back off the camera Brightness until I was just able to see the outlines of the hemocytometer. Then I was able to visualize the glowing bacteria perfectly. This wasn't possible before because cameras didn't have these options on them in real-time, and because scientists were using the eyepieces, and this method would be temporarily blinding if they attempted that. The lesson here is to try things, to experiment, and not accept the past work that's been done as set in stone.

Fungi

Different from bacteria, fungi can be harder to calculate. Fungi also contain bacteria as part of their nutrient acquisition pathways – the hyphae behave like plant roots doing rhizophagy. In fact, if a hyphae is filled with bacteria, we can assume it is non-pathogenic because it has its own food source and production internally. I've yet to see folks counting the bacteria inside fungi as an indicator of health or safety for their fungal ferments yet, but I imagine they will sooner than later, once we can get regular expressions of this type. It's not just do we have fungal strands – it's who they are, it's how well they are growing, and how favorable the overall environment is for them to continue to grow. This is a space where folks have been regularly tricking themselves, so let's proceed with the holistic context in mind.

Because of the nature of fungal growth, it always establishes in a single location and spreads following foods: it's nonmotile, we can always assume that fungi is found in pockets of significantly greater and lesser expression in the soil, and these pockets are found in relation to the location and abundance of fungal foods and the sporulation reach of the given fungi present in the soil unless it's no-till and there's been fungal foods throughout and sufficient time for the entire area to be inoculated more evenly, and even then it'll have variation in expression because of the legacy of the initial pattern of growth. This could also be an argument for amending and then tilling in those fungal amendments (inoculants and fungal foods), but once established, we want to let things flourish on their own. The delicate nature of hyphae means we are always destroying them when we disturb soil and especially when we mix our test tubes of diluted soil solution. We mince the fungi when we shake it up with sharp sand – it even looks like shards of glass under the microscope! You can imagine how much that could change things. This is why it's especially important to assume that there's always more fungi than we can see if we find any fungi because the process to prepare samples itself is destructive to fungi. This is why measuring the length of chunks can be misleading – the lengths will differ depending on the person doing the mixing even with the same exact test tube sample. They will also differ with the exact same person doing the exact same motions if the ratios of the minerals are more sandy – it's like adding in grit but boulder-sized sharp glass grit instead, so it always has an effect.

It is better to establish that you have fungi present, identify which kinds, and then to feed those types and to diversify the fungal populations and foods than it is to lump all fungi together as a group and counting or measuring them simply because calculating the biomass with a microscope manually has been shown to be misleading when used as a lead

100x

This is a composite image of a single hyphal strand of fungi from my friend Emin's backyard garden in Austin, TX – it's the longest hyphae I've ever seen under the microscope. This is a great example why we don't just count fungal strands – we look at the total length of hyphal strands per drop. These can be rounded to the nearest 10 microns.

If we're gentle, we can preserve large hyphal strands that can show us more accurately how they behave in the soil. Here we can see that this hyphae is connected to several pieces of organic matter.

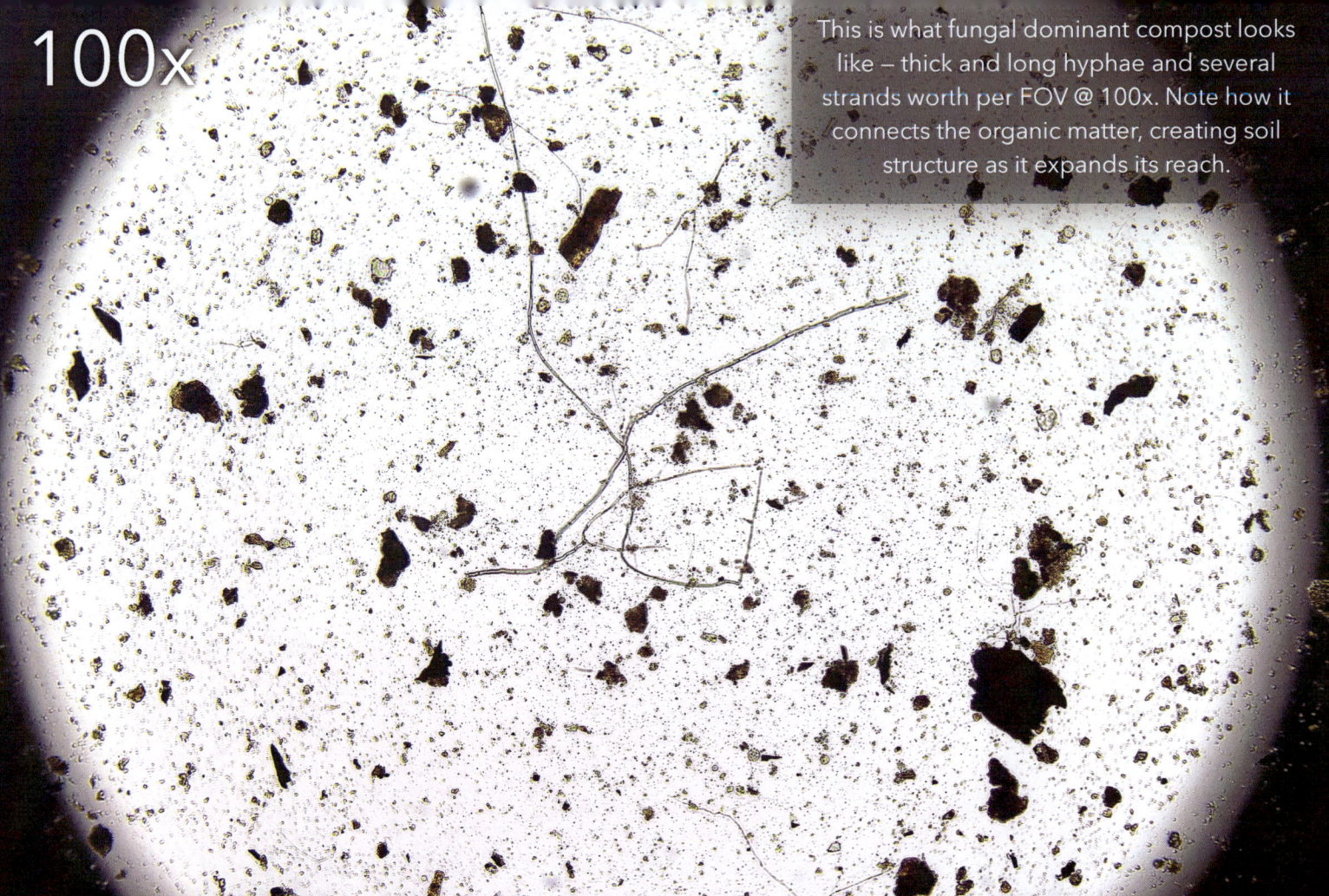

100x

This is what fungal dominant compost looks like – thick and long hyphae and several strands worth per FOV @ 100x. Note how it connects the organic matter, creating soil structure as it expands its reach.

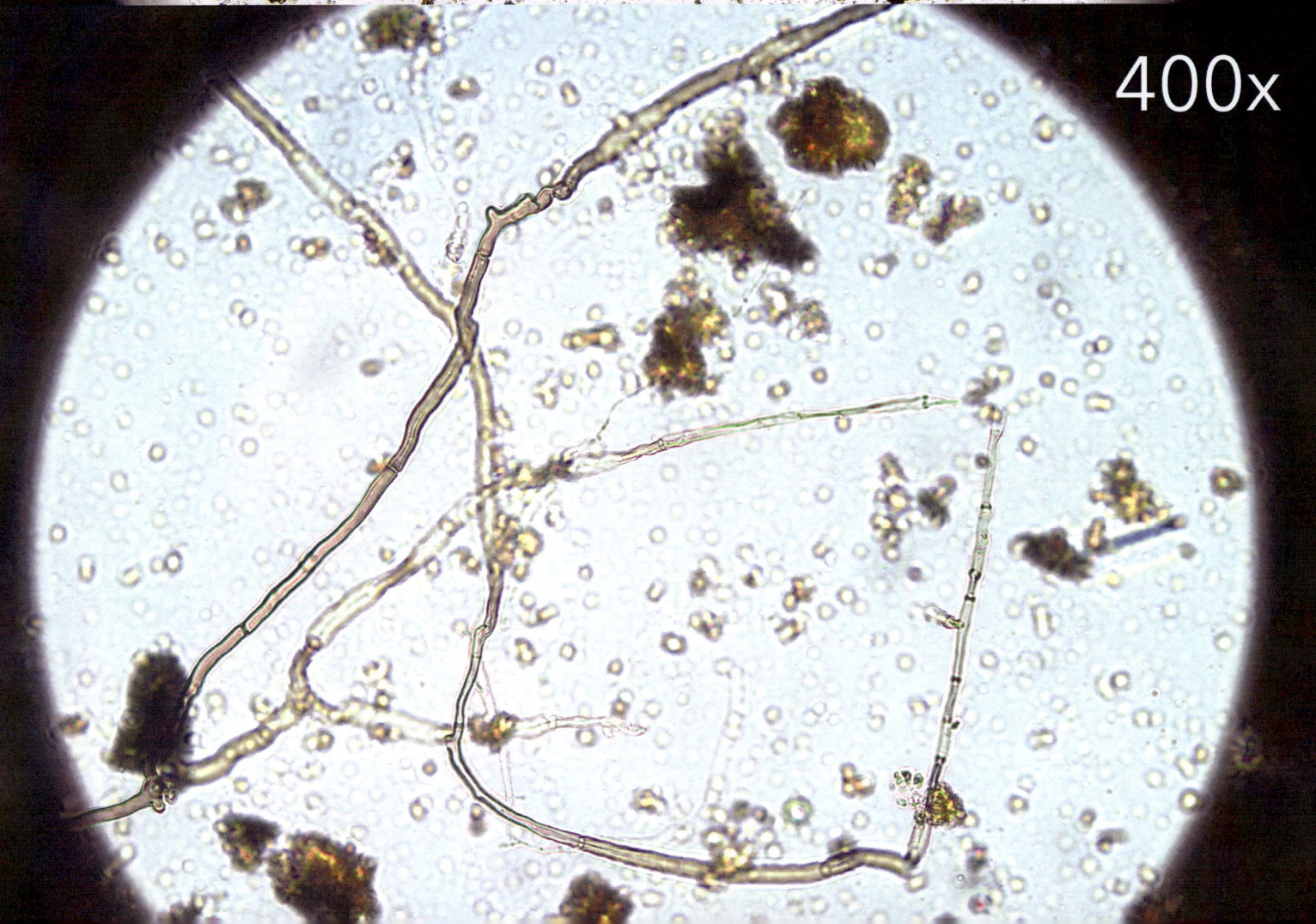

400x

indicator for fungal health and abundance in the soil in correlation with plant benefits. It is a hipshot testing method – that's why using something like the Microbiometer is useful as it saves us time and energy, gives us a generalized snapshot, and we don't over emphasize those numbers as they are extremely generalized and can be inaccurate as well. It's a better use of our time and effort to look at 40x over the entire slide first, seeking out the larger fungal strands and then move up in magnification to get a closer look. You can often see them holding sections of organic matter together. You may need to examine the slide at 100x to see hyphal fragments – the more broken up the hyphae is, the more work we have to do to see the fungi as it was before we disturbed the soil or compost, diluted it, and then shook it into a formless uniform solution. If we do things, extra *"thorough"* we can just end up making more work for ourselves and making the test less effective.

Our objective is to verify whether or not there is fungi in that sample, to identify what kinds of fungi are present, to determine if we have a diversity of fungi, and to identify its relative abundance, such that, we can then regeneratively respond by providing perhaps a greater diversity of fungi, more fungal foods, specific fungal foods, and if we have a minimum viable and beneficial population to promote. Sufficient levels of fungi are often dependent on a variety of factors like pH, Eh, moisture, temperature, competition, and the food sources. If we can shift our pH lower, most often we'll see a huge flush of indigenous fungi – this is often what folks are doing with fermentation of LAB and biofertilizers like EM. They are very acidic. This stimulates the fungi. This is also why folks like *Wormies* composting have noticed bokashi stimulates growth and a rise in fungal numbers.

What if I don't have any fungi in my soil or compost?
Then you craft or buy fungal dominant compost like a Johnson-Su compost or a bokashi compost – add that to your soil directly or as an extract. There's an art to making fungal dominant compost – because fungi like the stillness and thermophilic is usually all about disturbance, it can be a challenge raising those counts, BUT it's possible if you let it rest, if you feed it fungal foods, and if you inoculate it with fungal compost after it initially cools.

Which fungi are we talking about again?
Good question – if it's compost, if it's hyphae we are seeing, it's likely to be either streptomyces or another actinobacteria if it's earlier in it's static maturation process (i.e. bacteria, not fungi) and then basidiomycota if it's later in that same process which can take months – ascomycota has more specialized roles and so doesn't show up regularly (unless we are

looking at Korean natural farming IMO collections). So look for those fungi in those contexts. If it's garden soil in development, you are likely to see just actinobacteria and as things mature, as in a no-till garden, then you'll see basidiomycota, more often, BUT this is also true if you have alkaline soil you are trying to bring into more neutral or slightly acidic conditions. If you are in a pine forest, you may find a huge expression of fungi, acidity, and all the rest, but the tannins and the aggressive roots are hostile to growing a garden there, so keep your context in mind: this is why **Regenerative Soil**, the first book in this trilogy, is so vital. If it's the yard soil, it could be devoid of all fungi. If we are looking at roots, we are looking at mycorrhizal fungi and potentially yeasts if we are looking inside the root at endophytes and rhizophagy.

EPIFLUORESCENCE & FUNGI

Not all fungi autofluoresce, I've seen many basidiomycota strands that look like the poster-child of an ideal compost fungal hyphae, but they are invisible in the same lighting we use to see fungi inside and on roots, digesting organic matter, or growing our medicinal and edible mushrooms. Their high phosphorus content lights them up like halloween

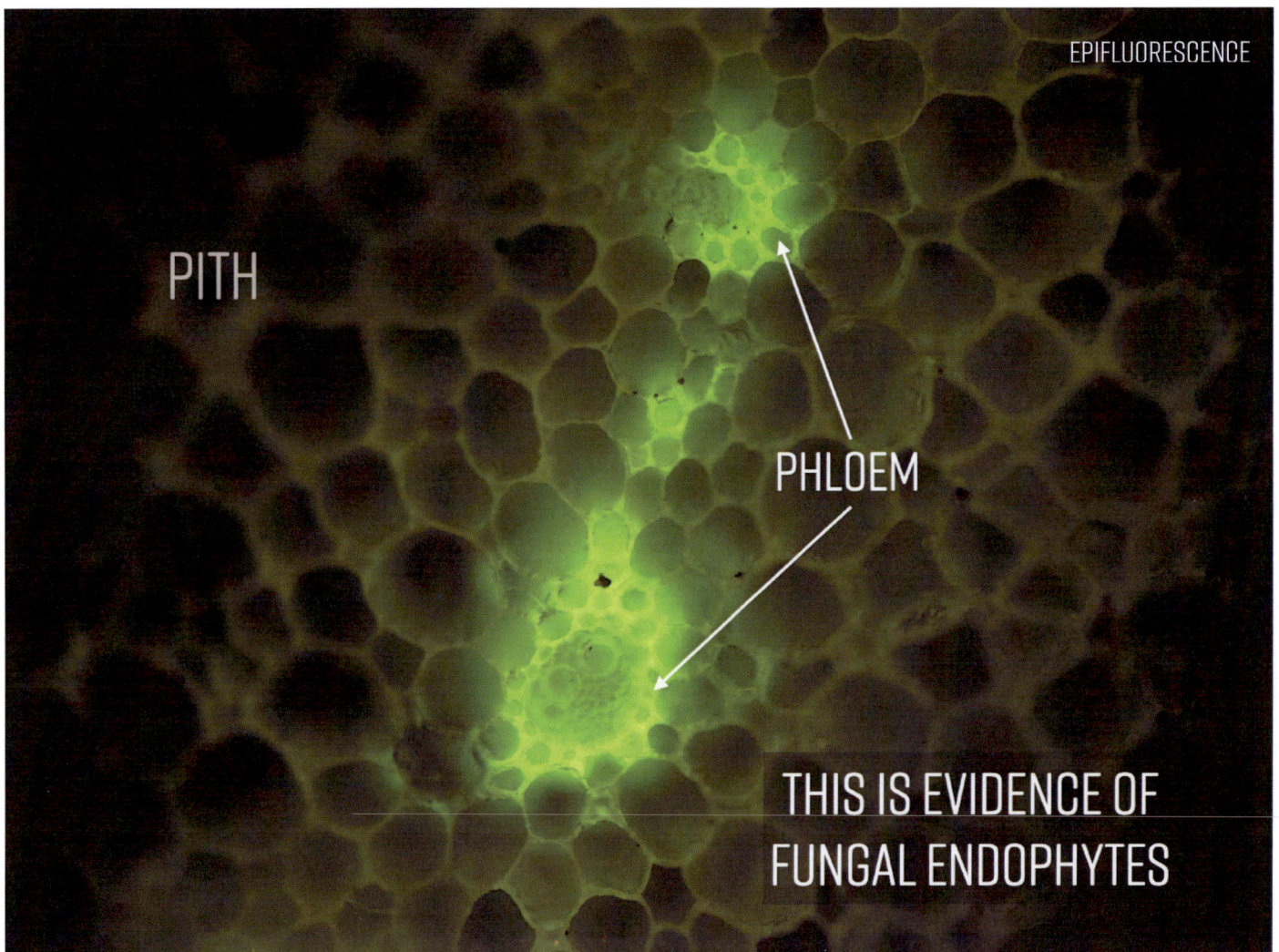

EPIFLUORESCENCE

PITH

PHLOEM

THIS IS EVIDENCE OF
FUNGAL ENDOPHYTES

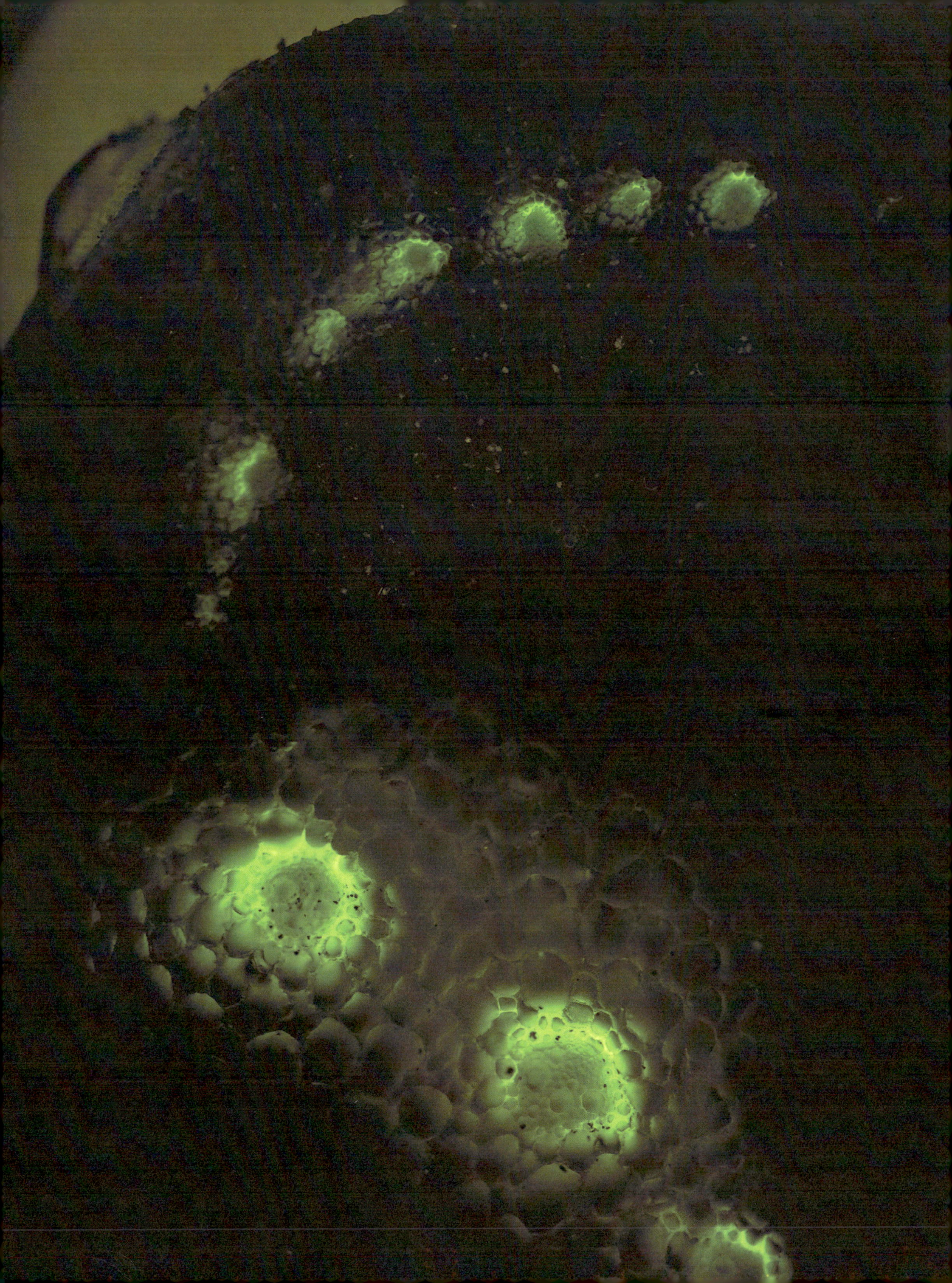

This is piece of organic matter inoculated by saprophytic fungi – it is digesting it.

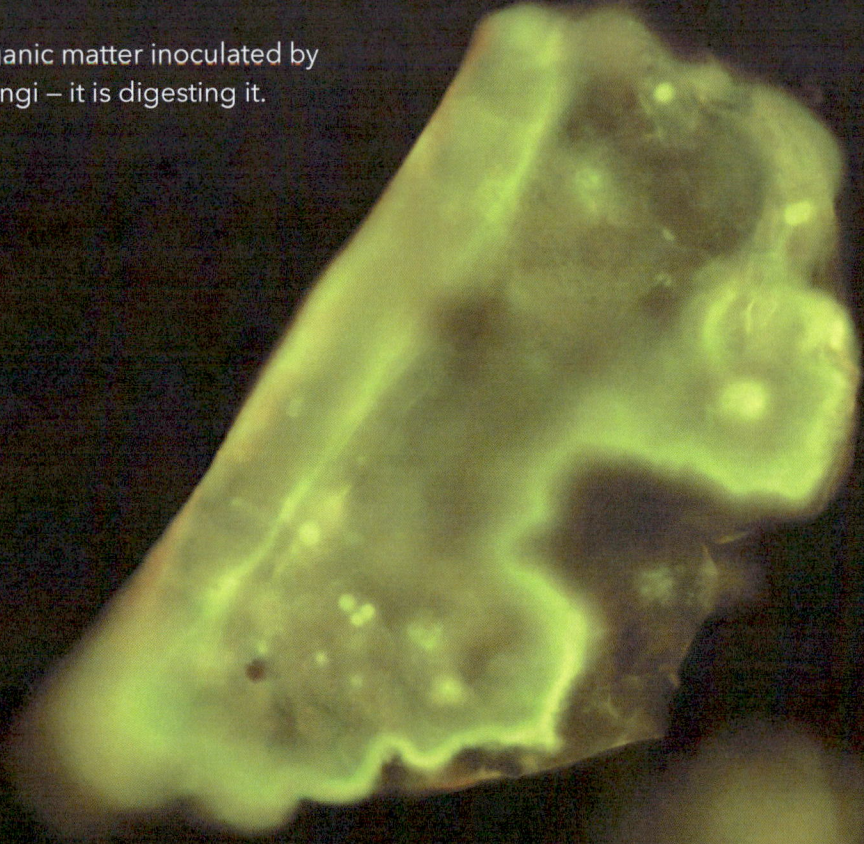

These are yeast cells from an EM-1 Pro sample from Teraganix – notice how some autofluoresce but some do not.

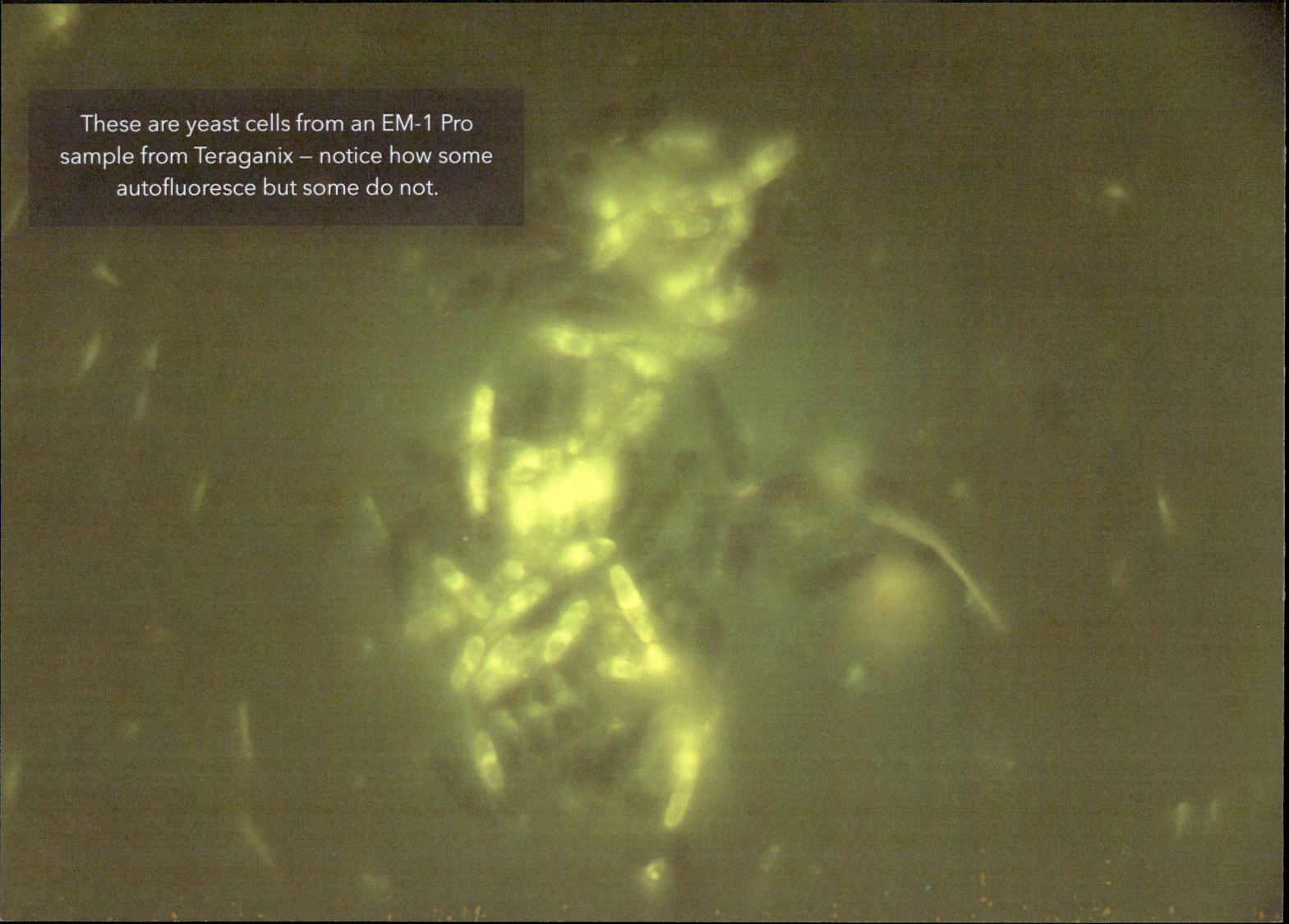

decorations. AMF fungi, at times yeast, phosphorus bearing minerals, some endophytes, and crystals made by the digestive activities of fungi – all show up brightly when excited with a 490nm cyan blue light and then when the reflecting light is viewed back through a 510nm filter cube, so only light at and above 510nm is released. What we are doing is shining a focused bandwidth of light at the soil and then filtering the light that comes back – animals and insects can have a wide range of visual experiences that range from UV to natural lighting to infrared to full color to color blind. We are each adapted to pay attention to a subset of the whole – we cannot perceive the world as it fully is with our eyes.

Organic Matter & Minerals

The types of organic matter and minerals we have present in our sample make a massive difference in how we treat and examine that sample. It also depends on our context. Compost is gathered organic matter in a controlled decomposition process – it's all organic matter (though some commercial composts may gather up some of the minerals from the pad they are turning their windrows on.) Soils can have very little soil organic matter, so if you are seeing mineral based soils with mostly bacteria: you need organic matter. If your organic matter is pale yellow, sparse, and thinly populated by bacteria and fungi, you need to add more reduced, fungal organic matter (compost) to that soil. The darker the organic matter, the more carbon it is holding – that's why there's such a thing as carbon black: it's just pure carbon. The more carbon, the longer the carbon chains. The longer the carbon chains, the more spaces there are for minerals and water to be bound to the organic matter. It's really just that simple.

Flip on the epifluorescence lamp and you can instantly see how fungal the decomposition process is in your compost or soil (check out the image in the fungi section in the pages prior to this). We can mix our samples extra gently and then we can stain 1 ml of that 1:10 or 1:5 solution with 2 – 4 drops of acridine orange to view the bacterial behavior in relation to the organic matter as well (see image to the right). We want to see organic matter that is infused with life, fungi, spores, and carbon.

I've considered trying to get more specific with the grading organic matter but attaching a metric to color can sometimes get us in trouble: anaerobic composts made at extremely high heats have a microbial soot that is black and actually screws up color-based test methods. The entire sample can be

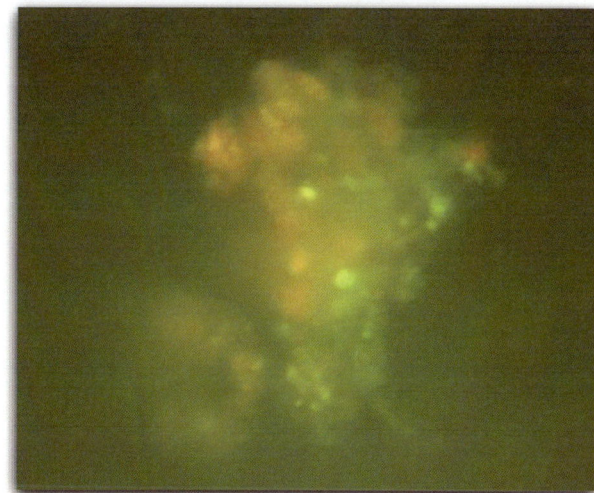

Organic Matter in the dark field has much more natural color to it – this is a fragment of partially burned and decomposed wood from a biochar-rich compost. The green and purple halos are artifacts of the objective lens and not really there, but they do indicate bacteria: the refraction halo is caused by their presence. On video, this image shimmers with movement. If you look at the middle and lower left of the fragment where it is more in focus, you can see how small the bacteria really are - they are small bright dots like stars in a galaxy.

murky due to this – stains can create a glowing fog-like effect that makes it very hard to evaluate and color-based mineral tests become meaningless as well.

Leaves & Roots

How much is too much? How much is just right? How much is too little?

With all things biological, it depends. Each plant has a different relationship with fungi and bacteria, in general and specific. I've heard some folks suggest too high an ectomycorrhizal connection turns pathogenic but that's only with specific mycorrhizae and specific plants from what I've read – meaning, there could easily be a 3rd factor at work that is not recognized in those studies: they could purely be correlative or the fungi is responding to the plants own signals to senesce. Because of the brief nature of many of these mycorrhizal relationships it can sound like they come and go, but

The Leaf Surface

RED = CLOROPHYLL

GLOWING GREEN = FUNGI

THIS IS A FUNGALLY
INOCULATED STOMA

Bright Field

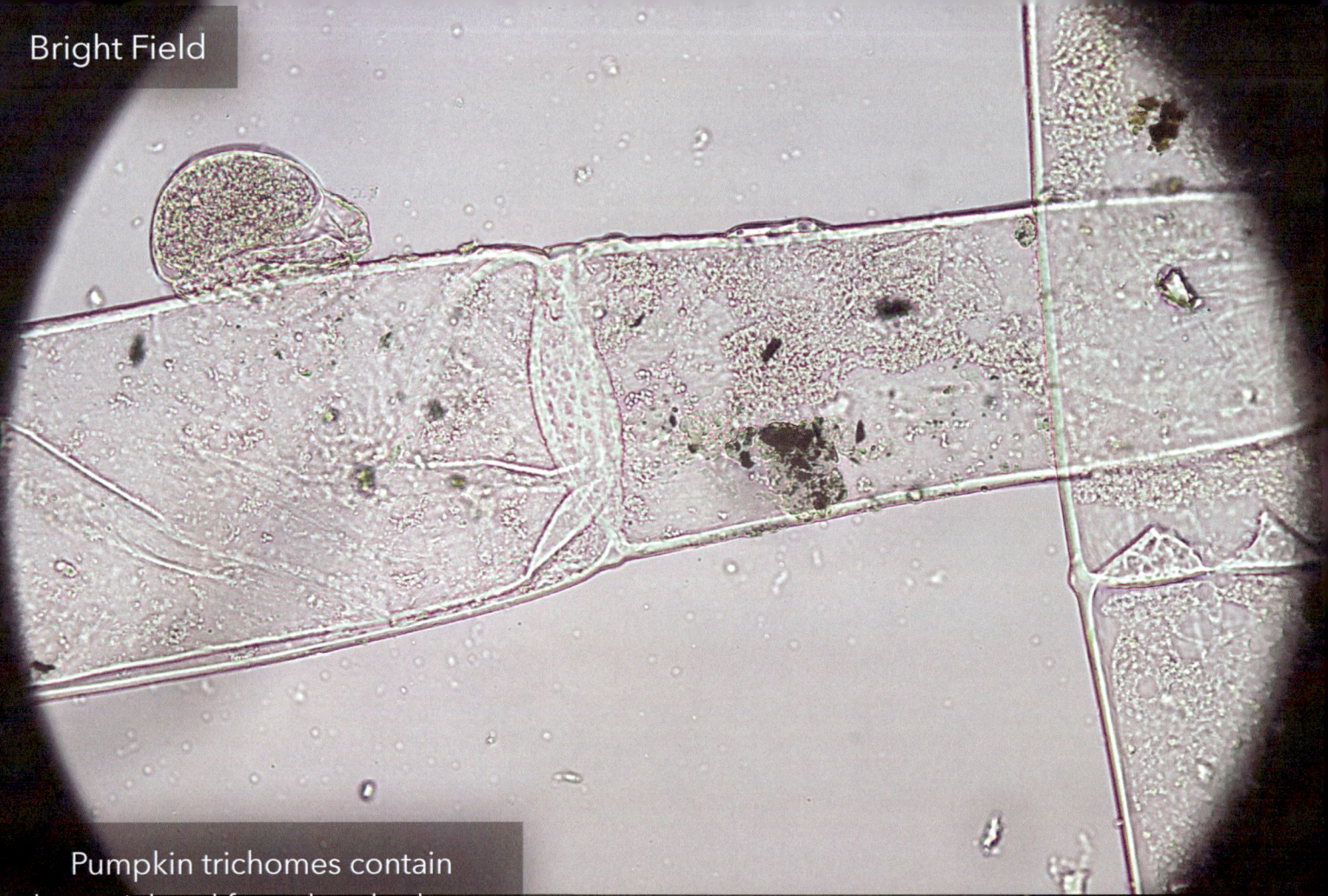

Pumpkin trichomes contain bacterial and fungal endophytes and likely algae (the red)

Epifluorescence

instead they are overlapping and chaotic with plants having many relationships with many different kinds of fungi all at once. The same is true for beneficial bacteria.

Healthy leaves and trichomes will be teaming with life. Their surfaces will have bacteria and fungi on them. Under epifluorescence, the leaf stomata will shine with endophytic fungal life, protecting and enhancing their function. We can see this easily by putting a leaf under the epifluorescence lighting and viewing it without any other preparation. This nondestructive viewing is the most natural (especially if as fresh as possible) and yields the most accurate and useful feedback (and it is indeed feedback as the light we send down excites and lights up the fungi in the sample that shines back to our cameras and our eyes). In the trichomes of cannabis, for example, the terpenes that are so sought after are actually communications between microbes in the trichomes (the plant hairs) themselves, so depending on the microbes we have present, we get different plant secondary metabolites, i.e. terpenes and essential oils. The future is going to be focused on using specific biofertilizers strategically timed and applied to create superior terpenes, essential oils, scents, flavors, colors, nutrient density, and more. Beyond plant immunity from pests, diseases, and viruses, these plants will be incredible to smell, taste, look upon, and experience.

For fungal inoculation levels, there is some literature out there, but it is primarily focused on documenting the inoculation rates of natural plants with some studies on what is an ideal percentage of the root to be inoculated. Endomycorrhizal fungi, including arbuscular mycorrhizal fungi (AMF), have an ideal range of inoculation of over 50% with some examples showing close to 100% inoculation. Ectomycorrhizal fungi have an ideal range of less than 50%, so don't expect to see a full inoculation or vigorous inoculation rate from ectomycorrhizal fungi. I've tried to talk with folks claiming to be experts in this area, and they avoid the question and then avoid me after I've asked it. I think this is because folks regularly aren't testing the roots of their plants while the plants are growing – to do a good job of this we basically would sacrifice a plant, but that's preposterous to a commercial grower who's site or situation requires them to use every square foot to grow product, BUT if we don't test, we don't know. If we don't look, we can't see. This is why I'm writing this book and creating the R-Soil Database.

EPIFLUORESCENCE + ROOTS

The best way to view plant roots for fungal activity is with epifluorescent lighting, specifically LED-based cyan blue 490nm wavelength light where the return light is filtered so only light at or above 510nm wavelength returns. LWscientific makes

In this picture, we can see the full life cycl
of arbuscular mycorrhizal fungi. Initiatio
of inoculation can be seen by the glowing
around the cells, the action of inoculation
indicated by the accumulation of
phosphorus (the extreme glow), and the fir
stage can be seen as they senesce, turn
opaquely reddish-orange, extend hyphae t
grow and release spores.

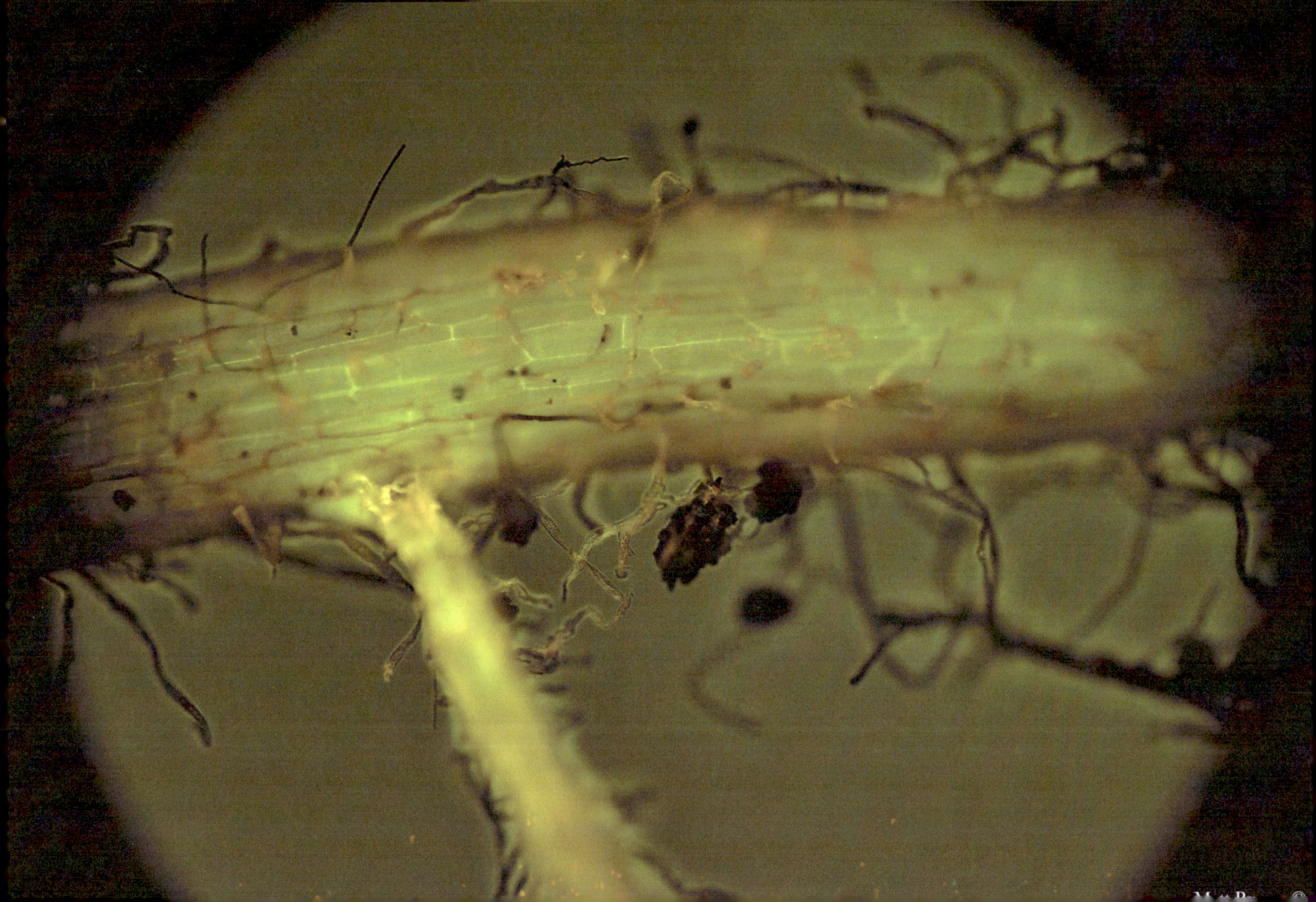

Though not as readily recognizable as reddish, we can see the senesced fungi does have a reddish cast here and there, and we can differentiate actively or recently inoculated cells from senesced. We can also note the extensive hyphae.

Initiation of inoculation can be very subtle at first, but then becomes very obvious as the relationships compound over time.

A Fully Inoculated Root

Partnering with Arbuscular Mycorrhizal Fungi (AMF)
most likely Glomus spp.

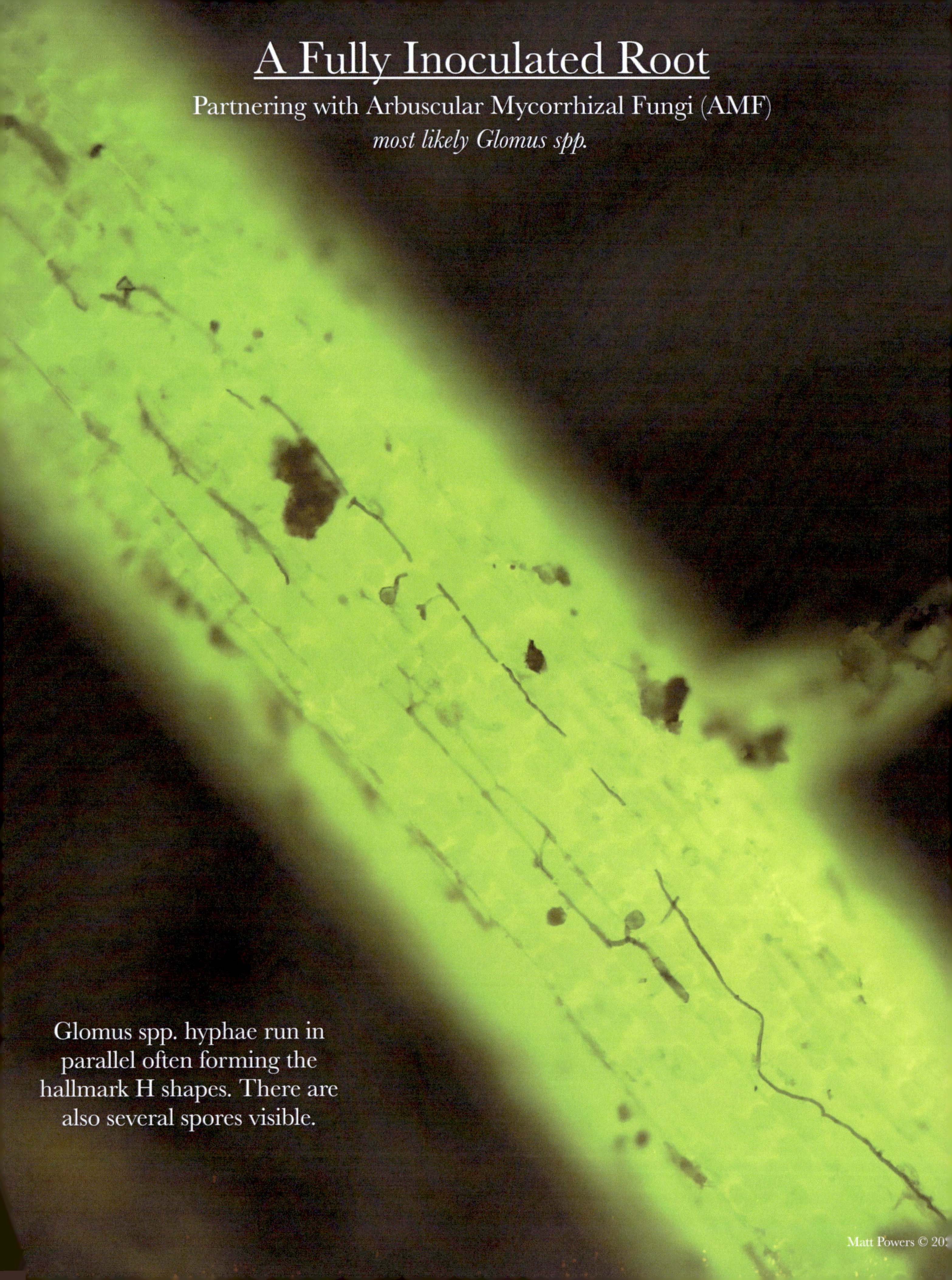

Glomus spp. hyphae run in
parallel often forming the
hallmark H shapes. There are
also several spores visible.

a LUMIN epifluorescence lamp that fits between the top of microscope and the base, so the BioVid 4k camera attaches above that and views through it. It's an incredible system that is nondestructive – I don't even add water or a cover slip!! Many people are astonished by the images I'm able to get with this camera and LUMIN lamp, but it's easy! It's doing less, not more. Anyone can view their roots using this lighting technique and instantly see the mycorrhizal relationships leap into the visible range. They are distinctly yellowish green, not the ambient green of the filtered blue light. With a little practice, anyone can recognize the presence of fungi in their roots.

Do I need a grid to calculate the fungal inoculation percentage?

Grids are useful, but most of us can judge half of something just by looking at it. Most of us are not working in native forestry nurseries, so we're looking for AMF only, and that means we are looking for a minimum of 50% inoculation rates.

What if it's less than 50%?

Test your phosphorus levels and pH. If you have high P and/or alkaline soils, those alone can inhibit mycorrhizal inoculation and growth (though there are some AMF that thrive in alkaline, arid conditions). For commercial farmers, this indicates that we need to address what we can, and then reinoculate, because we need to get above 50% minimum for the time and investment to be worth it.

Why do your images look different for AMF than other examples?

This is because those are destructive preparations that create that image temporarily before it is destroyed by the stains they use. I've reviewed their process and it's so far from natural that I think they have led us a bit astray: the arbuscules are stubs and knubs of what they are in reality. The process burns away the edges and finer points of distinction. This is why I prefer nondestructive methods. Even when I'm cutting a root in half to view it, by not adding a stain, I avoid that influence. I want things to be as natural as possible in our viewings to understand how they work in the actual soil as much as possible.

THE MANUAL LIGHTING METHOD + ROOTS

As I began to work more consistently with roots, I began to realize how strange it was that we were having to cut up the root to see it. The lighting coming from below (bright field) was a hinderance and only generated silhouettes or light drenched images. Using the epifluorescence lamp opened a new door: lighting from above! This changed everything.

Bright Field

Manual Lighting

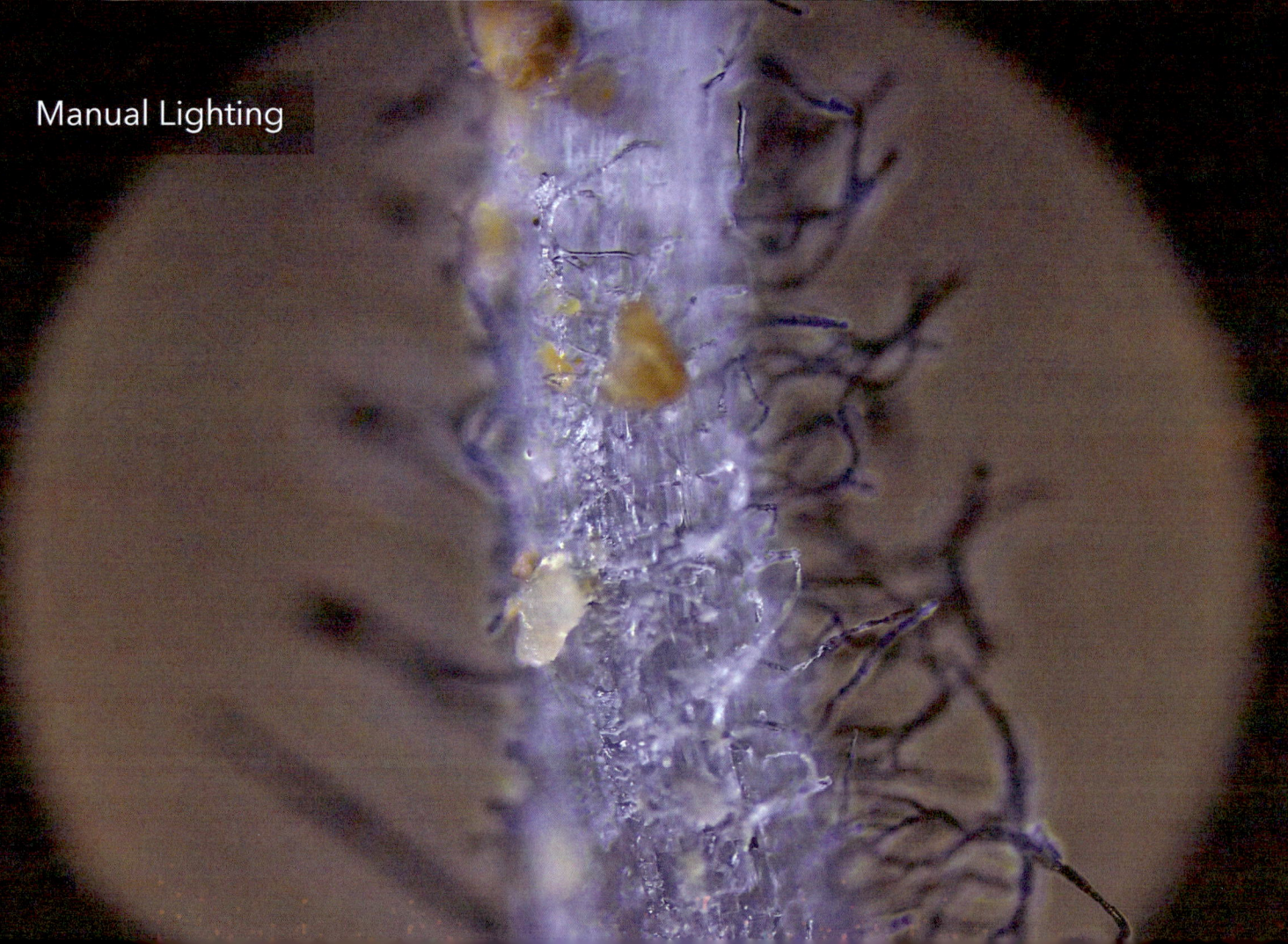

Bright Field

Manual Lighting

Bright Field

Manual Lighting

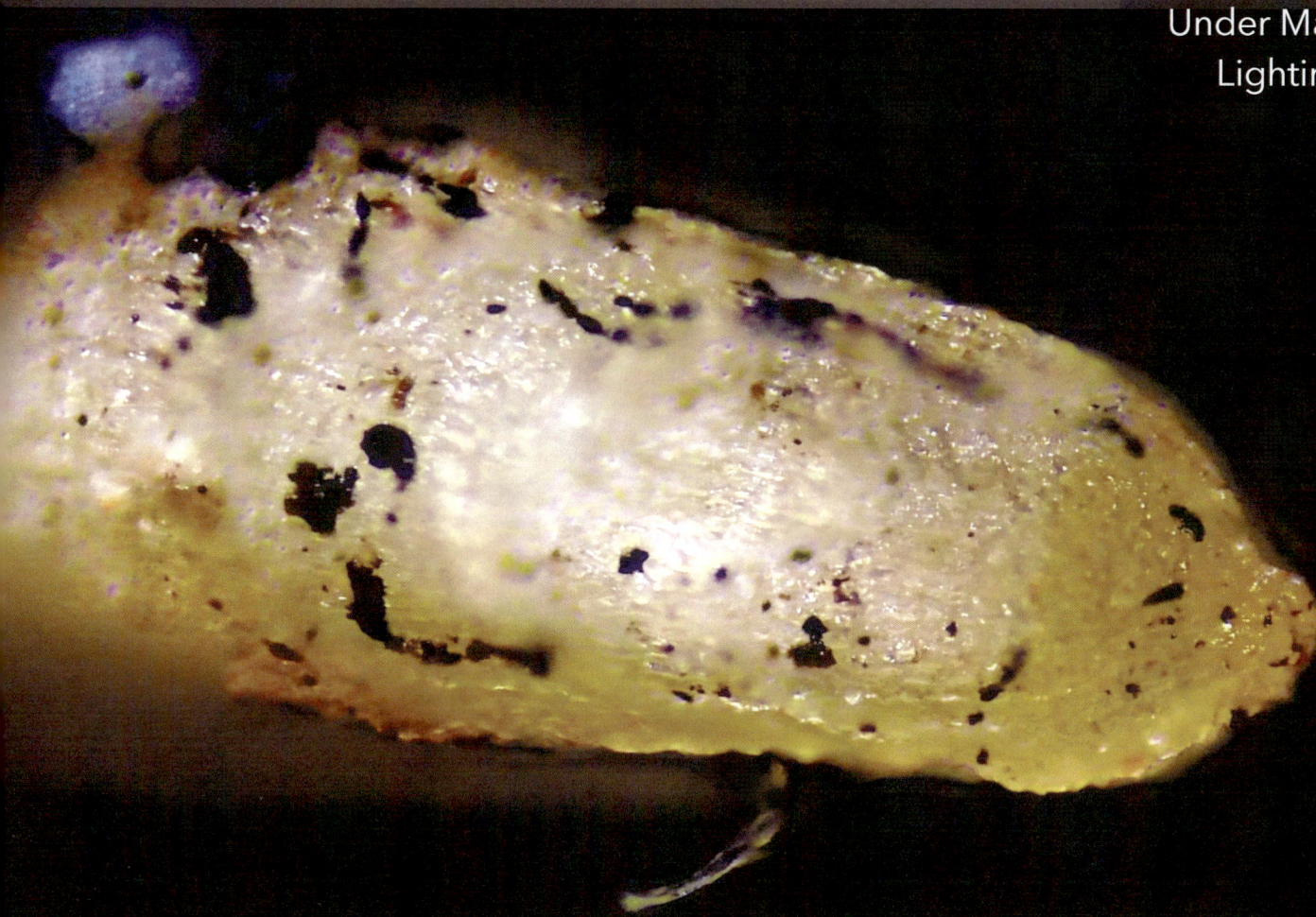

While dark field was diffuse light from below, it was still from below. I could see the *surface* for the first time. It was

revelatory – this is what it really looked like… but it was through the lamp only. I wanted to see things without the

epifluorescence lighting but from above, so I grabbed a flashlight, and at first, it seemed like it wasn't working or effective.

I then discovered there's a specific angle, a sweet spot (likely a birefringence point), where the root leaps into view.

Though this is such an easy thing to do, I've never seen anyone else do it! It also yields the most stunning images – we get

to see roots as they appear in natural light but UP CLOSE without water distorting the image, without being smashed

between glass cover and slide, without removing all the soil particles, and without washing away all the microbes. We can

get to 400x or beyond with this method because we'd have the objective pressed directly against the root which is not

ideal – we want to keep our objectives in pristine condition and we want to avoid getting moisture or oil in them. They are

designed to flex inward as we press down against the slide as we magnify things (keep in mind it's only a small amount of give), but when we do this, we can get moisture or oil into the objective which may require professional cleaning if it's oil in a 400x for instance. For this technique stick to 40x – 100x range, just so you keep your objective lens from touching it.

I, at first, called this the light field technique but it's now called the Manual Lighting method as a light field objective already does exist. You can even do this with epifluorescent flashlights – the angles transform the image in a startling way.

WHAT ABOUT METHYLENE BLUE?

Used for all sorts of purposes, this amazing metabolic drug/industrial blue dye/mycological stain allows us to see microbes and fungi and bacteria in and around roots more distinctly. You can find it at most aquarium stores as it is used in larger doses to kill fungi and prevent fungi from infecting fish eggs. It is superb for visualizing the rhizophagy process (though to visualize all the steps in the rhizophagy process requires a long list of stains, some of which are difficult to obtain like tetrazolium blue – please refer to the work of Dr. James Francis White for more details on that). He's created a way for every stage of the rhizophagy process to be seen and verified as well as the endophytic relationships. We don't nee

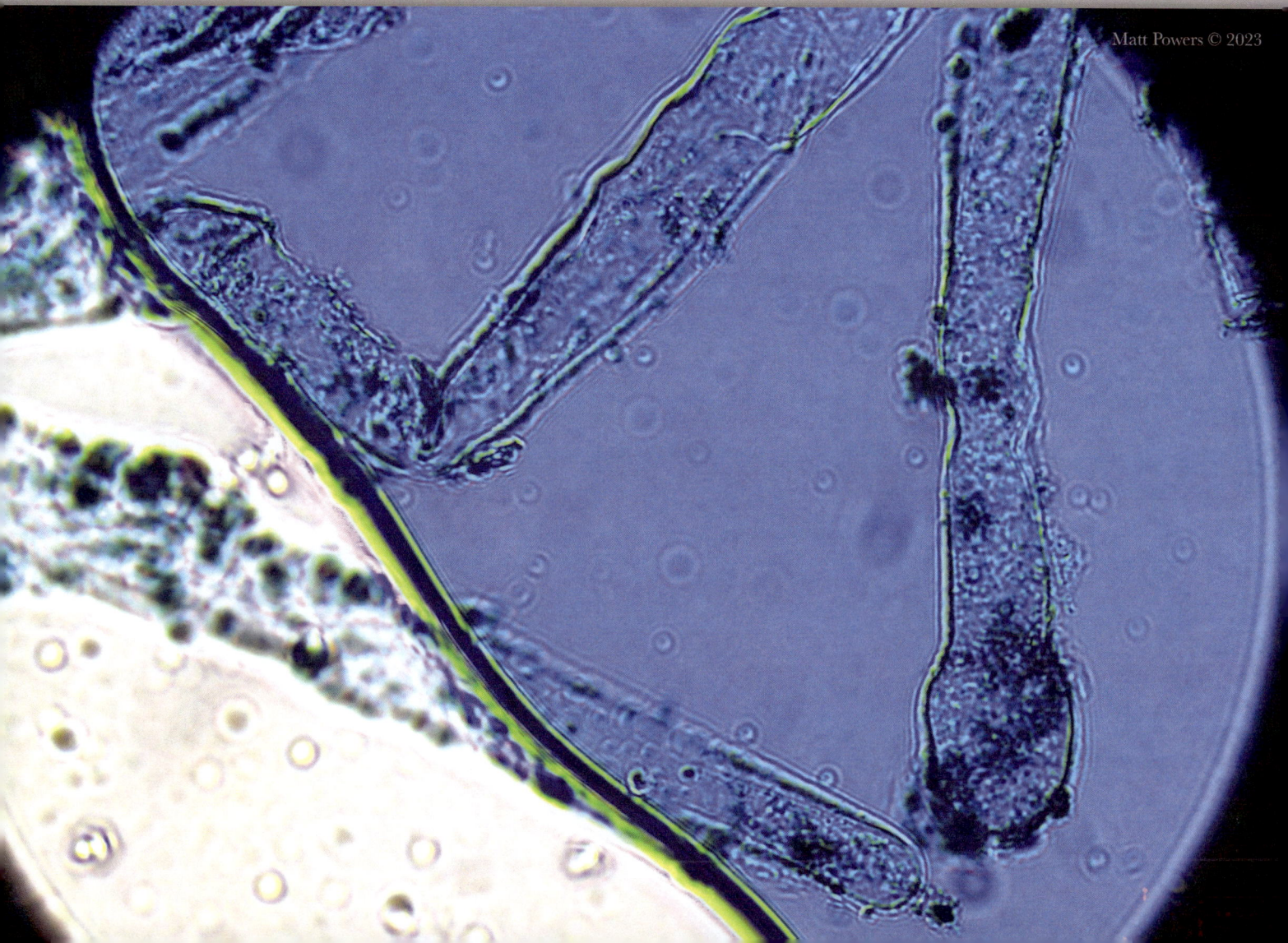

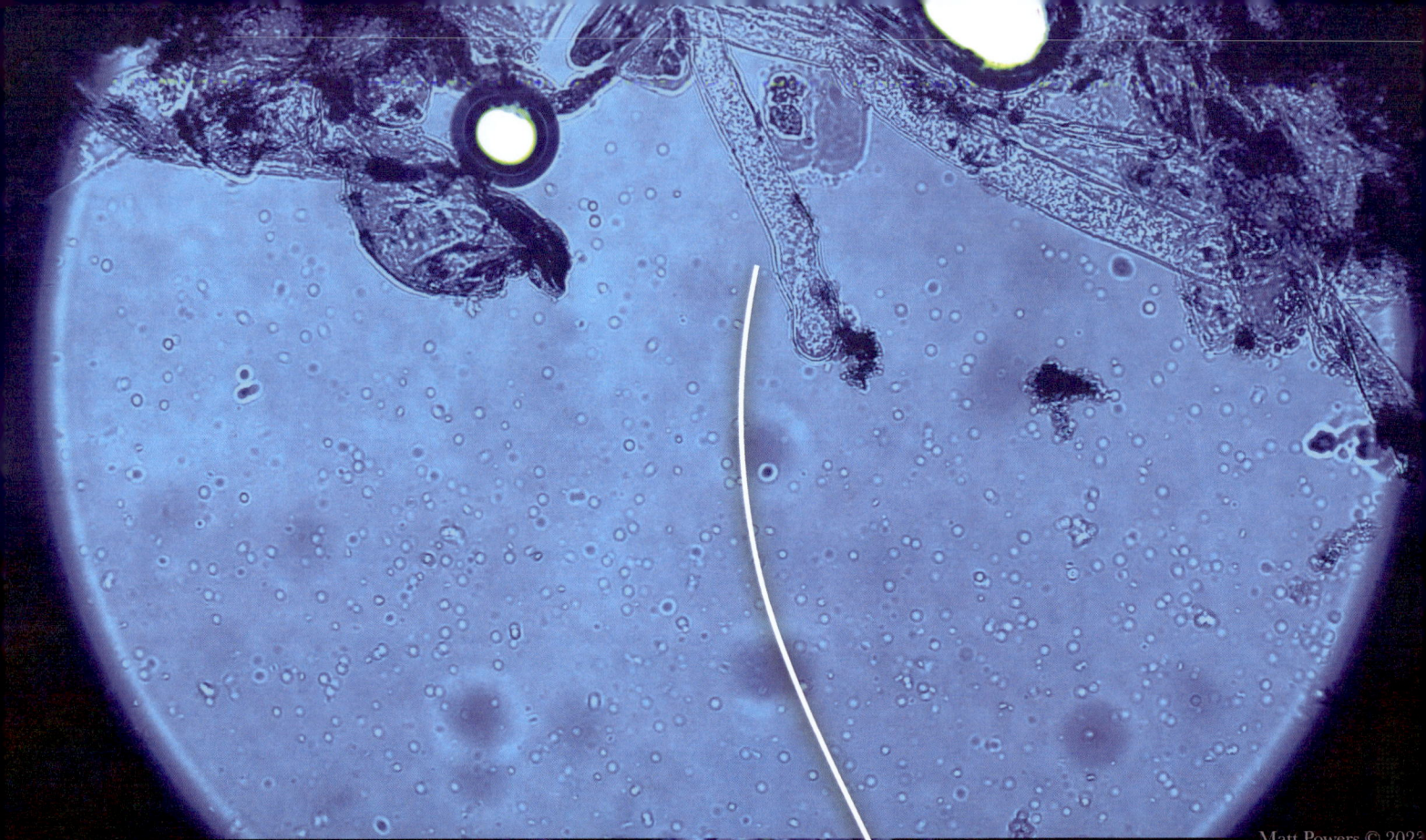

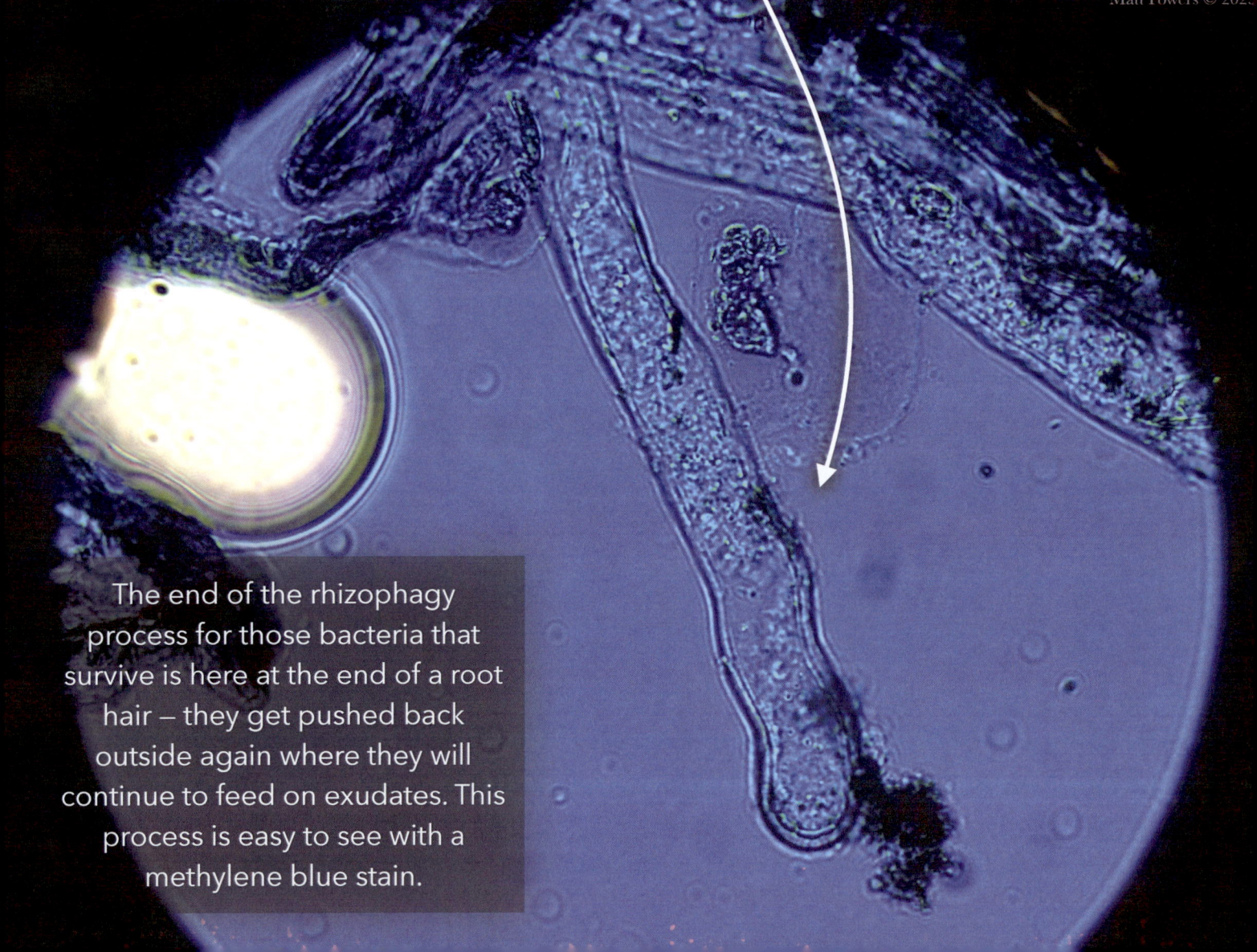

The end of the rhizophagy process for those bacteria that survive is here at the end of a root hair – they get pushed back outside again where they will continue to feed on exudates. This process is easy to see with a methylene blue stain.

to verify each and every step ourselves – if we have root hairs, we know our roots are utilizing rhizophagy to gain their nutrition. A few drops of methylene blue in 1 ml of water and we can stain the root with that solution. Some methods recommend stains and then washing the sample. If we washed the sample, we'd lose much of what's of interest in the sample! Instead, we dilute the stain, and add it to the root, let it adhere and soak in for a few minutes, and then view it. I crushed the root under a cover slip awkwardly for these photos. You may have to do the same. Being creative with our methods is the key to having new insights into our soil, plants, roots, mycorrhizae, and more.

OTHER TESTS

- **Clay/Sand/Silt** – if you aren't looking at the ratios of clay/sand/silt, you may be missing a fundamental aspect of your soil, OR if you are buying compost in bulk, this is a quick way to see if you are being ripped off.

- **NPK** – plants are primarily composed of NPK + C, so it makes sense that compost can be very high in nitrates and phosphorus (both can overwhelm and/or inhibit a system or soil if improperly applied at the wrong time or the wrong rates). Having a home NPK test kit like a LaMotte NPK test kit can be a fast and easy way to get a general idea of how your compost amendment will affect your soil and plants.

- **pH** – if we know our pH, we can understand the forces and pressures at work on the biology in our samples.

- **EH** – if we know our REDOX potential and our pH, we can fully map out what's possibly available in our soils on a regular basis. Your soil mineral test report may list a mineral being there but if your Eh is in the wrong range, it won't be available to the plant roots.

- **Salinity** – salting is killing soils all over the world – knowing how salty your water is, your soil, your compost, etc. is important to maintaining the integrity of your soil and rhizosphere.

- **Mesofauna & Macrofauna Tests** – pitfall traps & burlese funnels can help us get a snapshot of some of the microarthropods and arthropods cycling nutrients and

This is why you test any compost or amendment you buy - this was recommended as the best compost in the Austin, TX area by a full-time soil food web consultant who is an awesome enthusiastic person who immerses herself in promoting and educating people on compost, but it never occurred to her to even test the compost.
Please make sure you test your compost.

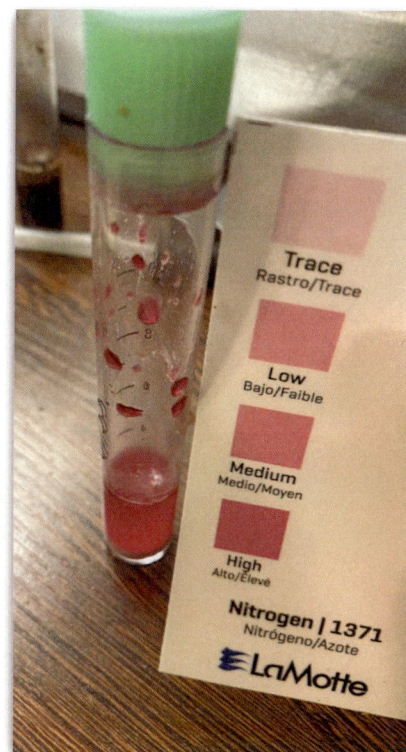

While folks may argue about the exactness of home tests, they are great generalized indicators, and since 80% of what plants absorb is NITROGEN, it shouldn't surprise us that our composts are high in nitrogen, and if we do thermophilic compost that we turn, we will have high nitrates - this is why LAB is useful to turning the nitrates into amino acids.

THIS IS A DIRTY SLIDE....
SEE THE STREAKS?

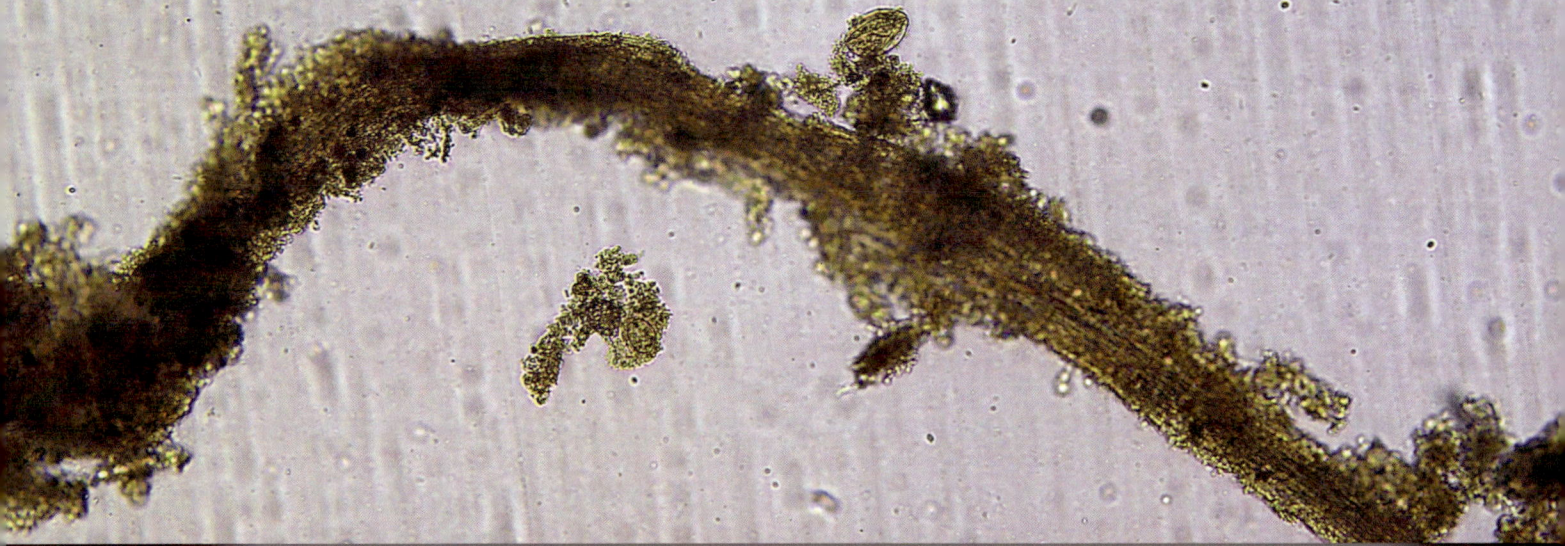

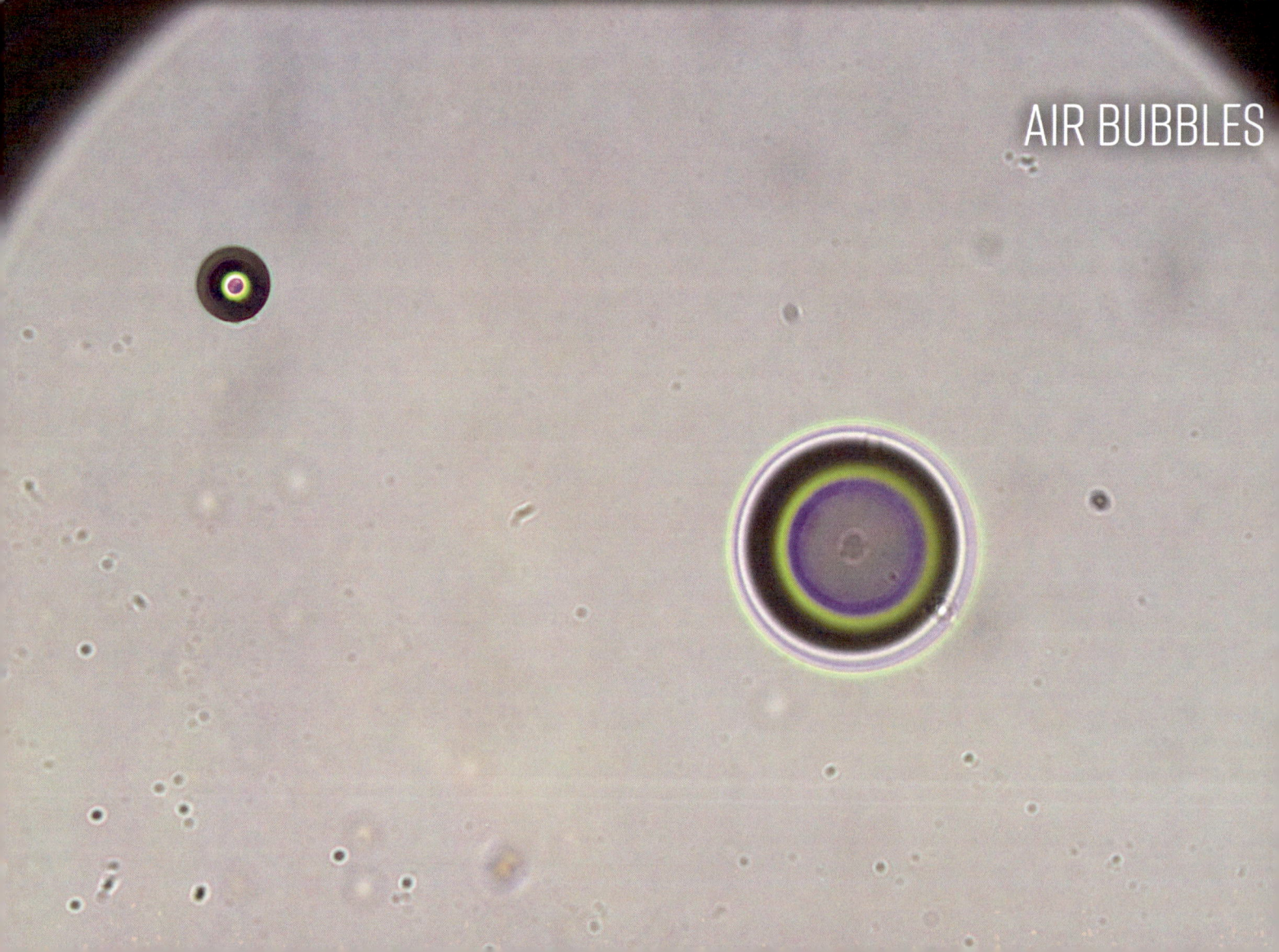

AIR BUBBLES

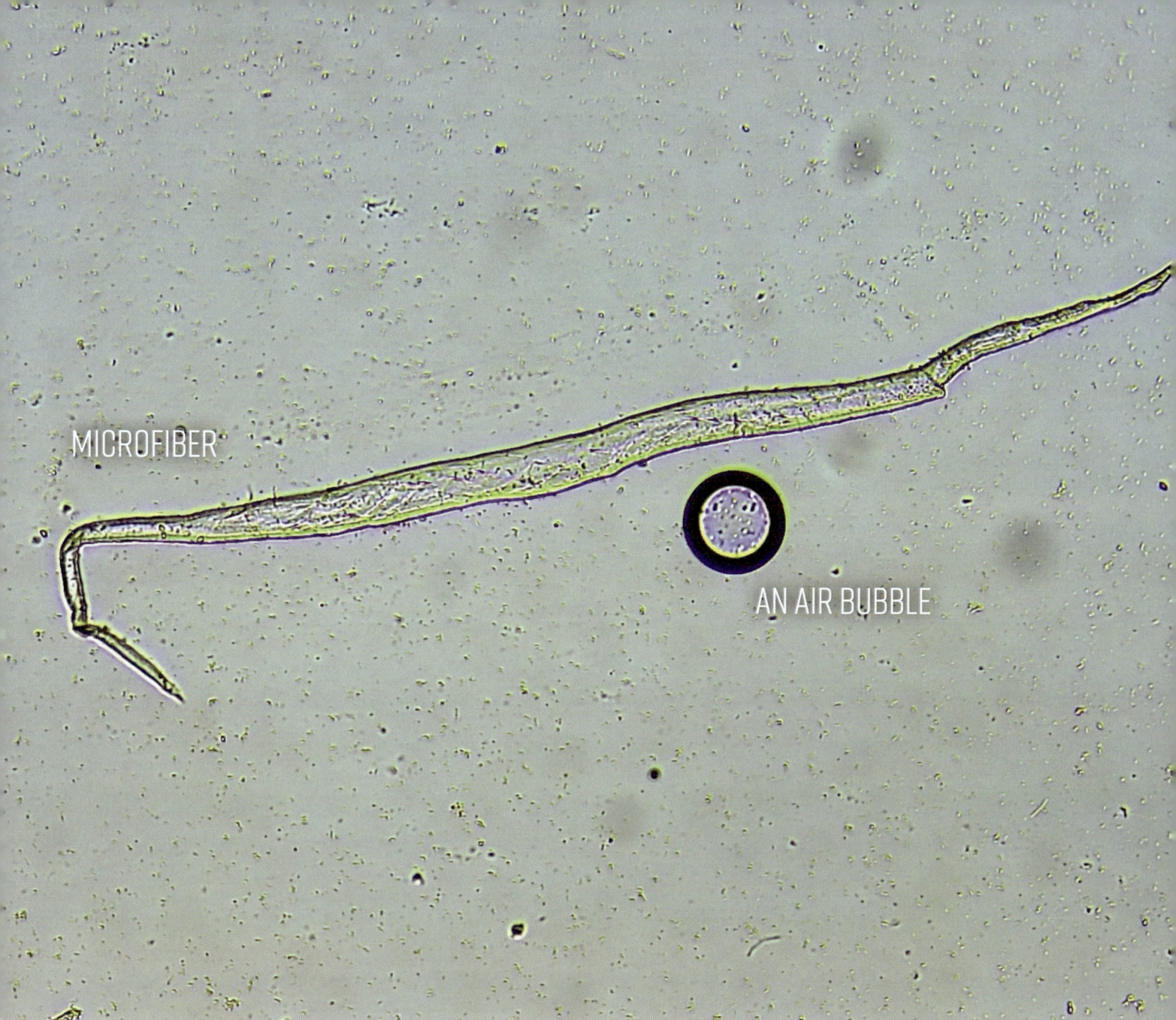

MICROFIBER

AN AIR BUBBLE

shredding organic matter at the higher levels of our soil.

- **MicroBiometer** – for automatically calculated F:B ratios especially for clients that desire this metric (this saves time and energy better spent on examining more samples and doing other tests). I don't think manual counting is a reliable method, so that is why I do not lean heavily upon it in this book. It's better to observe the foundational layer (bacteria) as its own metric, and then to view fungi independently especially because it depends on what fungi we are talking about: we want beneficial, desirable microbes, and specific biofertilizers and mycorrhizal fungi. The microbiometer provides a way to check our work and our thinking - it detects yeast, spores, and all forms of fungi.

Test your water always before you test any solution with soil or compost.

Always account for as many variables as possible.

Common problems

At first you're going to encounter challenges, but as your eye develops and your microscope skills improve, you will overcome most if not all of these challenges. I provide these pages and some images as guides to help you avoid the most obvious errors.

WHAT IF...

- You can't see ANYTHING but specs and tiny black dots – you are likely focused on the top of the slide cover, above all of the action which is between the sections of glass, or you are focused on the light lens below it all. Either way, you've gone too far. Start over at 40x magnification, get in focus with the coarse focus knob first, then dial it in with the fine focus. Then move up again in magnification.
- You keep smashing the objective into the slide, breaking slides, breaking the cover glass, or there's no room to even put the objective down in place: the microscope stage is too high. Start over at 40x and try again OR remove debris from under the cover slip so it can be lower.
- There's no bacteria at 400x – 1000x – slowly adjust the height of your condenser and close your condenser shutter to see if you can see them better, but know that some samples have little to no life. They also may be dehydrated and extra small – letting samples sit and rehydrate at a microbial level may be necessary.
- You see strange orbs? Thick black iridescent, perfect circles? These are air bubbles – often caused by oils still on the slide.
- Large, clear corkscrew spirals? It's likely cellulose released as a single fiber in the heat of a decomposition event (composting) though if they are small spirals and motile, they are likely spirilla bacteria.
- You see straight lines or streaks across the FOV – that's either a dirty slide or scratched glass
- You see strange geometric ephemeral shapes and refractions that appear and disappear: that's light going through some organism or mineral
- It's total CHAOS – dilute your sample more!
- I see GIANT THINGS – it could be seed coats, insect parts, insect eggs, earthworm casting, microplastics, etc.

The Regenerative Soil Ratios

These are numbers I've generated from viewing actual compost, soil, and biofertilizers over time with my microscope using a combination of testing methods – I learned Dr. Elaine Ingham's methods through her Microscope Intensive online, but I also learned what veterinarians and ranchers are doing to verify the efficacy of their dewormer treatments – calculating up from a grid or hemocytometer is widely practiced and standardized. Depending on your size of measurement, you'll find the same exact calculation or formula everywhere, so I've verified these methods in my home lab, acquired a medical-grade hemocytometer, and used the most commonly accepted and regularly utilized measurements and formulas to generate my numbers. I have seen the recommendations by other soil microscopy teachers, and I do not see 600,000,000 bacteria per ml (or gram to be more imprecise) to be a reasonable number. I instead see in most composts 100 – 1000x less that amount using the commonly accepted formulas and counting methods – perhaps there is a calculation error or an error in the method in which they are counting. The power of the hemocytometer is you can verify the formulas for a 3mm by 3mm grid in any major language through the internet at any time – *it won't change* because so many branches of science rely upon it being a standardized testing method.

WHAT ARE THE IDEAL RANGES & RATIOS?

Just as there are many variable inputs that go into the ongoing open air experiments we do in our gardens or fields, the outputs are just as variable, so we will find good soils with a wide range of biological expression. What we are looking for also depends on our goals – good rangeland soil is not the same as garden soil nor is it like orchard soil. The plant types, minerals, climate, pH – all of these factors will affect what is "ideal" for us individually on case by case, site by site, area by area basis. What if you are growing non-mycorrhizal plants like amaranth or swiss chard? You're not going to see the fungal images or numbers shown in the examples in this book. If you are growing a no-till orchard in a humid temperate climate, you should be seeing high fungal counts and diverse expressions of fungi in your orchard soil though if you do not, that's a clear signal of what actions to take. Context, once again, is critical to keep in mind.

Given all that, what is ideal? Having minimum desirable levels as a guide can help us find a starting place that is common to all. The "perfect" ratios and numbers do not exist in reality – instead we see a wide range of expression dependent on several factors. As we support the nutrient cycles, fill in the gaps in the microbial trophic layers, and build our soil's structure biologically, the ranges of expression will change and develop new patterns as the soil passes through its own form of succession. We've learned so much by observing the best soils in the organic farming movement – it's what led to the regenerative soil science breakthroughs we have today, but it came out of observation and practice. We need to be open to what nature is teaching us, we need to catalog it thoroughly, and we need to share it so it can be compared across samples, bioregions, climates, soil types, plants types, management methods, etc. It's the only way we can map the next layer of nuance in soil science – again, this is why the R-Soil Database is so vital. Out of that process of community communication, observation, and learning, we will develop ideal ranges by soil type, bioregion, plant type, biology present, etc. beyond what is currently available or even possible to know.

THE MINIMUM DESIRABLE RATIOS

	ORGANIC MATTER	BACTERIA	FUNGI	PROTOZOA	NEMATODES
SOIL	~5 – 45% or higher *(Peatlands are ~100% organic matter).*	~1 – 3 billion per ml/g of soil	• Avg 1 – 5 Fungal Hyphae per FOV @ 100x • Total lengths of 150 – 1000 µm per drop @ an average width between 2 – 3 µm	• 1 – 5 Beneficial Protozoa per FOV @ 400x • Less than 1 – 3 ciliates total per drop	• 2 – 5 Beneficial Nematodes total per drop • More Bacterial than Fungal, & More of Both than Predatory
COMPOST	~80 – 100 % *(Pit composts can get a lot of soil mixed in)*	~4 – 10 billion per ml/g of compost	• Avg 1 – 5 Fungal Hyphae per FOV @ 100x	• 3 – 12 Beneficial Protozoa per FOV @ 400x • Less than 3 – 5 ciliates total per drop	• 3 – 7 Beneficial Nematodes total per drop • More Bacterial than Fungal, & More of Both than Predatory

Extremes Require or Create Balance

What if you have compost or a biofertilizer that is imbalanced? It's all bacteria, or it's predominantly predators. Sometimes, as in some BEAM composts, it's all saprophytic fungi and spores with a lot of diversity of bacteria, specifically no rhizobia (which is common in thermophilic composts), so we can find ourselves in all sorts of situations where our piles or brews lean one way or the other. This is an opportunity to use them as a tool – if we have bacterial dominant soils lacking predators, we can add compost or extracts from that compost to the soil to balance and cycle the nutrients trapped in the bacteria. Creating something truly neutral in pH, whether in a large pile or an open field, can be difficult – it's much easier to land on one or two extremes out of 4 biological groups we are trying to balance, and then to use them strategically. This is exactly the case with compost itself – we must use the kind of compost we have (fungal, bacterial, oxidized, reduced) for the purpose it lends itself to, OR we have to modify it before we apply it in any way. Maybe you add EM-1 Pro® or a homemade equivalent to your predator-rich pile to give them something to eat and to grow and stimulate the bacterial foundation of the food web. Maybe you add watered down molasses to your pile when you add the EM-1 Pro®. Or maybe it's the other way around and you need to boost your fungal foods: combine your hot thermophilic compost pile on day 3 or 4 at the peak of heat with a pile of equal size but purely ramial wood chips (twig and new growth chipped). Let that sit for a year, and you'll have incredible fungal compost. I highly suggest getting **Regenerative Soil** if you don't already have it for the recipes and nutrient cycle diagrams.

What about Mycorrhizal Inoculation %?

Arbuscular mycorrhizal fungi (AMF), and all endomycorrhizal fungi, inoculate plant roots at 50 – 100% of the root in healthy symbiosis. The longer the AMF inoculated roots are on a plant, the more abundant AMF is in the soil profile. This can be difficult to gauge on living plants, but we can gently excavate the surface soil until a root is exposed and we can harvest that root and cover the area back up again. The plant might not benefit from this treatment but it will give us significant data that we can compare with other tests and plants for greater insight.

Endomycorrhizal fungi inoculate plant roots at less than 50% – if we know which plant type we are examining the roots of and we can examine the roots with a microscope, we can see the sheathing of the root cells that this fungi performs: it is starkly different from AMF, and only occurs with a small amount of trees, specifically cold temperate and high altitude native and timber species like pine, fir, beech, manzanita, and chestnut, though there is a small group of ecto- endo-mycorrhizal plants like eucalyptus, willow, aspen, poplar, cottonwood, and alder. Notice how we aren't going to see these in our orchards nor in our fields or gardens.

REGENERATIVE SOIL MICROSCOPY
Soil & Compost Food Web Analysis Worksheet & Guide

BACTERIA

Count at 400 – 600x
& Calculate per ml/g
@ 1:100 dilution
Hemocytometer + AO & Epifluorescence

Use 400x - 600x to diagnose "types" & behavior

Bacteria types & # per ml/gram of soil
= Food Chain Ratios, Defines the Limits,
Relative Abundance, & Imbalances

FUNGI

Locate at 40x – 100x
& Calculate per FOV @ 400x (or total micron length per drop)
@ 1:10 dilution
Bright Field, Dark Field, & Epifluorescence

Use 400x – 1000x to diagnose "types" & behavior

Fungal types & # per drop of soil or compost dilution
= Food Chain Ratios, Defines the Limits,
Relative Abundance, Disturbance Levels,
& Imbalances

PROTOZOA

Count at 400x
& Calculate the average per FOV
@ 1:10 dilution
Bright Field, Dark Field, & Epifluorescence

Use 400x – 1000x to diagnose "types" & behavior

Protozoa types & # per FOV
= Food Chain Ratios, defines the limits of the soil,
Relative Abundance, Disturbance Levels,
& Imbalances

NEMATODES

Count at 40 –100x
& Calculate per drop
@ 1:10 dilution
Bright Field, Dark Field, & Epifluorescence

Use 400x - 600x to diagnose "type" & behavior

Nematode types & # per drop of soil or compost dilution
= Food Chain Ratios, Types of Cycling,
Relative Abundance, & Noticeable Gaps

Soil/Compost Food Web Ratios:
BACTERIA:FUNGI:PROTOZOA:NEMATODES

ORGANIC MATTER & MINERAL CHARACTERIZATION

THE AVAILABLE NUTRIENT CYCLING PATHWAYS

GAPS, DEFICIENCIES, & IMBALANCES:

REGENERATIVE SOIL MICROSCOPY
Root & Leaf Analysis Worksheet Guide

ROOT INOCULATION %

Mycorrhizal Type

Plant Type

Inoculation % of Root

Average Length of Root

TRICHOME INOCULATION %

Inoculation Type

Plant Type

Inoculation % of Trichome

Average Length of Trichome

STOMATA INOCULATION %

Inoculation Type

Plant Type

Inoculation % of Stomata

THE CASE STUDIES

Now that you know what's minimally required, what to look for, and how to look for them, let's look at how things compare across a broad range of examples from biofertilizers to different kinds of compost and soil. Please feel free to replicate this exact work with your own compost or biofertilizer brews and post your results on the R-Soil Database.

JOHNSON-SU STUDY

In the pursuit of ever more fungal compost, the Johnson-Su style bioreactor composting has emerged a leading method – whether by microscopic means or through the microBIOMETER, it is clear that homemade Johnson-Su composts have the highest fungal counts (D.B.'s for example). A commercial lawn care business, students of my Regenerative Soil program, and BEAM have all sent me samples with varying expressions. Stangl's lawn care business in Ontario, Canada (stangls.com) has amazing Johnson-Su compost that looks like a near cousin to Catalyst Bioamendments thermophilic windrow turned commercial compost (their team was trained by Dr. Elaine Ingham). The static aeration of JS compost is a powerfully balanced method that naturally promotes fungi through the lack of disturbance, consistent moisture, and long duration of decomposition. Catalyst Bioamendments gets similar looking results but in a fraction of the time by combining methods: hot composting with aeration and following that a lack of aeration for an extended period of time for fermentation and facultative anaerobes and fungi to do their work. (There's a reason the facultative microbes of EM® promote flushes of fungi). BEAM compost lacked nematodes entirely while in the case of Stangl's Enviro Lawn Care, a Canadian company, their JS compost had a great balance of all levels of the food web and a surprising number of testate amoebae (similar to Catalyst). Just looking at the microscope images without counting or testing, we can see a stark difference. The student's Johnson-Su compost is much more similar to commercial lawn care company's compost (though technically they're both students of my program).

I have seen JS composts with high nitrate and variable phosphorus levels which implies that the

Omnivore Nematode from an Ontario
Johnson-Su Compost Pile

185

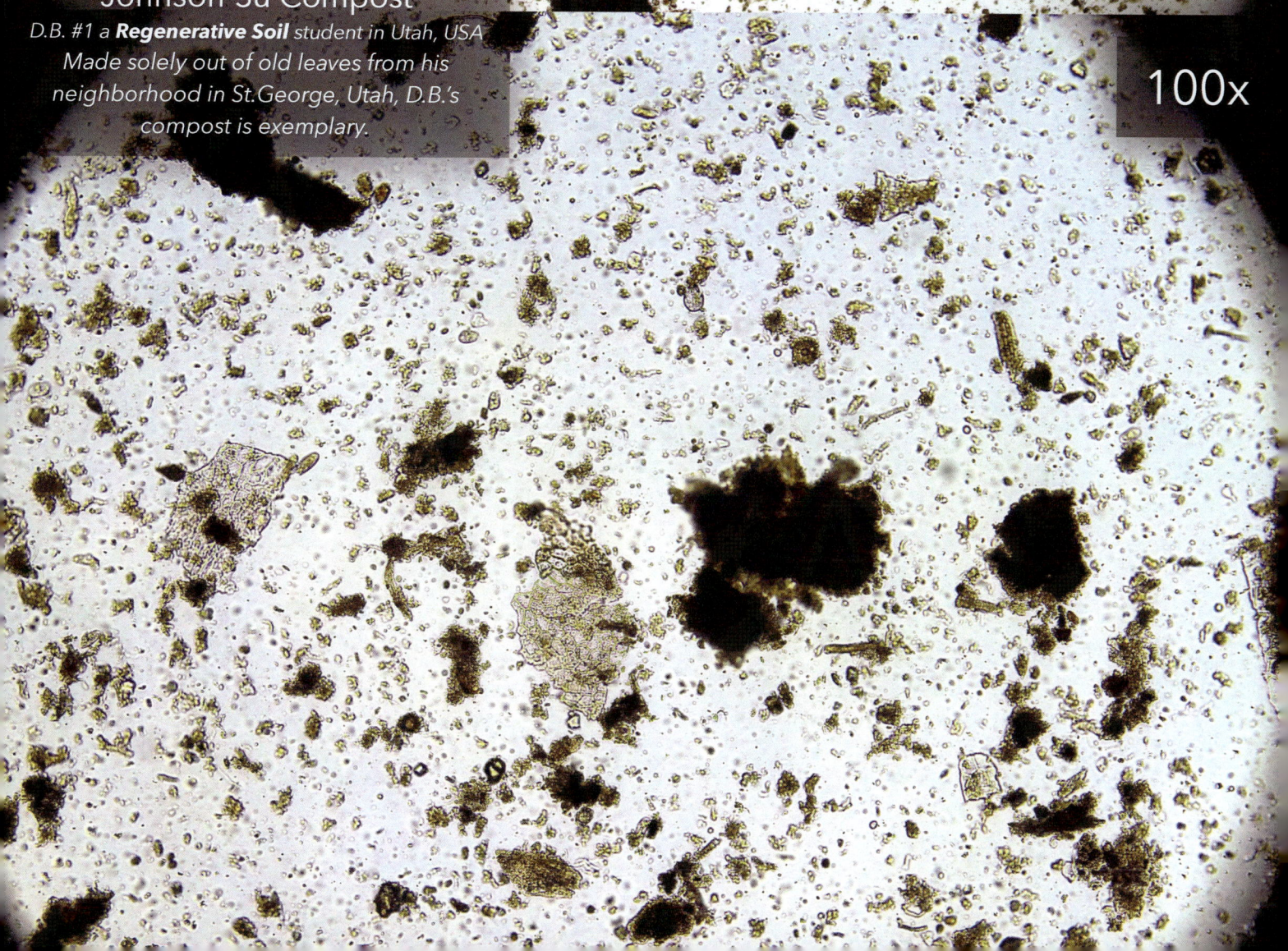

40x

100x

Johnson-Su Compost
D.B. #1 a **Regenerative Soil** student in Utah, USA
Made solely out of old leaves from his
neighborhood in St.George, Utah, D.B.'s
compost is exemplary.

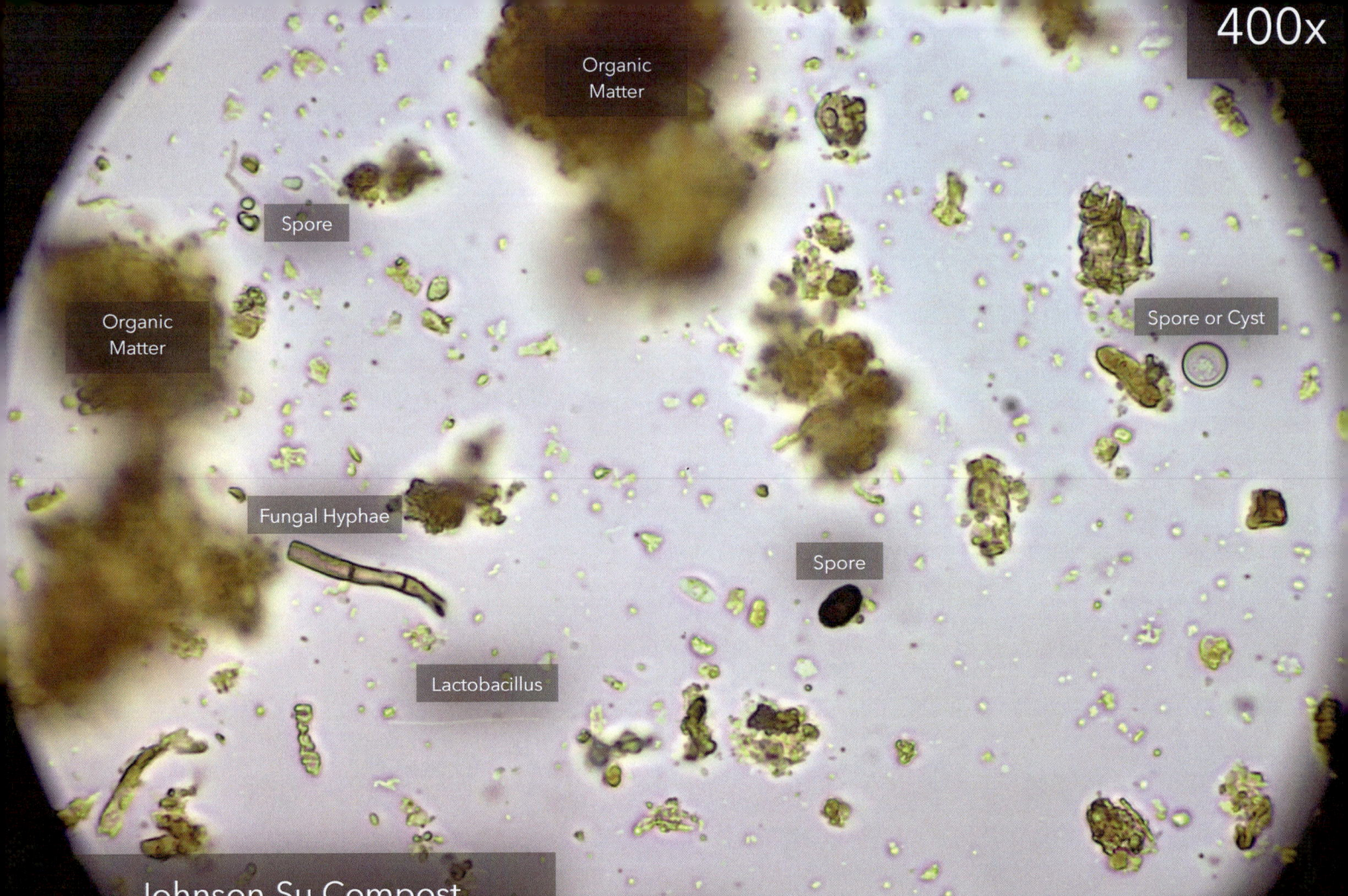

400x

Organic Matter

Spore

Organic Matter

Spore or Cyst

Fungal Hyphae

Spore

Lactobacillus

Johnson-Su Compost

D.B. #1 a **Regenerative Soil** student in Utah, USA

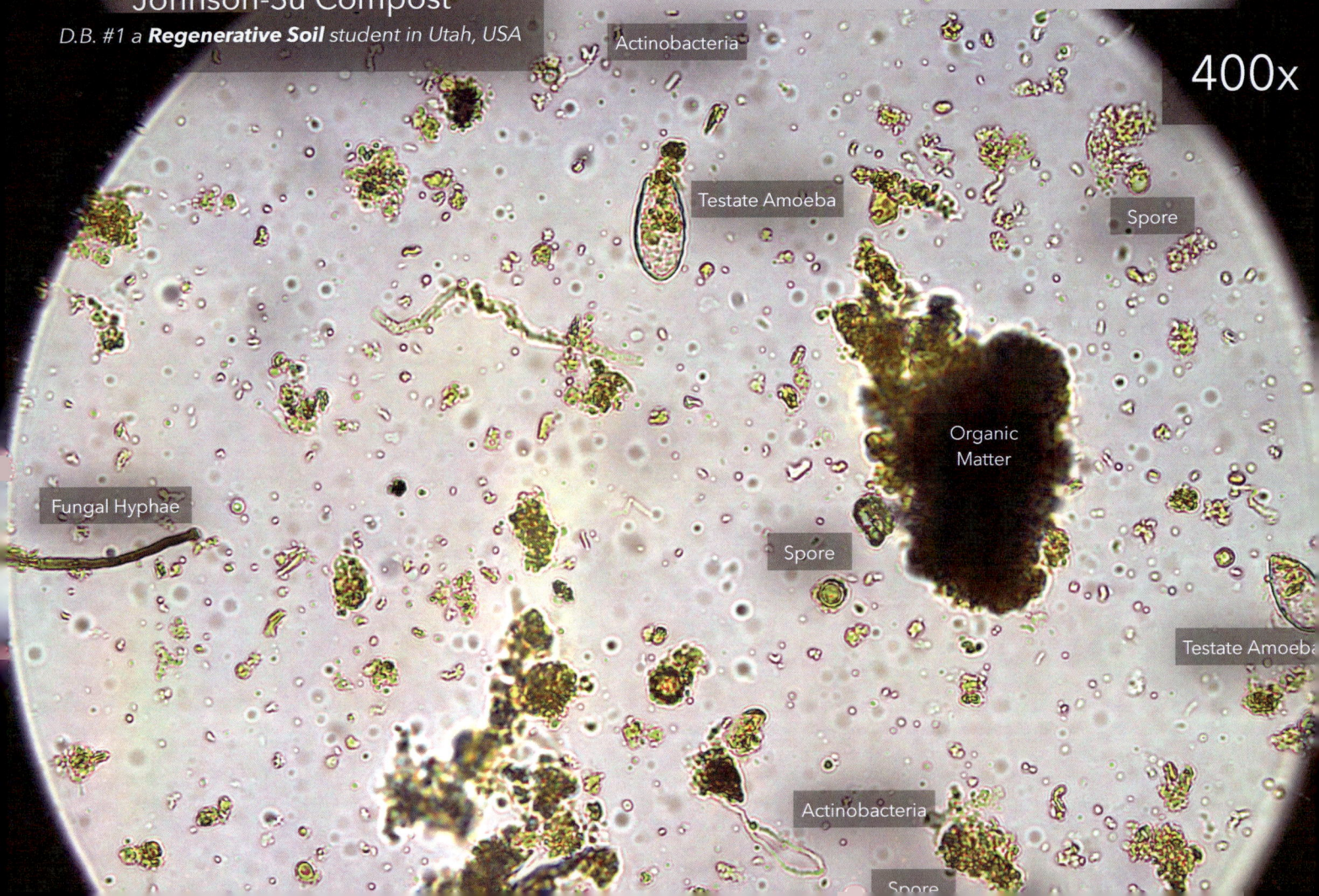

Actinobacteria

400x

Testate Amoeba

Spore

Organic Matter

Fungal Hyphae

Spore

Testate Amoeba

Actinobacteria

Spore

400x

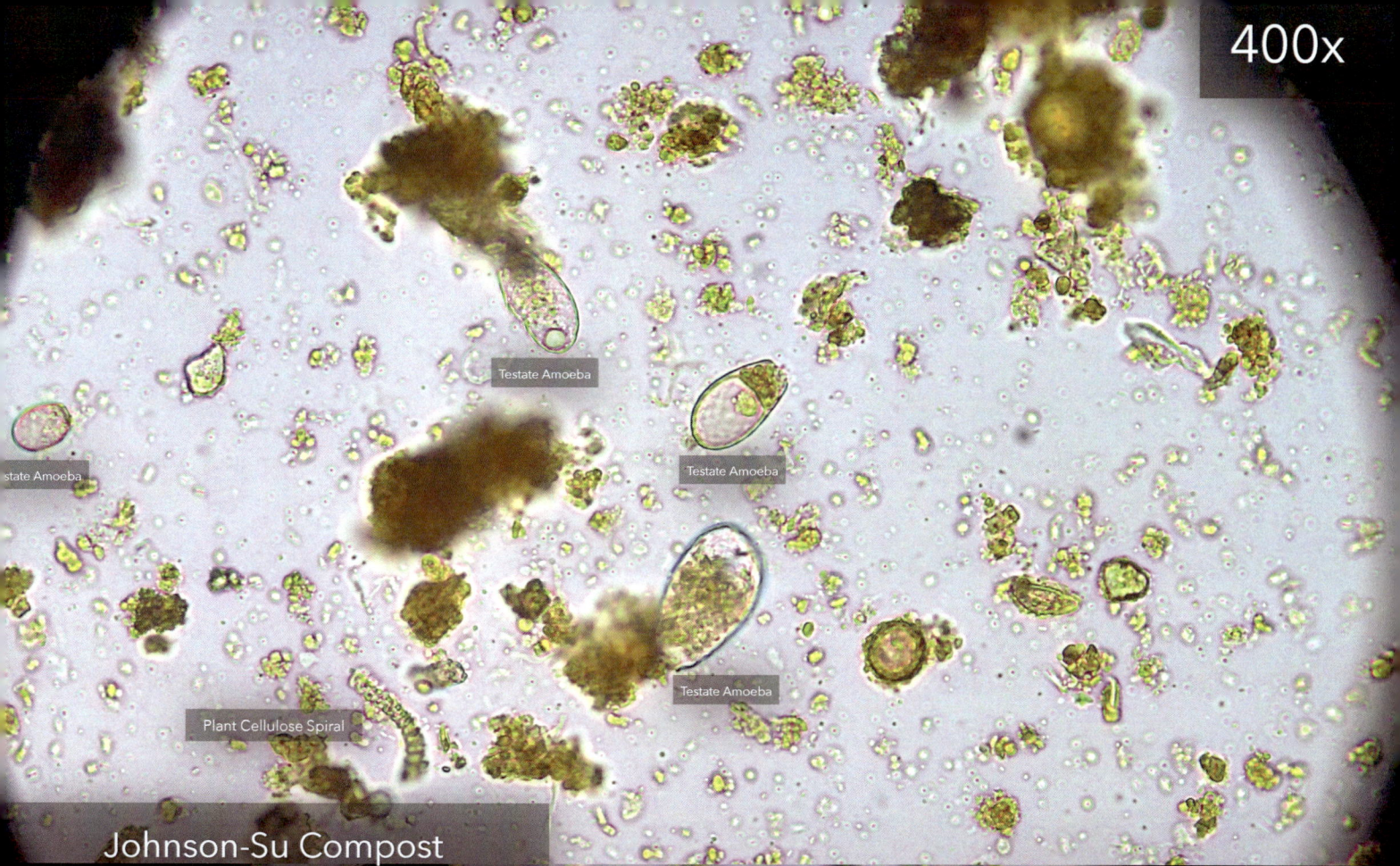

Testate Amoeba

state Amoeba

Testate Amoeba

Testate Amoeba

Plant Cellulose Spiral

Johnson-Su Compost

*D.B. #1 a **Regenerative Soil** student in Utah, USA*

400x

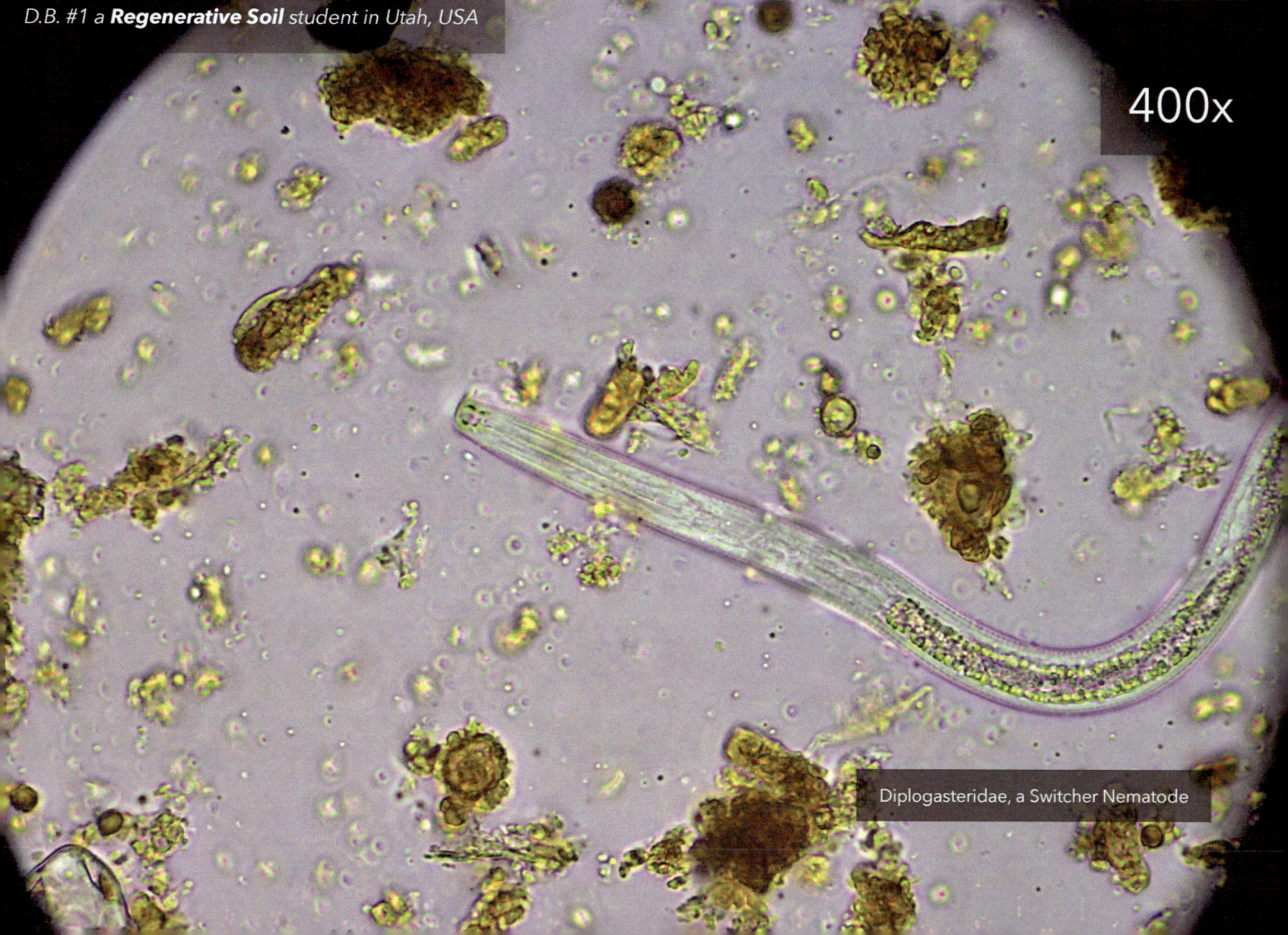

Diplogasteridae, a Switcher Nematode

DECOMPOSING WOOD CHIP FRAGMENT

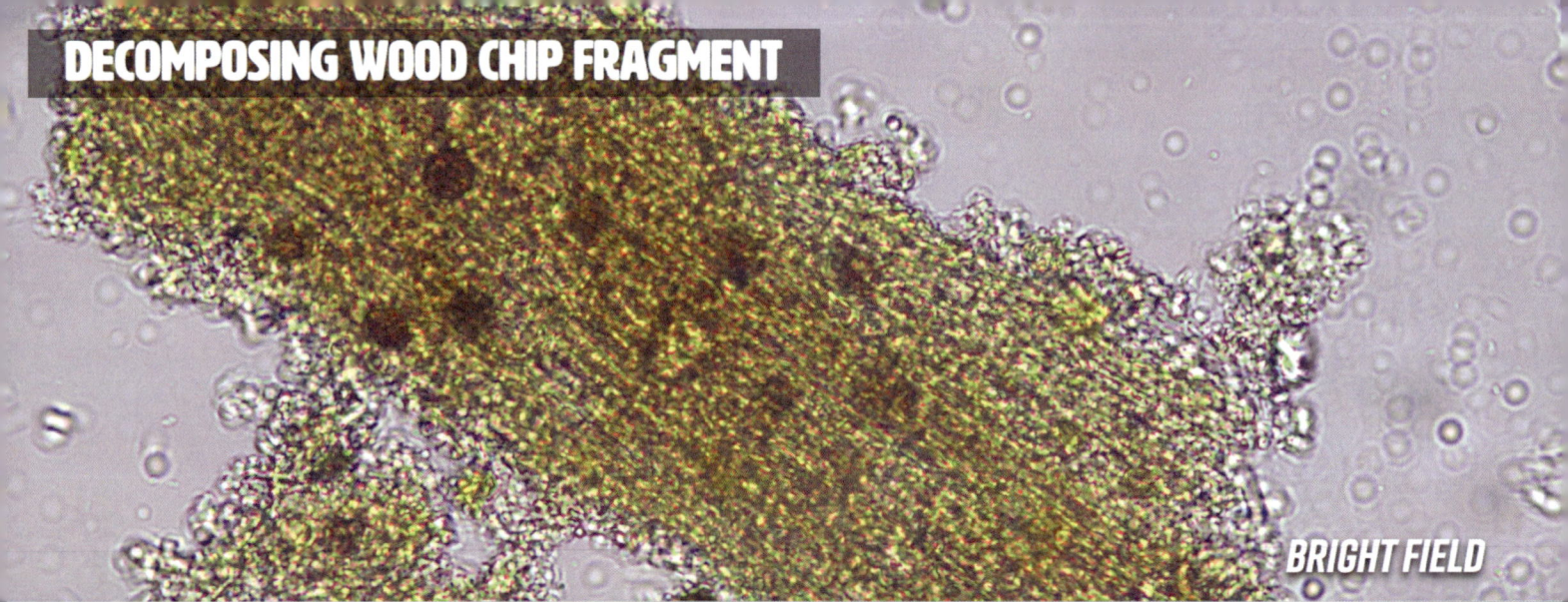

BRIGHT FIELD

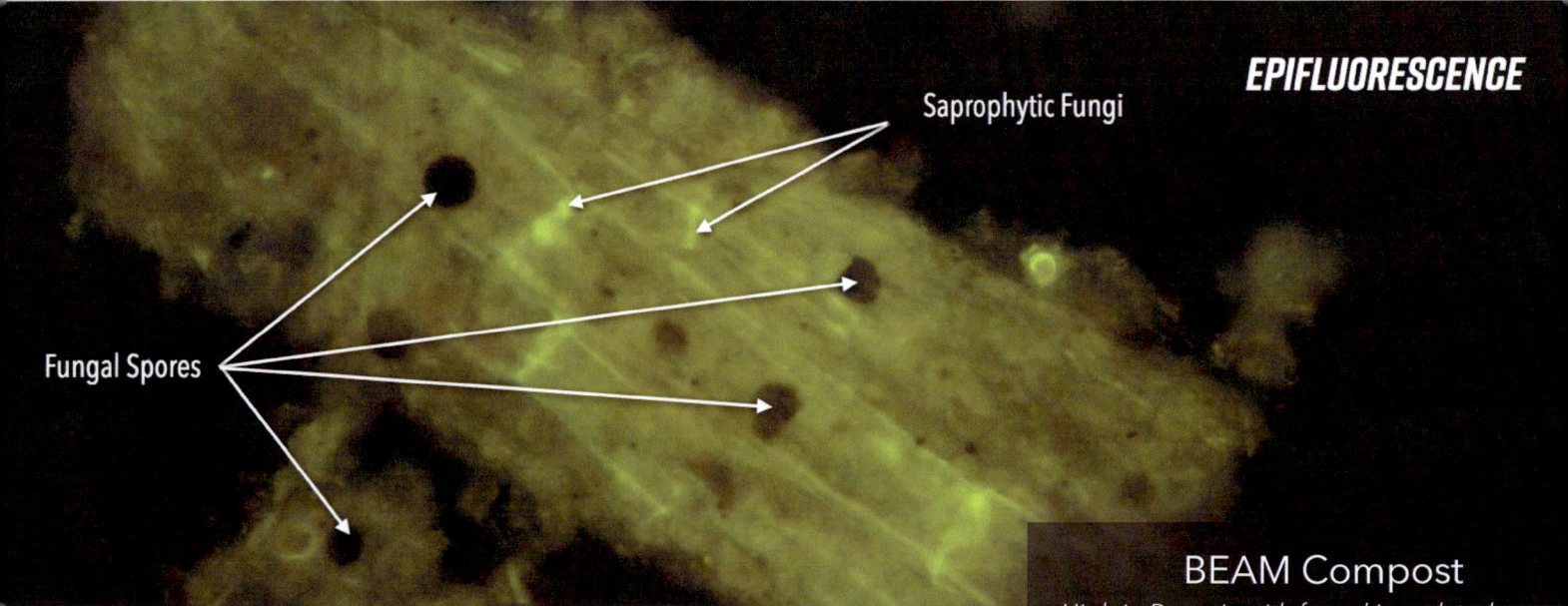

EPIFLUORESCENCE

Saprophytic Fungi

Fungal Spores

BEAM Compost
High in Bacteria with fungal inoculated organic matter & purple testate amoebae.

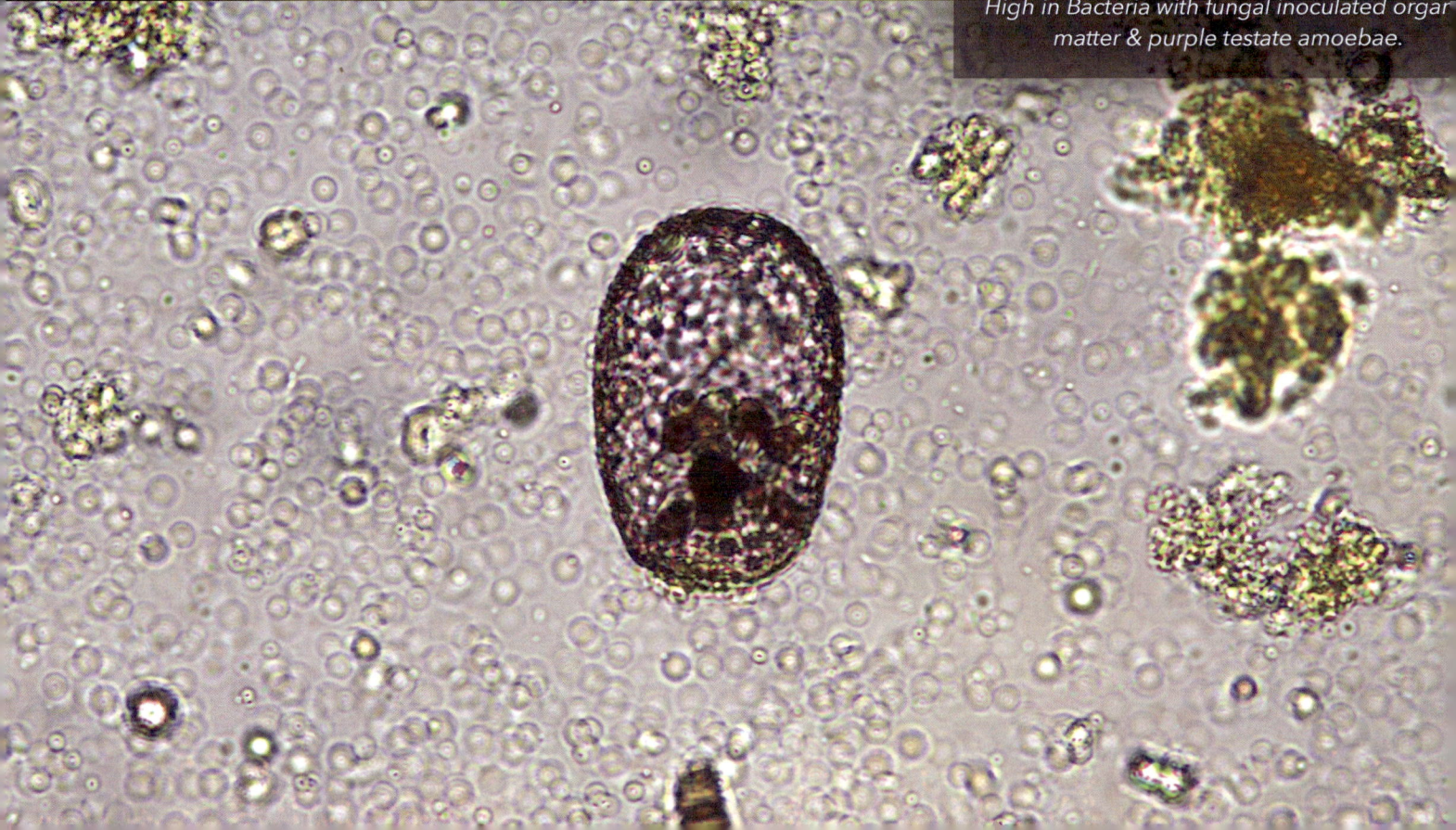

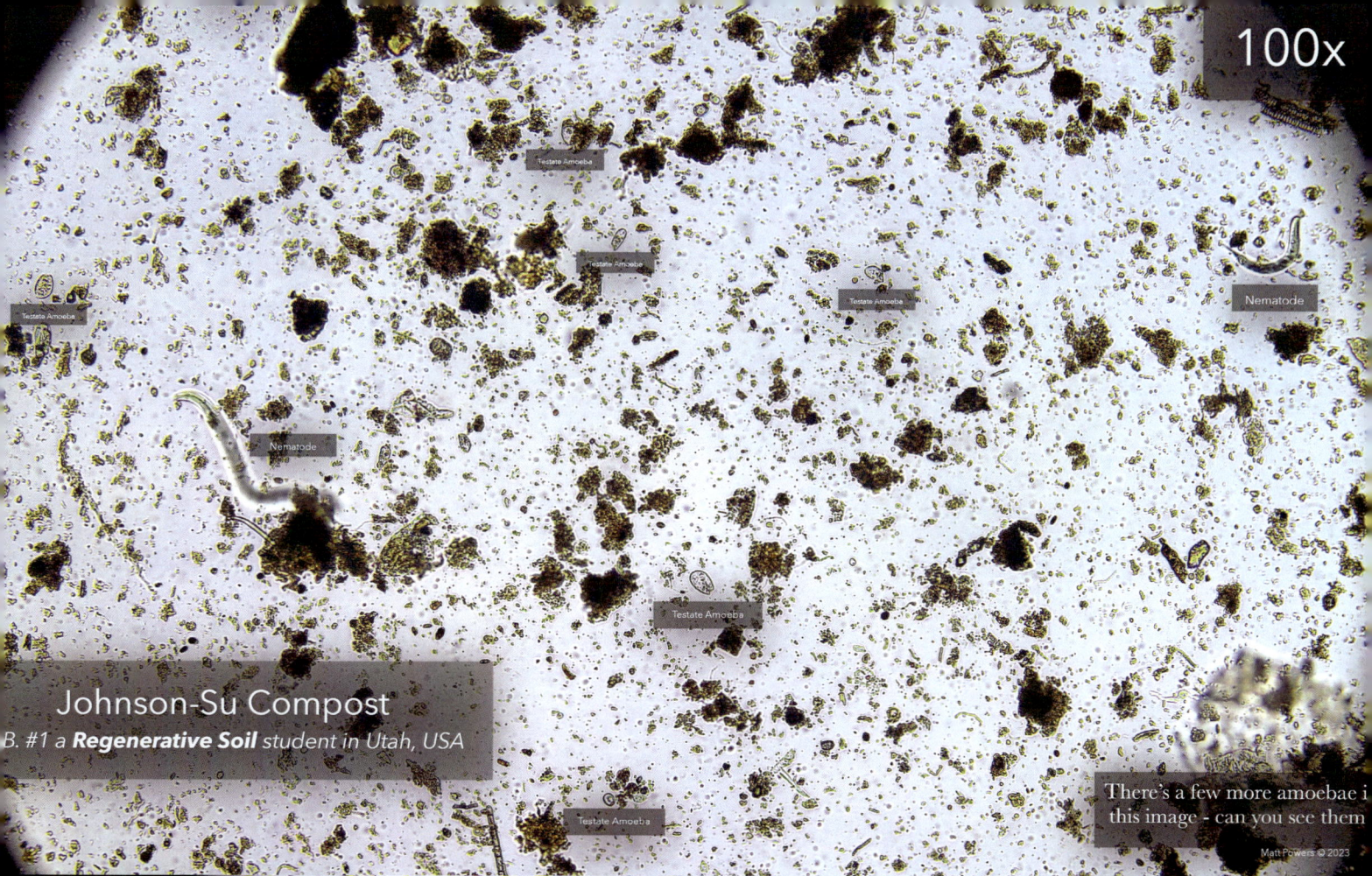

100x

Testate Amoeba

Testate Amoeba

Testate Amoeba

Nematode

Testate Amoeba

Nematode

Testate Amoeba

Johnson-Su Compost
*B. #1 a **Regenerative Soil** student in Utah, USA*

Testate Amoeba

There's a few more amoebae in this image - can you see them

Matt Powers © 2023

BEAM Compost
from BeamCompost.com

100x

Notice how big this testate amoeba is in contrast to the above compost - it's the same magnification (100x)

Testate Amoeba

Testate Amoeba

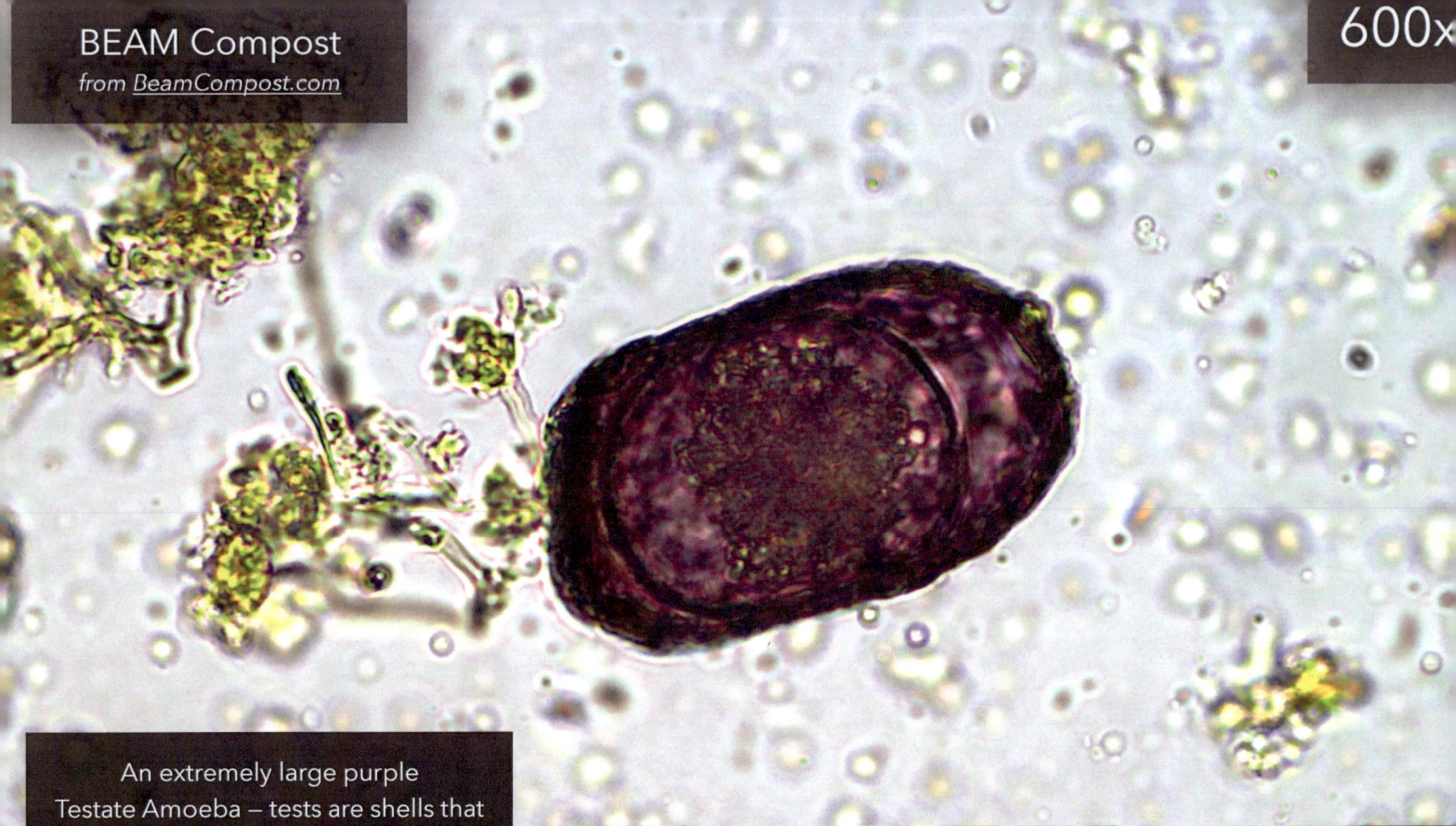

An extremely large purple Testate Amoeba — tests are shells that protect testate amoebae. The darker the color of their test, the less light changes influence them. The hardier the test, the more durable the shelter it is against times of stress: watering, drying, thermophilic heat, and pH/Eh swings.

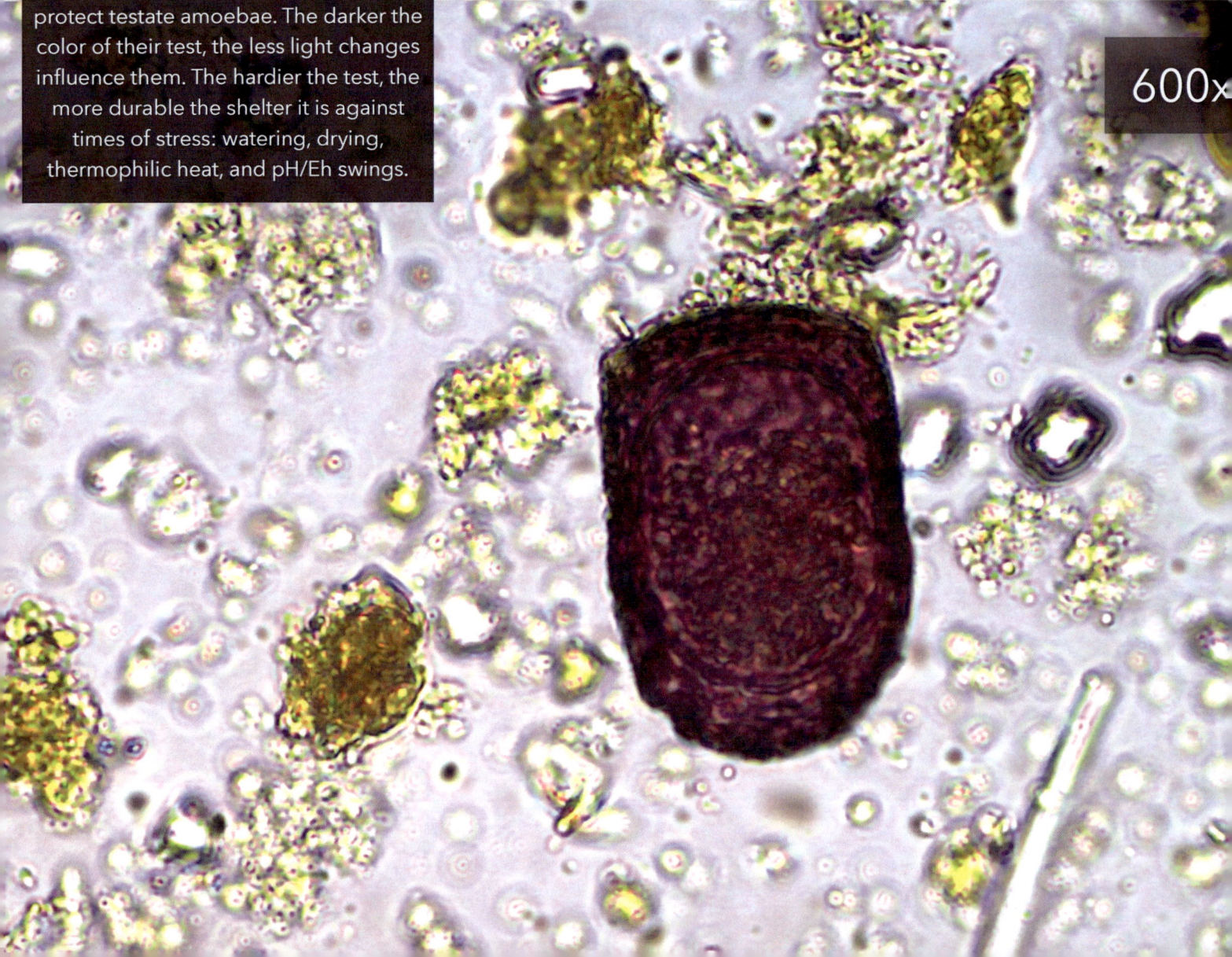

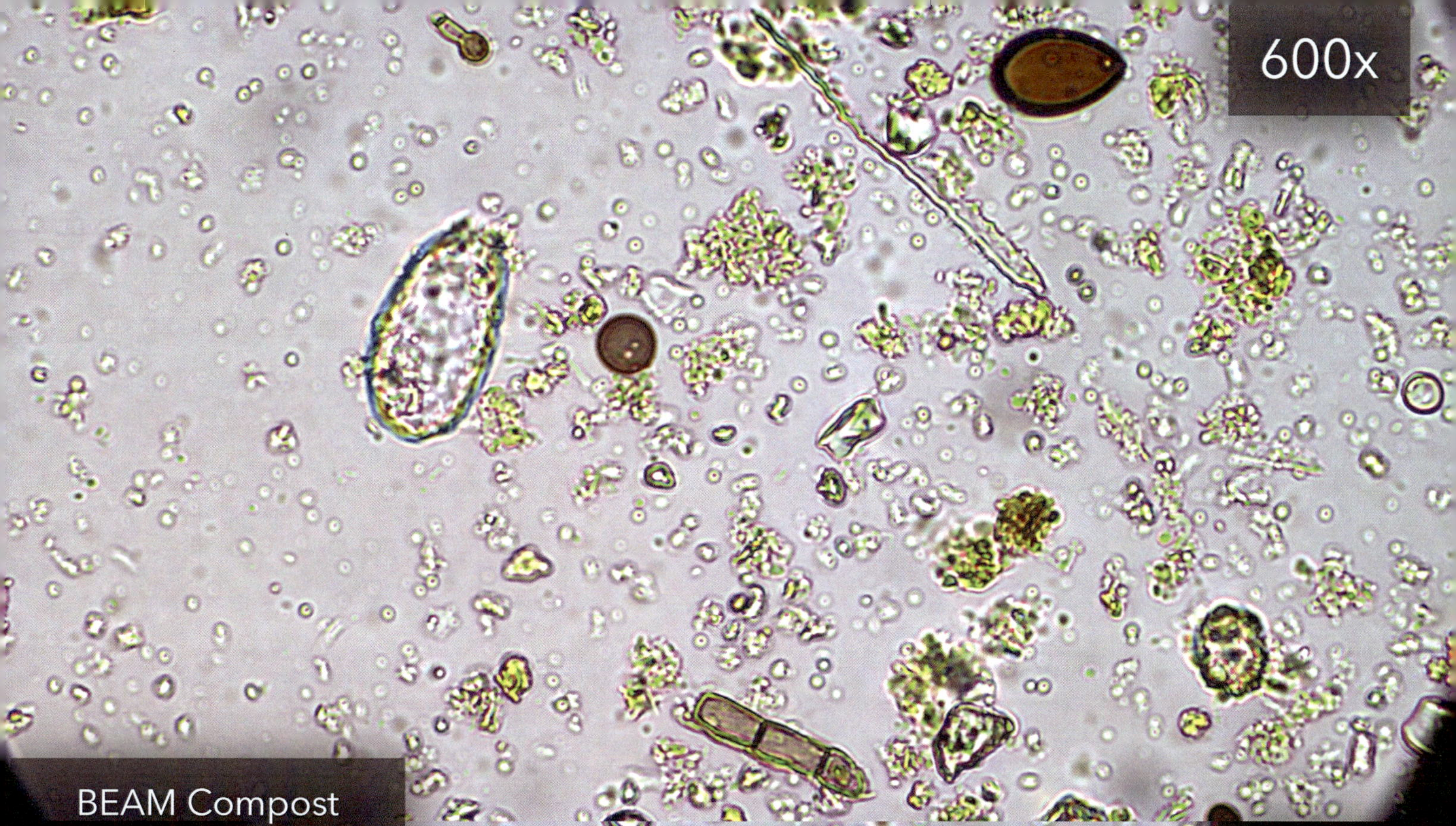

600x

600x

BEAM Compost
from BeamCompost.com

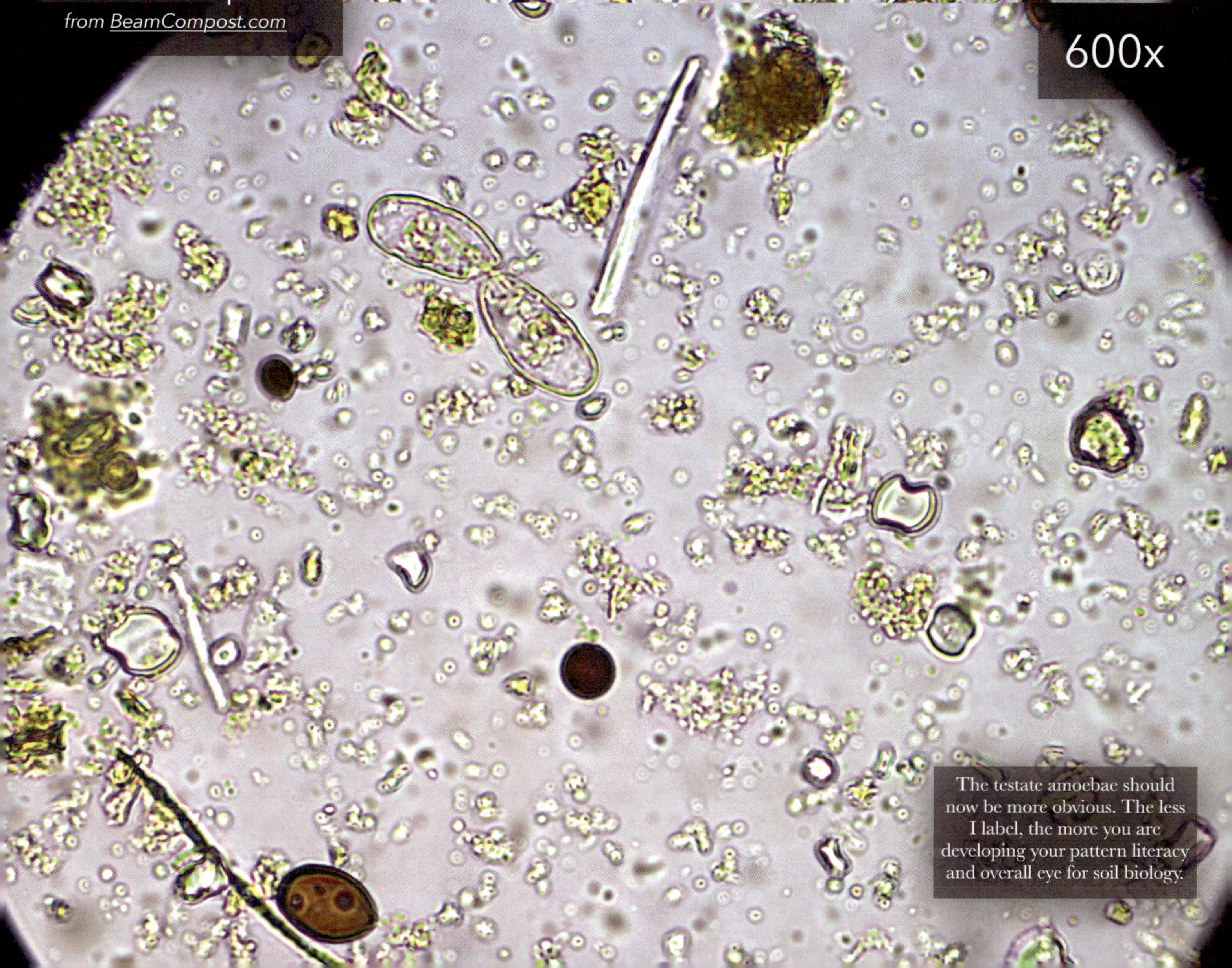

The testate amoebae should
now be more obvious. The less
I label, the more you are
developing your pattern literacy
and overall eye for soil biology.

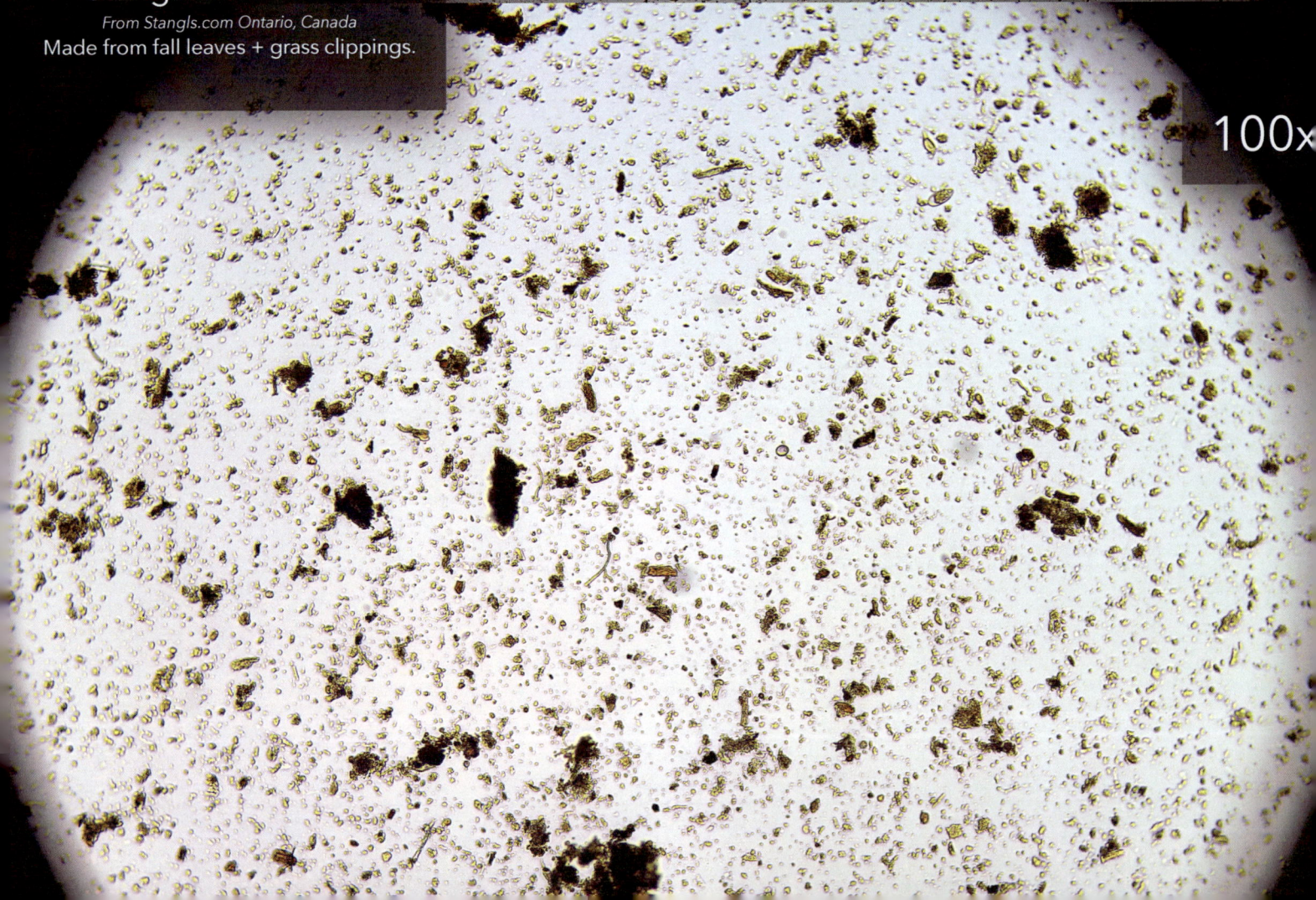

40x

100x

Stangls Johnson-Su
From Stangls.com Ontario, Canada
Made from fall leaves + grass clippings.

400x

400x

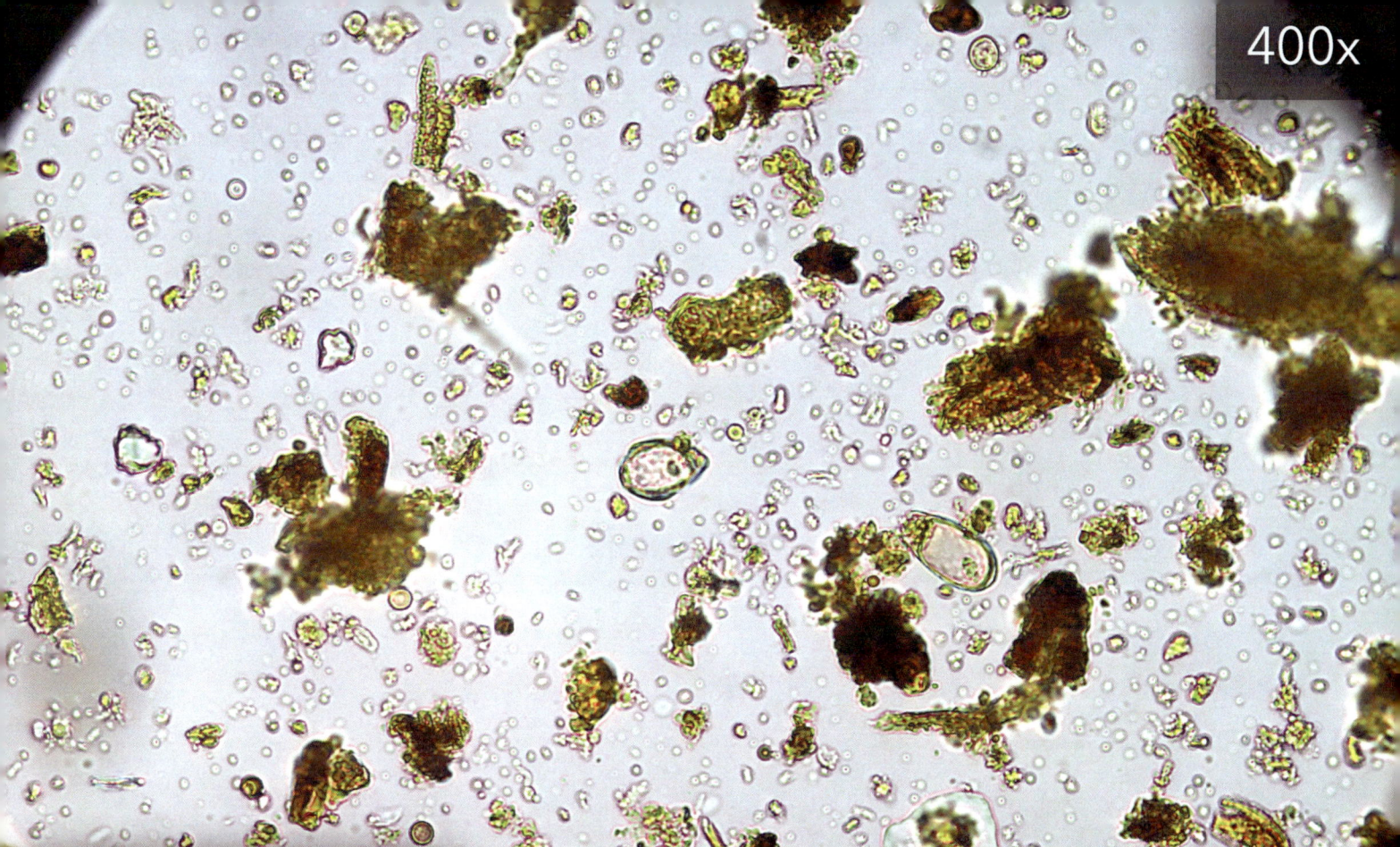

Stangls Johnson-Su
From Stangls.com Ontario, Canada

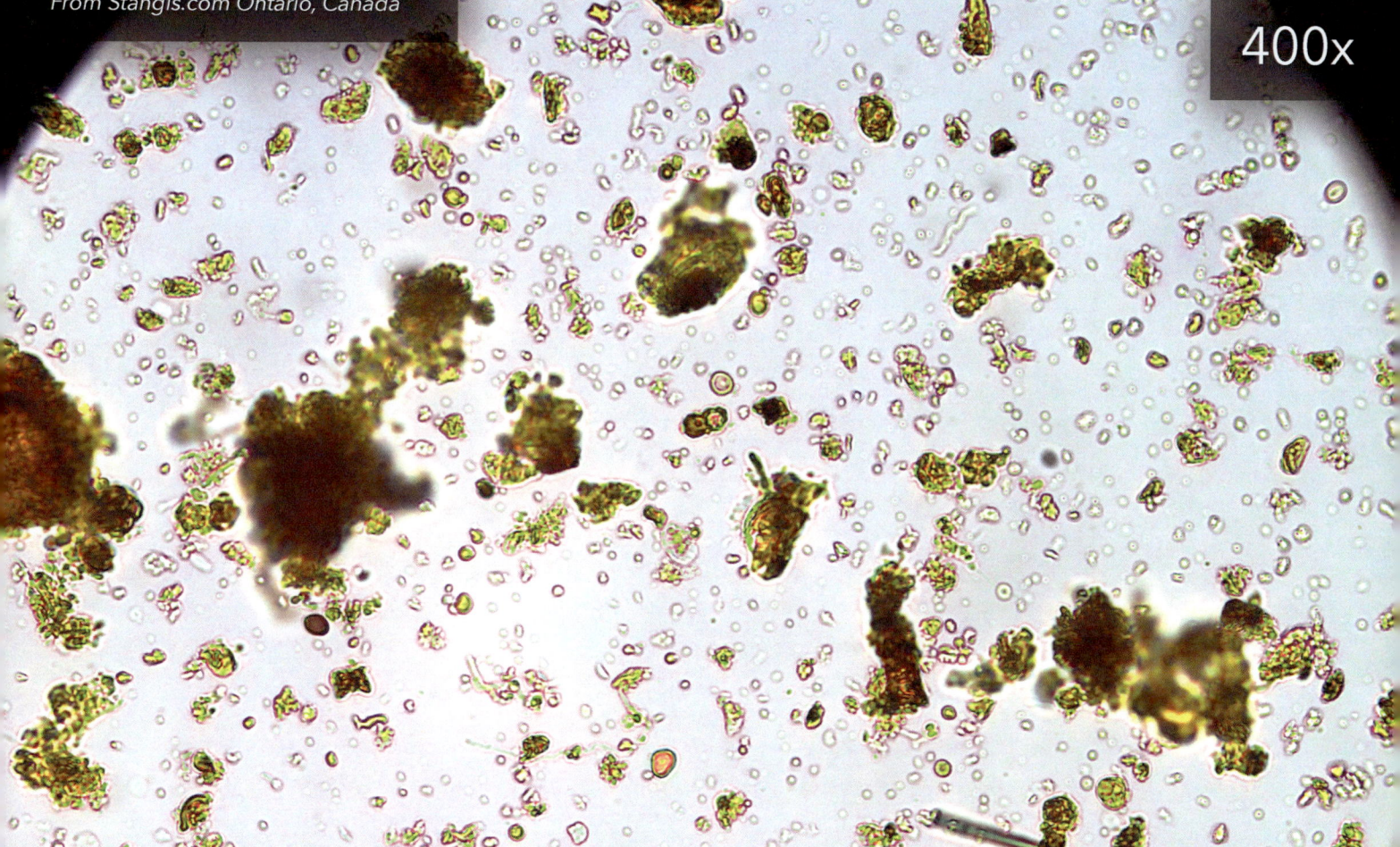

Fungal Hyphae

nitrogen and sometimes the phosphorus aren't being gassed off, and since the top 3 macronutrients plants all need are: NPK, our plants are all naturally higher in those minerals (their required macronutrients) than other minerals, so when we compost them properly and retain those nutrients, we'd expect to see them in high numbers and in the forms that the compost's pH/Eh levels would dictate – it's important to know these levels because high phosphorus would inhibit the growth of mycorrhizal fungi if you added it around the roots when transplanting. Excessively high nitrate levels in anything we add to our soils or water onto plants will create plants prone to fungal infection, disease, virus, and pest attack (and require 4x the water as other forms of nitrogen), but all plants require nitrates for vegetative growth primarily and then in a very small amount during fruiting. In compost, nitrates can be high, so test for N as well, and then if you have high nitrates and want to convert them, use LAB or EM (check out **Regenerative Soil** for more details on how to fix things + all the recipes). The fungi I have seen have been purely saprophytic, and the endophytes have been more abundant in some JS composts than other. I've heard of piles that were heavily populated with protozoa, sometimes ciliates. Sometimes there are no nematodes, and sometimes I find beneficial nematodes and sometimes omnivore nematodes – the broad range of expression coupled with the consistent success many have experienced using this type of compost suggests that more study is needed on the inputs, climate, water, and management methods and their effects.

What is clear is Johnson-Su composting methods aren't going anywhere anytime soon – they generate a prolific biology that transforms soils in field trials despite there being a broad variation of expression of that biology, and despite BEAM seeming the least biologically active, the high numbers of bacteria and spores spring into action in situ – even in high desert barren soils, anecdotal field trials are showing it is transformative. This is why it's so important to examine our samples and different types of soil, compost, and biofertilizers with an open mind: everything has something to teach us.

	ORGANIC MATTER	BACTERIA	FUNGI	PROTOZOA	NEMATODES	MICROARTHROPODS & WORMS	N	P	SALINITY	F:B	TOTAL MICROBIAL BIOMASS	PH
STANGLS JOHNSON-SU #1 ONTARIO, CANADA	Golden to Dark Organic Matter	10 per 0.025mm² ~4 billion bacteria per ml of soil 3 actinobacteria per FOV @ 400x	6 hyphae per drop but a great diversity of spores: 10 – 20 per FOV @ 400x	2 testate amoebae per FOV @ 400x: *giant pink ones & iridescent smaller ones* 1 ciliate per drop *Small & fast*	1 – 2 per drop	2 earthworms per cup	High	Low	2600 ppm 0.26%	**4.9:1** F: 83% B: 17%	2423 ug C/g	8.0
D.B.'S JOHNSON-SU #1 UTAH, USA	Lighter colored OM (more fulvic than humic), but covered with active bacteria, some sand present	15 per 0.025mm² ~6 billion bacteria per ml of soil 1 – 3 Actinobacteria per FOV @ 400x Lactobacilli 0 – 3 per FOV @ 400x	0 – 1 per FOV @ 400x Low Diversity, but TONS of Spores	5 testate amoebae per FOV @ 400x	3-5 per drop *Bacterial Feeding Omnivores*	Mites & several earthworms per cup	High	Trace	1820 ppm 0.18%	**3.0:1** F: 75% B: 25%	1340 ug C/g	8.0

	ORGANIC MATTER	BACTERIA	FUNGI	PROTOZOA	NEMATODES	MICROARTHROPODS & WORMS	N	P	SALINITY	F:B	TOTAL MICROBIAL BIOMASS	PH
BEAM ILLINOIS, USA	Fibrous Organic Matter and humic compounds – very fungal: almost all has spores embedded within	27 per 0.025mm² ~10.8 billion bacteria per ml of soil **Very High** Bacteria (cocci & bacilli) with 1 actinobacteria per FOV @ 400x	5 spores per FOV @ 400x - all organic matter is fungally inoculated, hyaline hyphae with clear	Ward Laboratories Inc.'s PLFA says 0 protozoa, but there's 2 – 5 testate amoeba per FOV @ 400x!! *Giant Purple Testate Amoebae!!*	0	0.25 enchytraeid or pot worms per drop	High	Med	2100 ppm 0.21%	**3.6:1** F:78% B: 22%	1572 ug C/g	4.5

THERMOPHILIC STUDY

Thermophilic composting is the classic and fast-acting hot composting method that so many of us rely upon to efficiently break down organic matter. It's what Dr. Elaine Ingham has taught for decades, what Catalyst Bioamendments follows, and so many others utilize as well because it works: it gets hot, the plant fibers split and burst apart (look for those cellulose spirals), the bacteria and fungi access more and more surface area – like a wedge in a crack, they break everything apart and then down, using the opening the heat's effect gives them. We can see the bacteria all over the organic matter if we are close enough – dark field and epifluorescence lighting can really help us see this clearly.

For years, I've had students seek to find the ultimate, "perfect" compost without realizing that all decomposition leads back to soil, and nature uses a variety of decomposition pathways. After finding the members of the EM biofertilizer blend in all the hot compost samples I could DNA sequence, I realized that there's a lot more common between these methods than different. After seeing them under the microscope, I'm only more certain of this fact. Good composts look similar even if we use different methods, but that doesn't mean they're all the same or will be the same every time: this is why a microscope is so vital, so you can see what kinds of nematodes you have, what kinds of fungi, etc. Whether it's something new from the ingredients, weather, season, or a change in the

The test tube on the left, the 1st one, is a sample from Catalyst Bioamendments. It's almost pure organic matter. It has good structure, leaves only a faint humic tint to the water - low turbidity. This is what compost should look like.

The black in the 2nd test tube is microbial soot from anaerobic decomposition. The 2nd test tube is nearly all sand with some organic matter - high turbidity due to the unstructured organic matter, soot, dust, & likely some clays (superfine mineral particles). Nature's Way compost is the 2nd test tube.

400x

100x

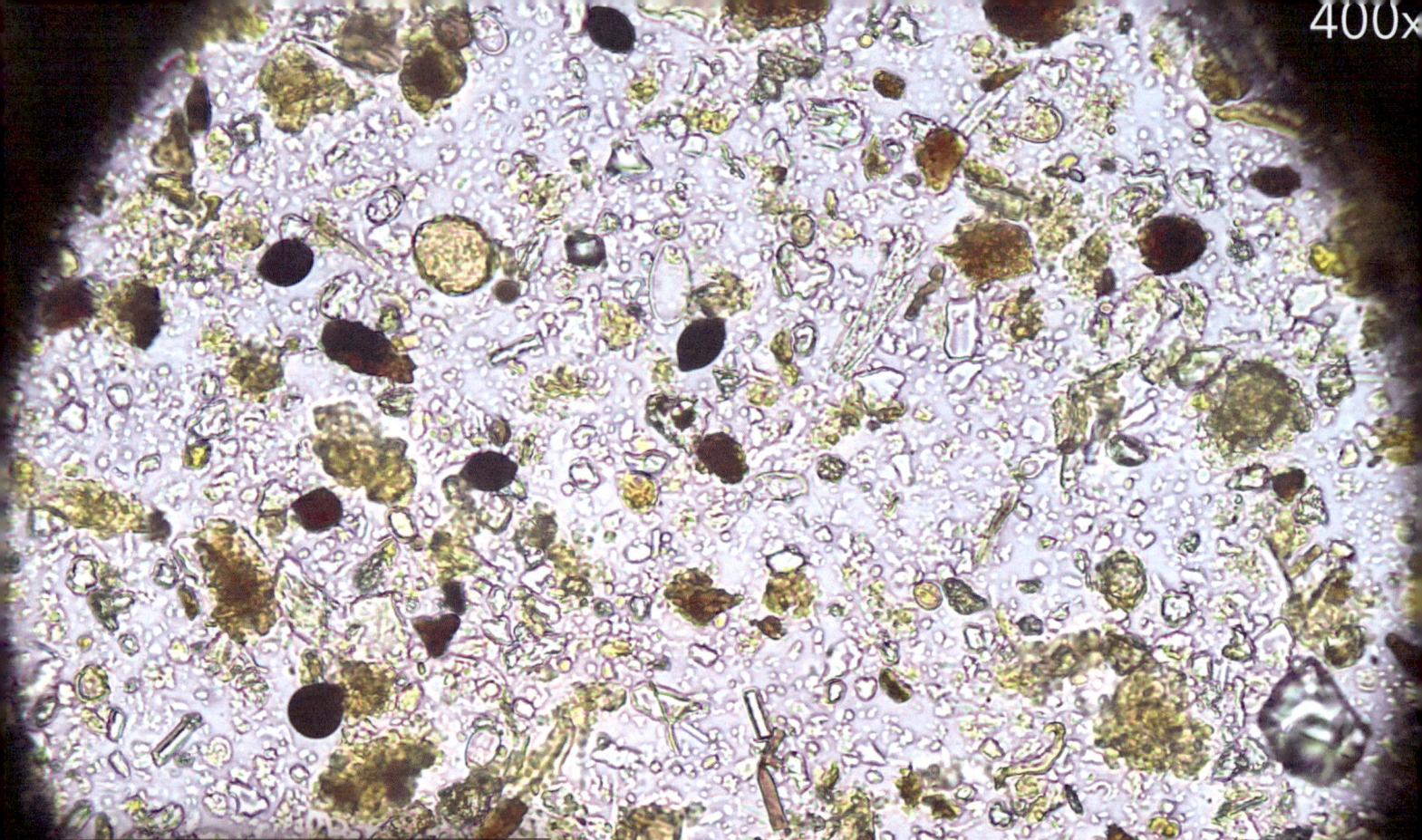

Thermophilic Compost

from **Catalyst Bioamendments** California, USA
Diluted at 1:5 to better showcase diversity — we can
better see what it would be like in the soil
environment, only it would be 5x more chaotic and
abundant with spores, testate amoebae, bacteria,
organic matter, fungi, and more.
What else do you see?

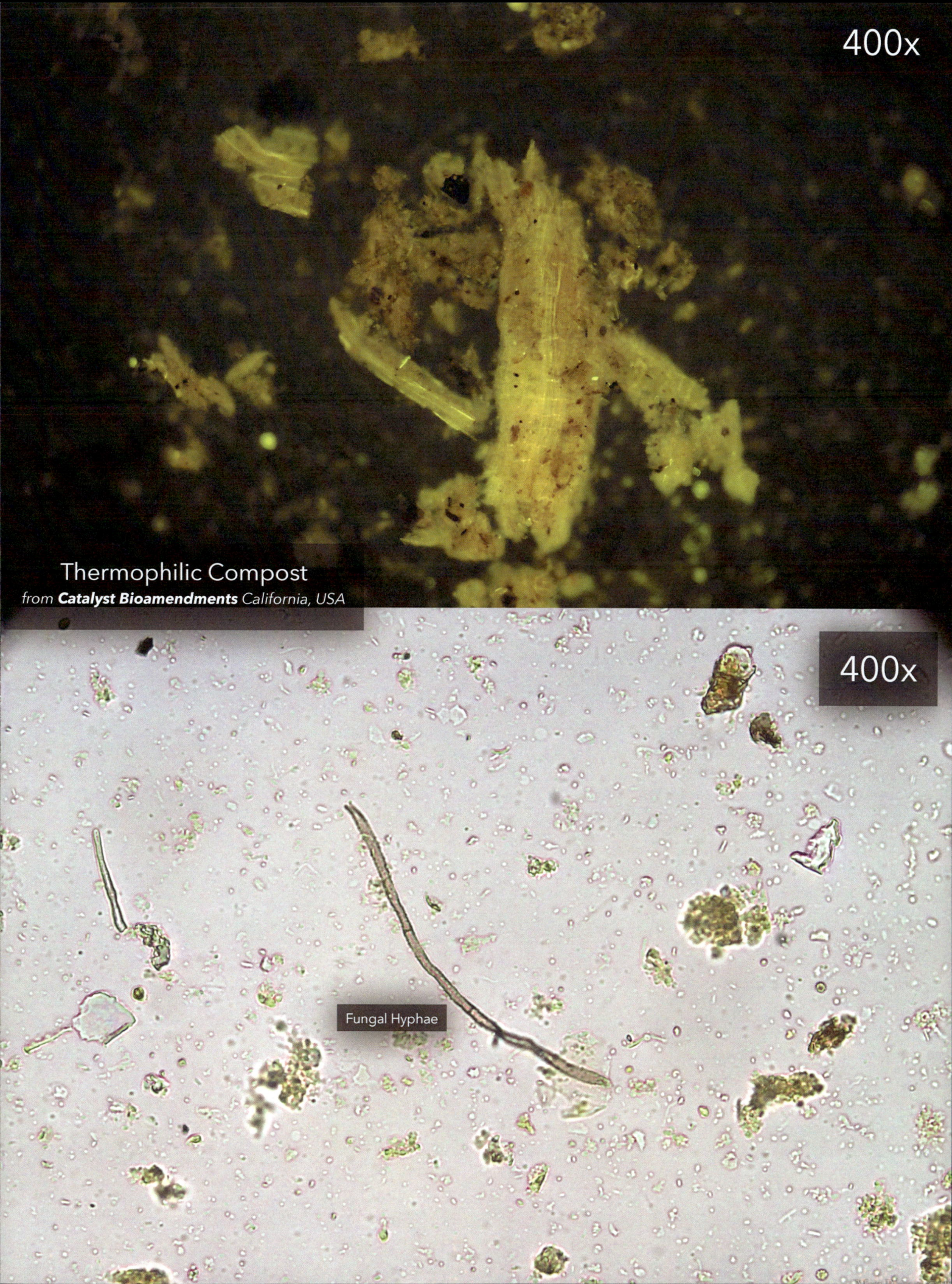

400x

Thermophilic Compost
from **Catalyst Bioamendments** *California, USA*

400x

Fungal Hyphae

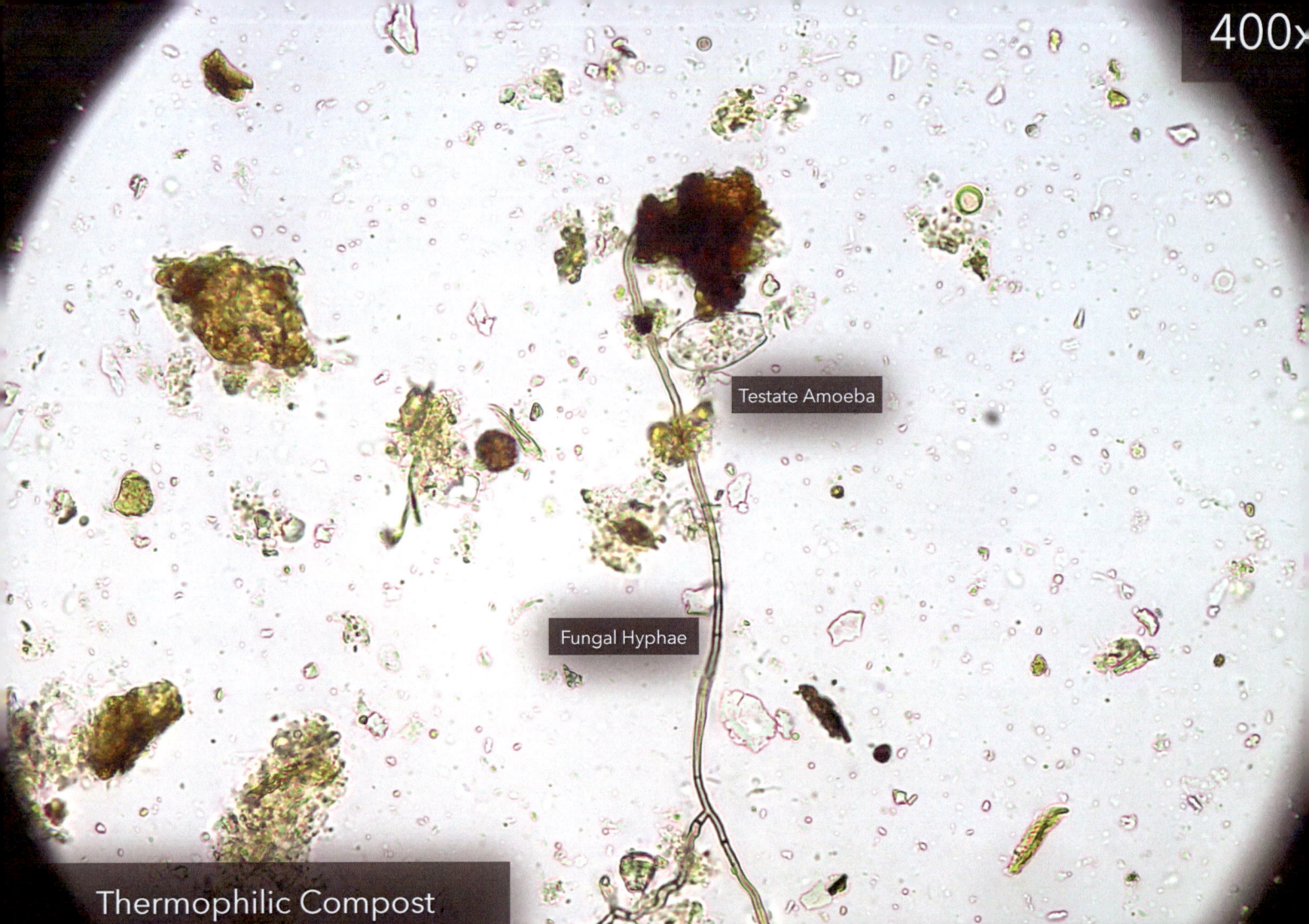

400×

Testate Amoeba

Fungal Hyphae

Thermophilic Compost
from **Catalyst Bioamendments** California, USA
Diluted at 1:10 as per normal protocols

40x

Nematode

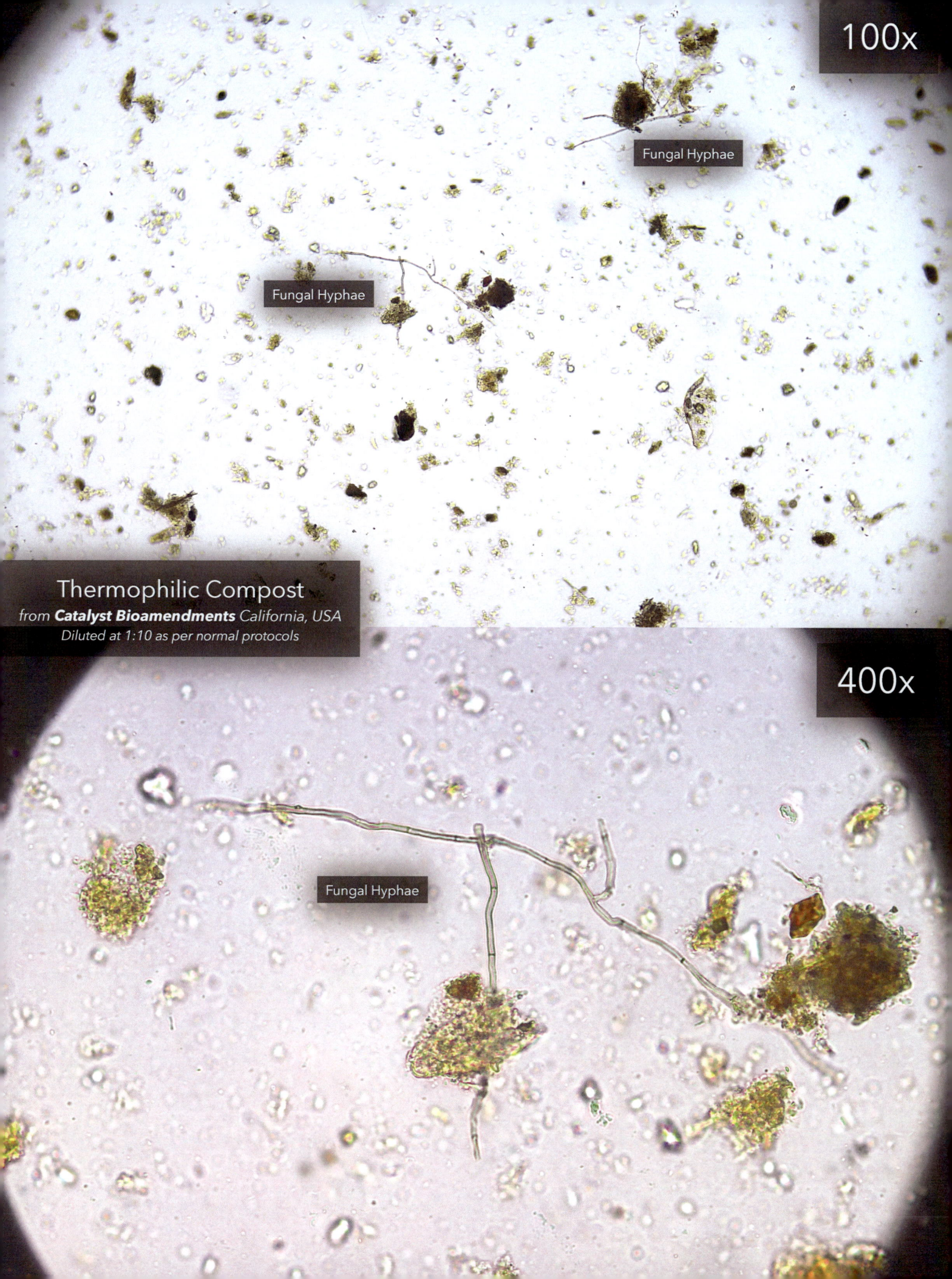

100x

Fungal Hyphae

Fungal Hyphae

Thermophilic Compost
from **Catalyst Bioamendments** *California, USA*
Diluted at 1:10 as per normal protocols

400x

Fungal Hyphae

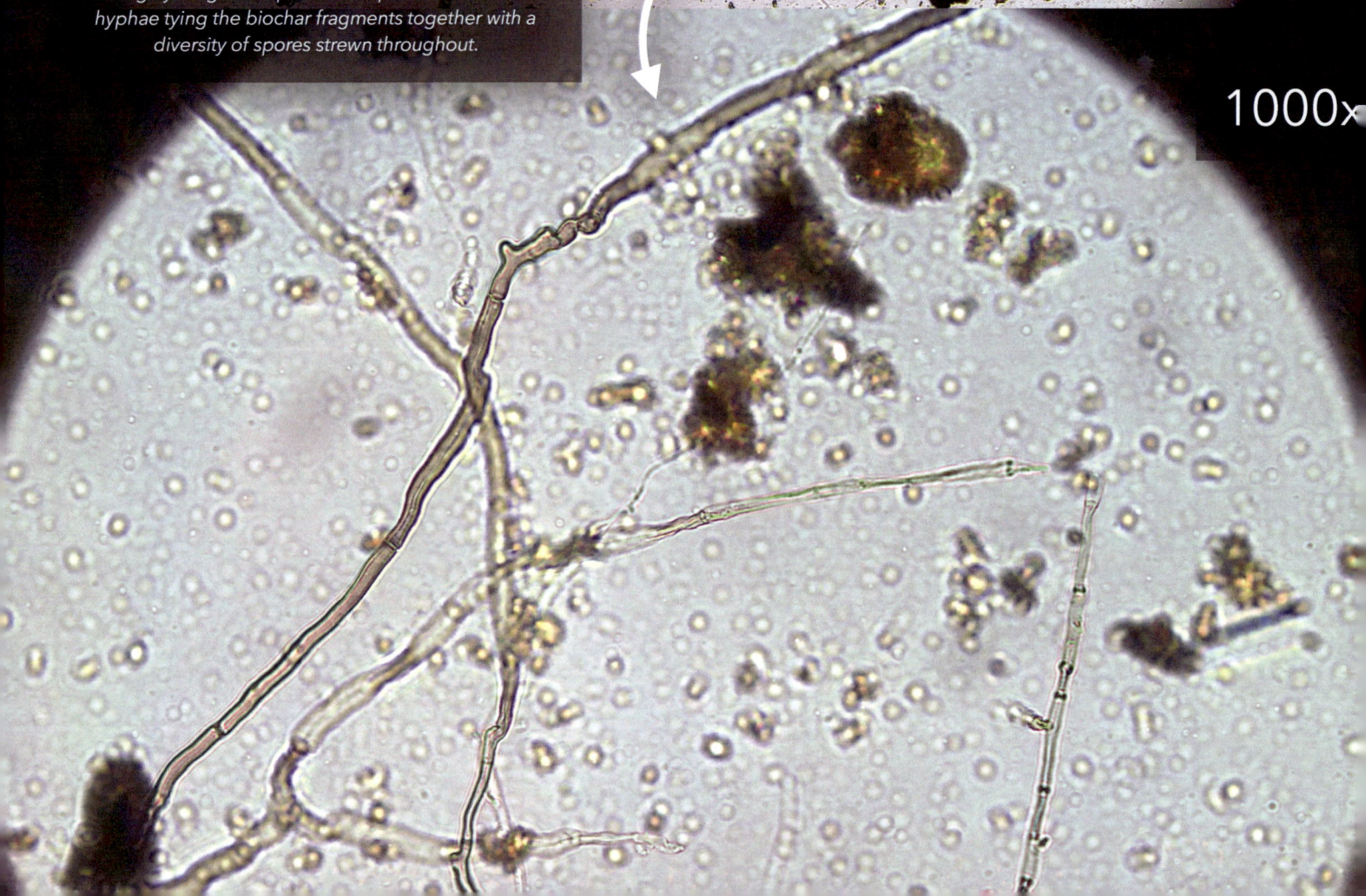

100x

Fungal Hyphae

Thermophilic Biochar Compost
from **Pacific Biochar** California, USA
This highly fungal compost had impressive amounts of hyphae tying the biochar fragments together with a diversity of spores strewn throughout.

1000x

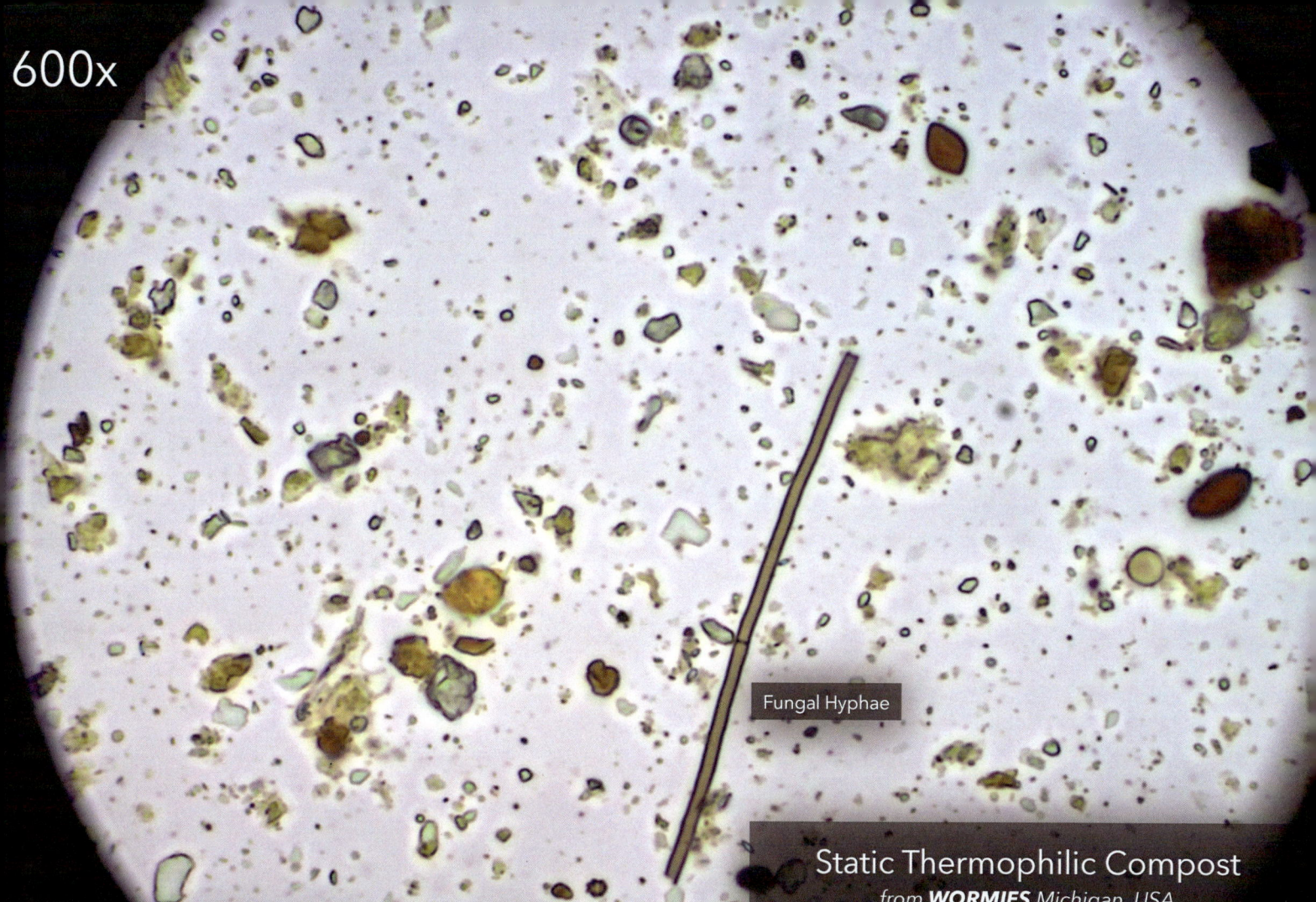

600x

400x

Fungal Hyphae

Fungal Hyphae

Static Thermophilic Compost
*from **WORMIES** Michigan, USA*
This highly fungal compost has an incredible amount of diversity of spores and fungi.

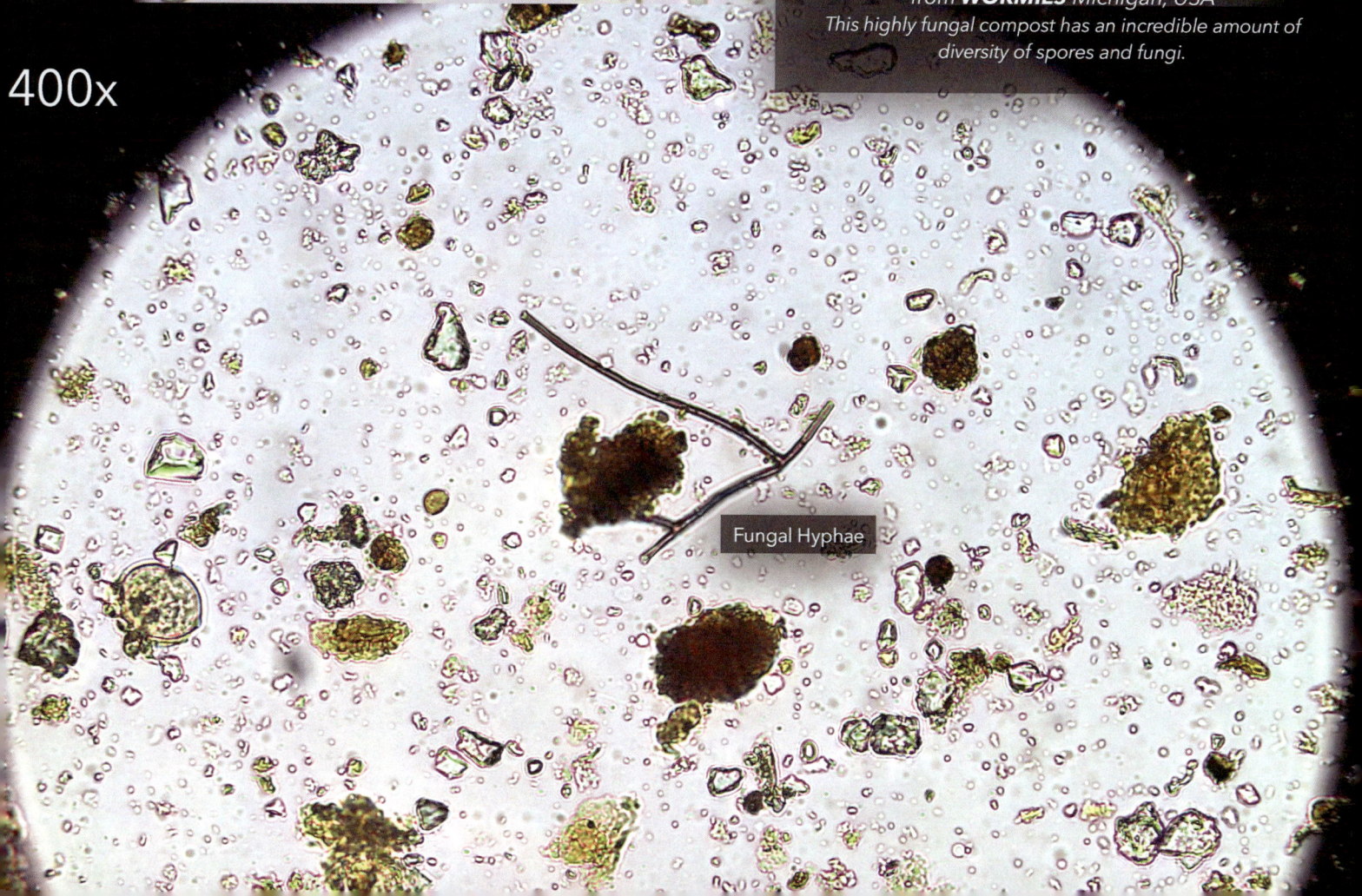

40x

100x

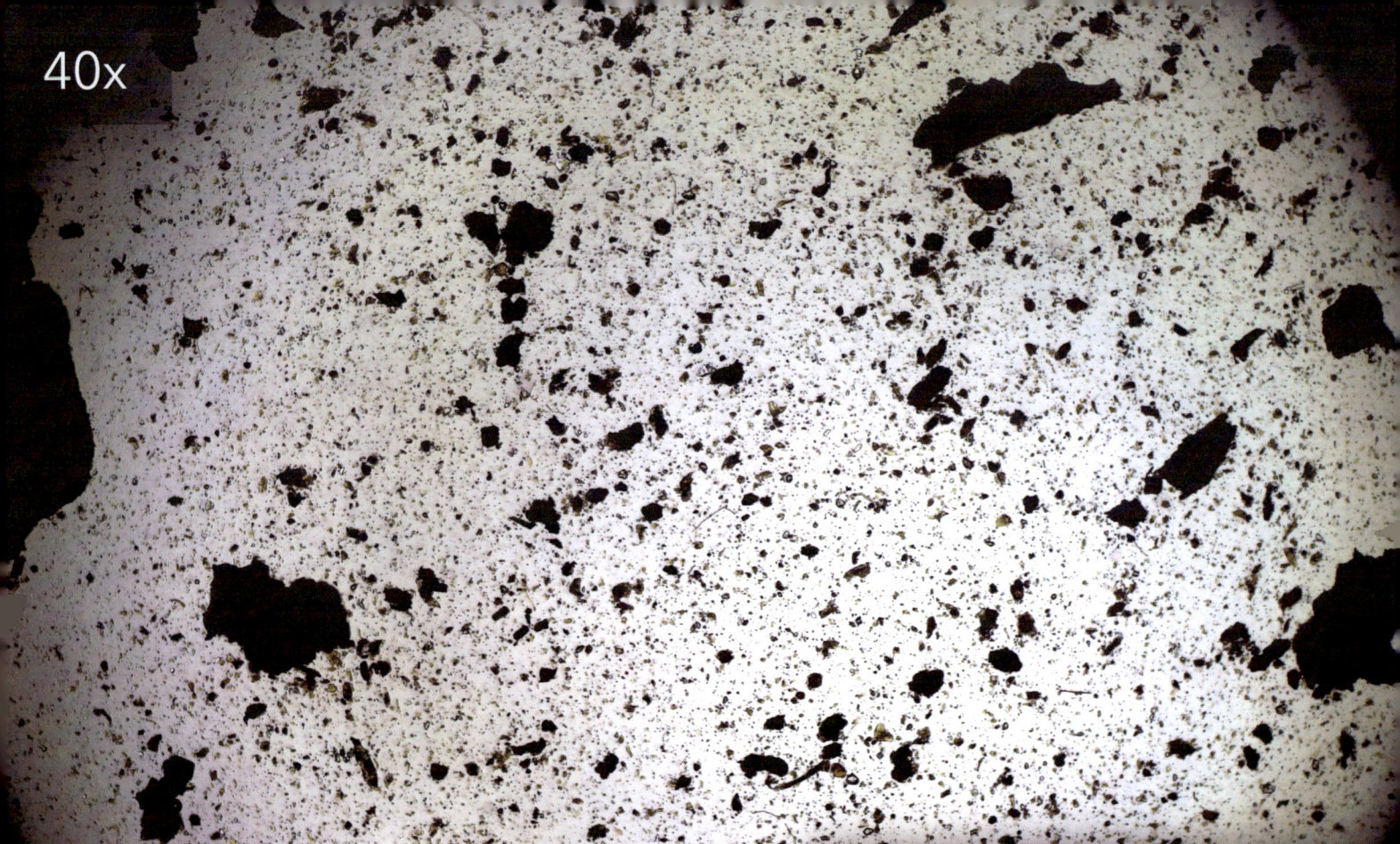

Static Thermophilic Compost
from **WORMIES** Michigan, USA
This compost is finished with bokashi, a cover crop, and worms.

Fungal Hyphae

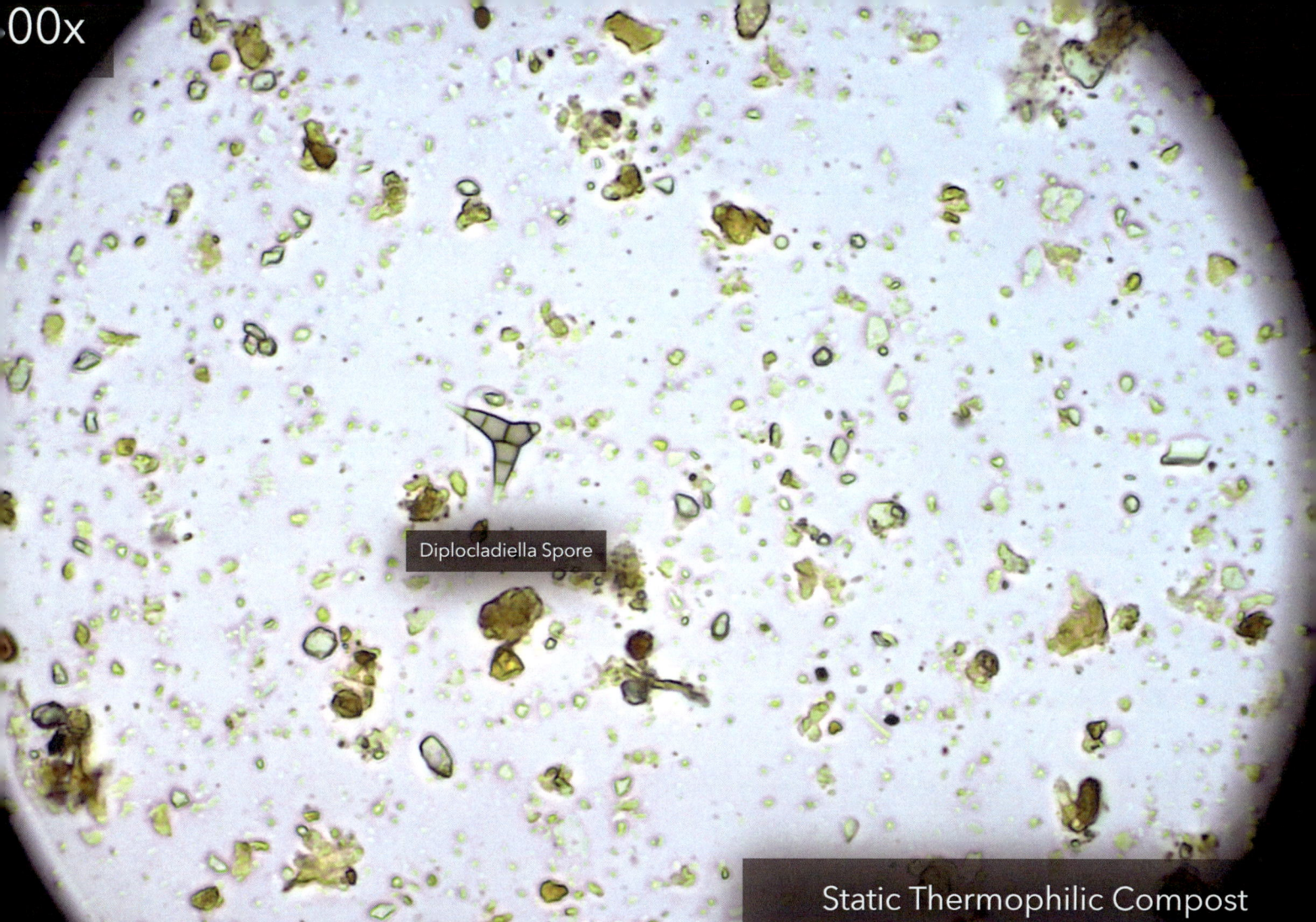

Diplocladiella Spore

Static Thermophilic Compost
from **WORMIES** Michigan, USA
Nutrients are cycled through fungi, bacteria, and protozoa

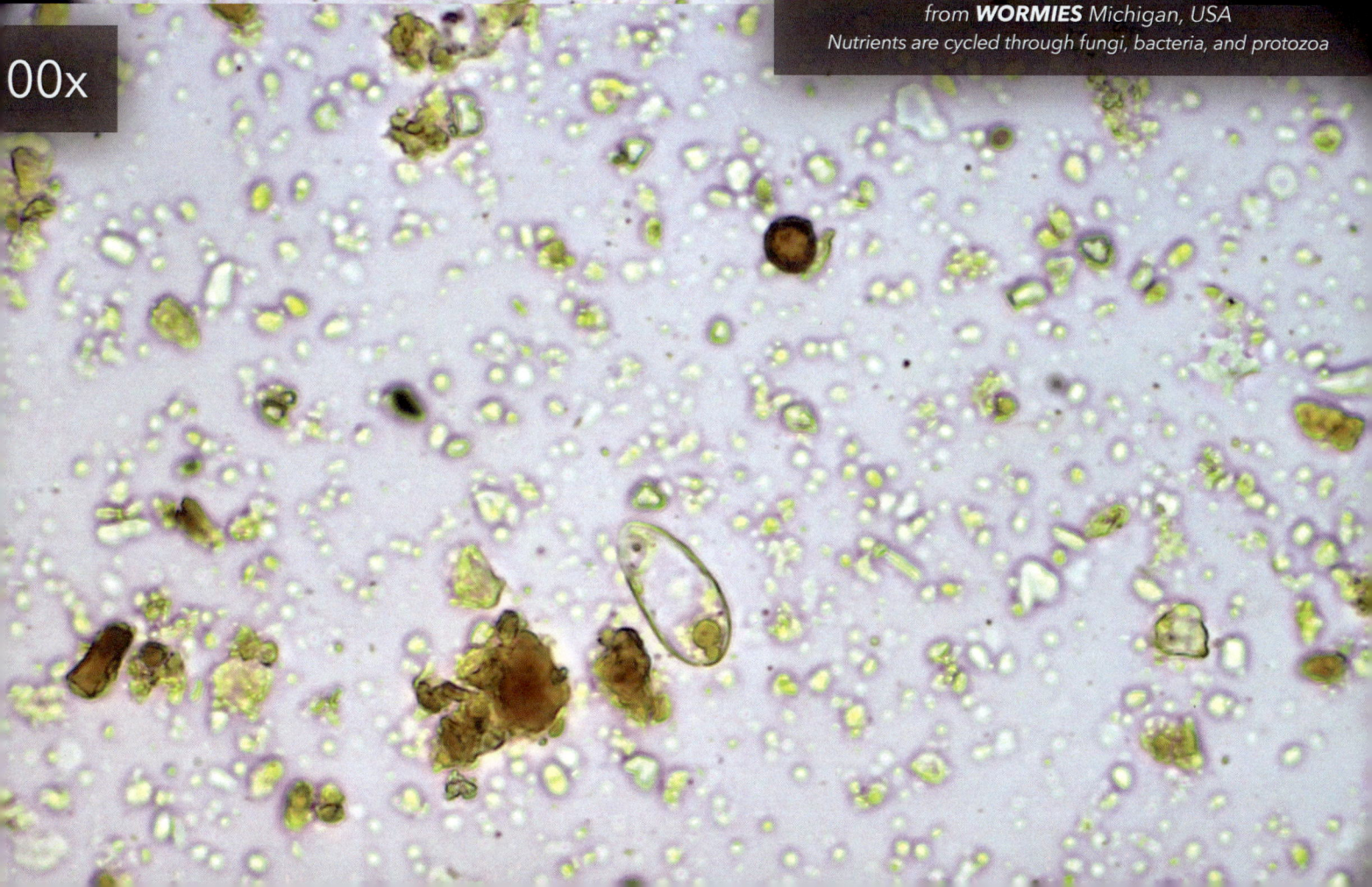

40x

100x

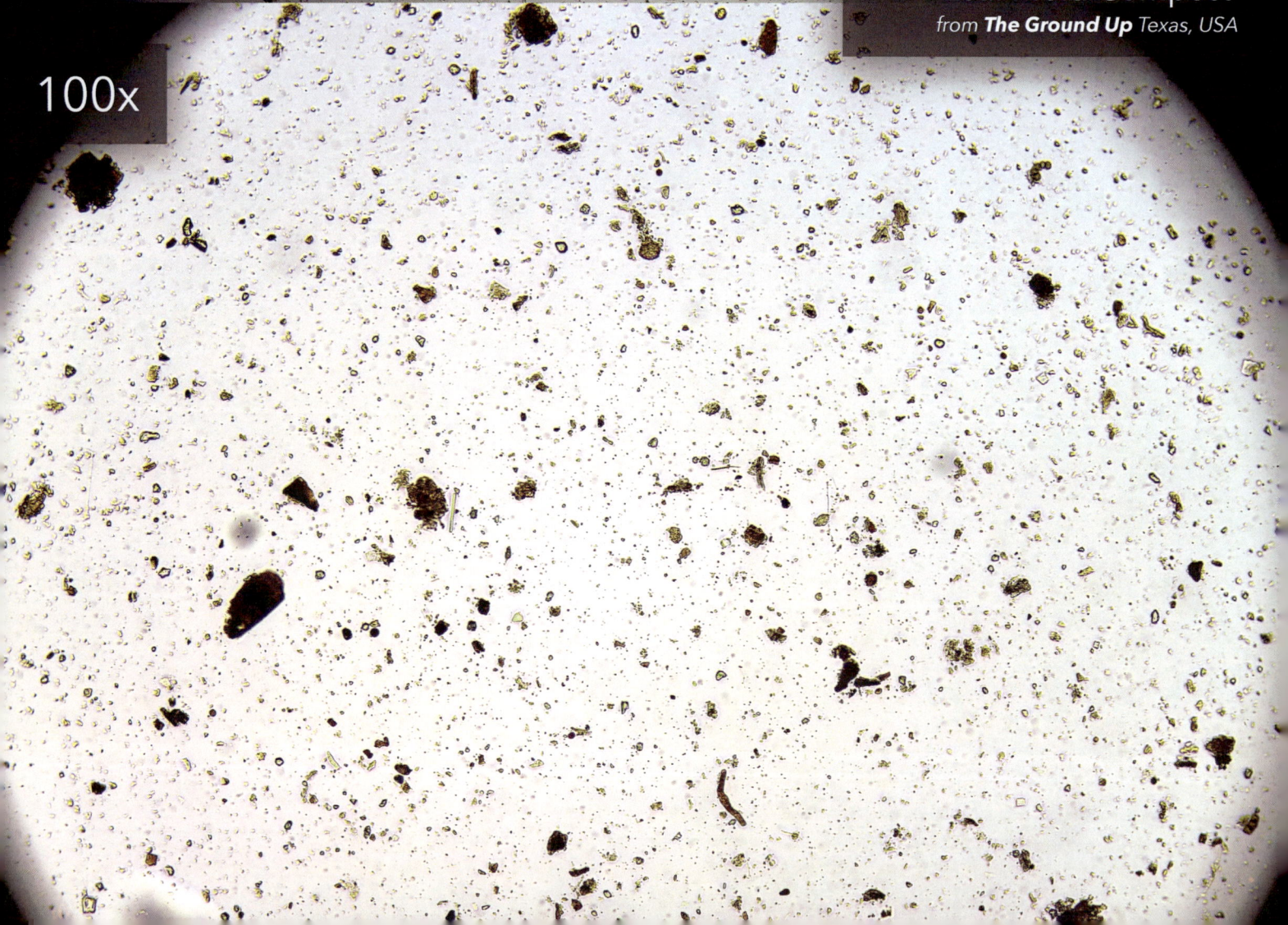

Leaf Mold Compost
from *The Ground Up* Texas, USA

400x

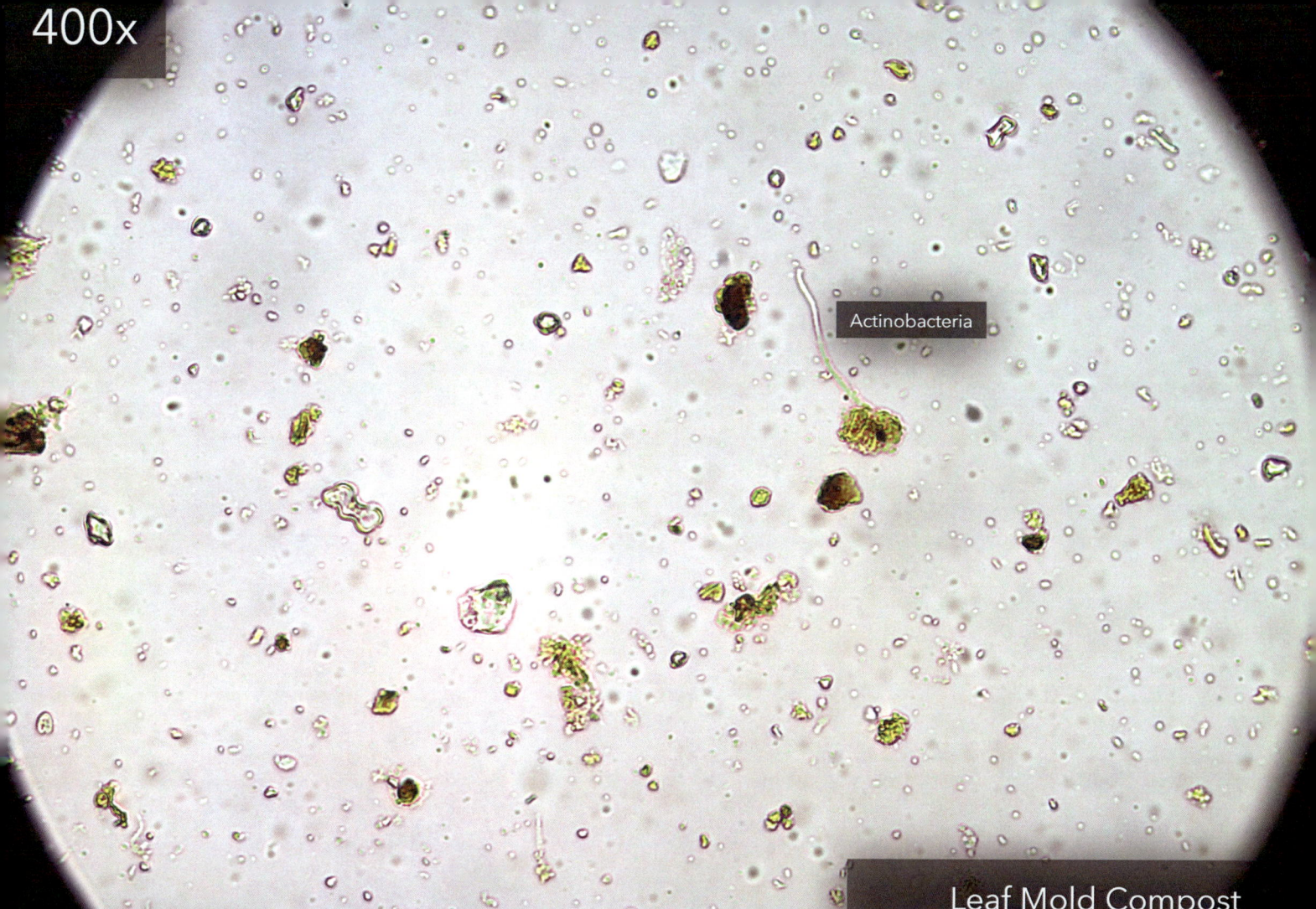

Actinobacteria

Leaf Mold Compost
from **The Ground Up** Texas, USA

400x

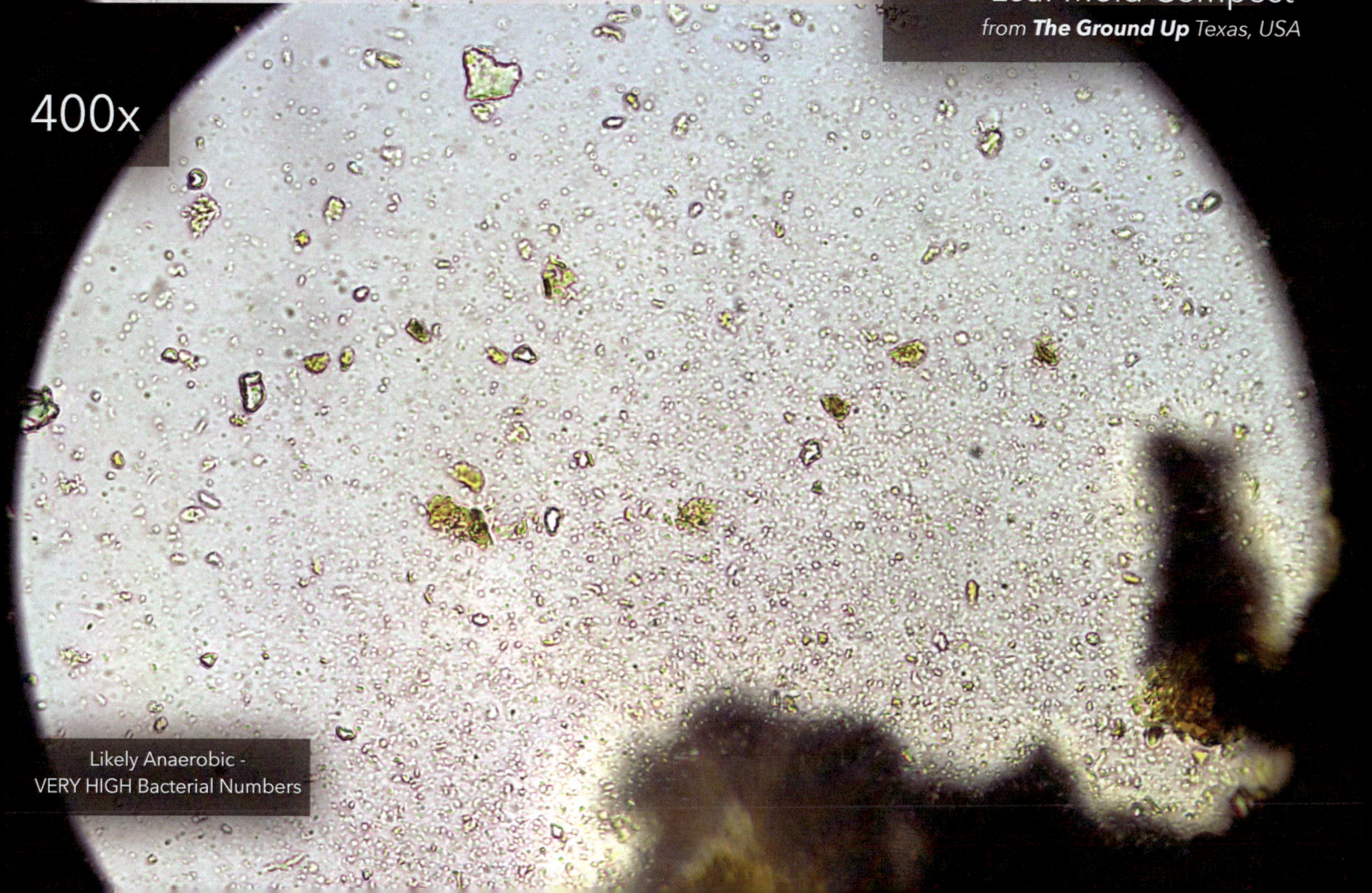

Likely Anaerobic -
VERY HIGH Bacterial Numbers

400x

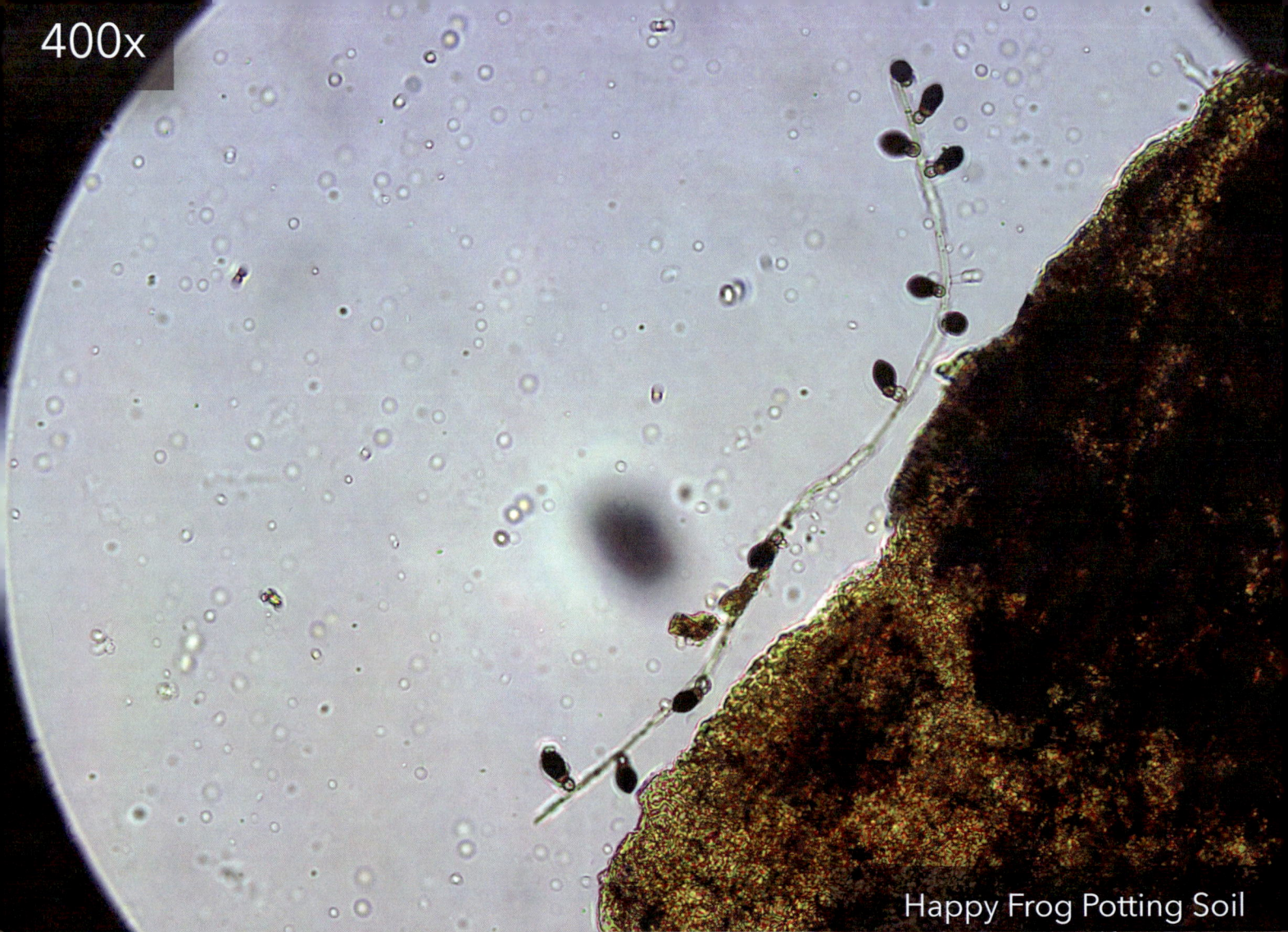

Happy Frog Potting Soil
from *FoxFarm* California, USA

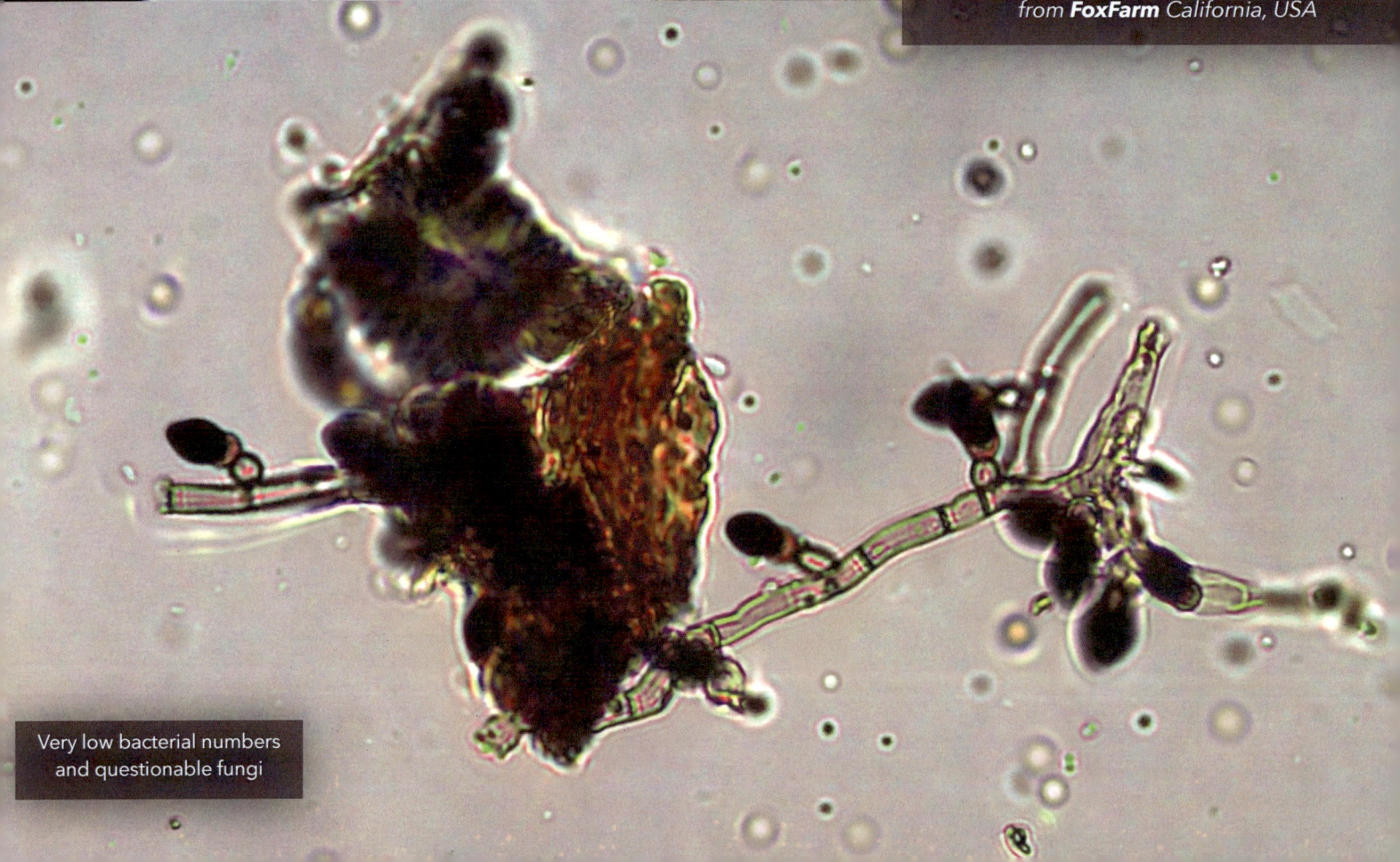

Very low bacterial numbers
and questionable fungi

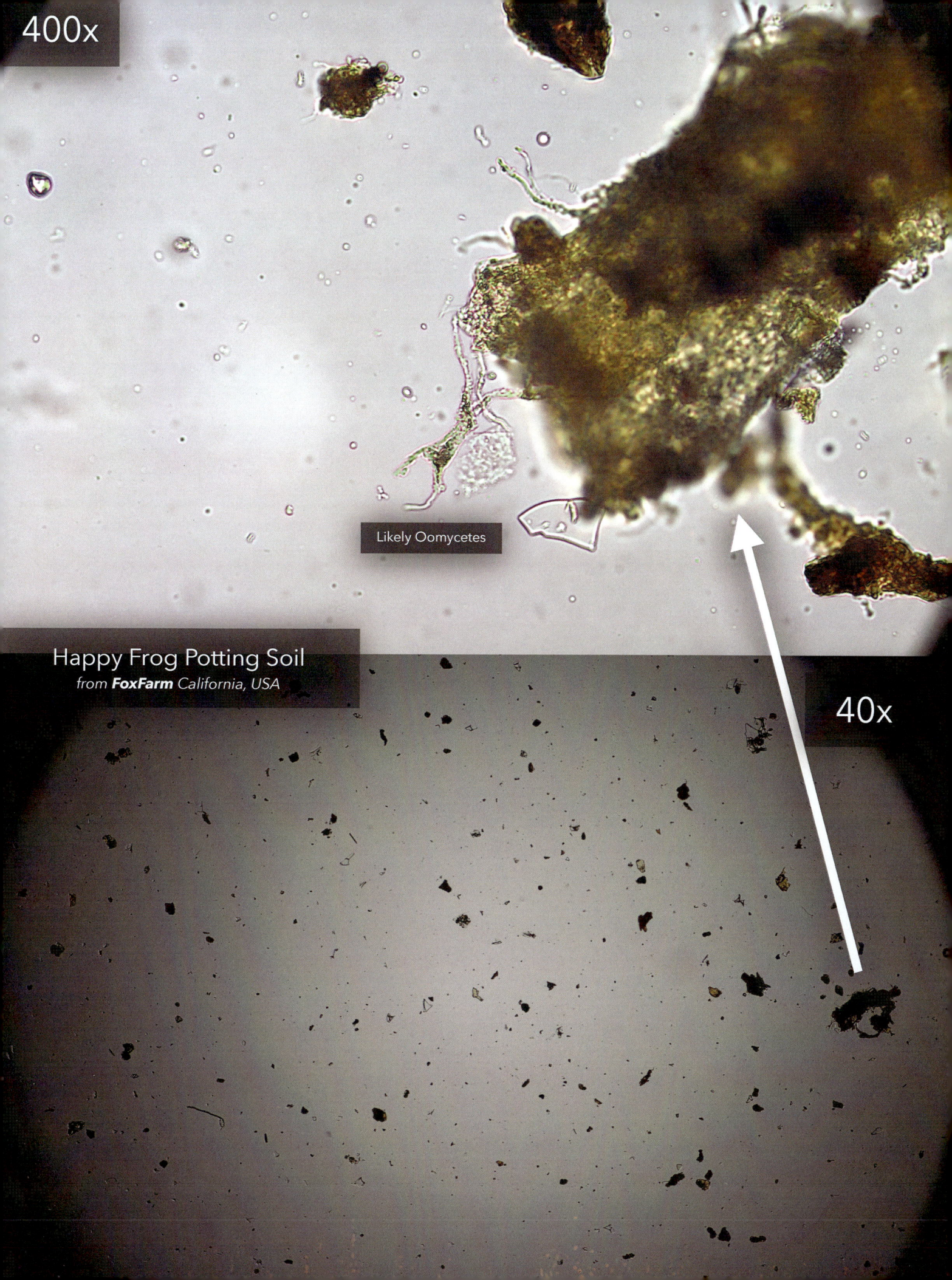

400x

Likely Oomycetes

Happy Frog Potting Soil
from **FoxFarm** *California, USA*

40x

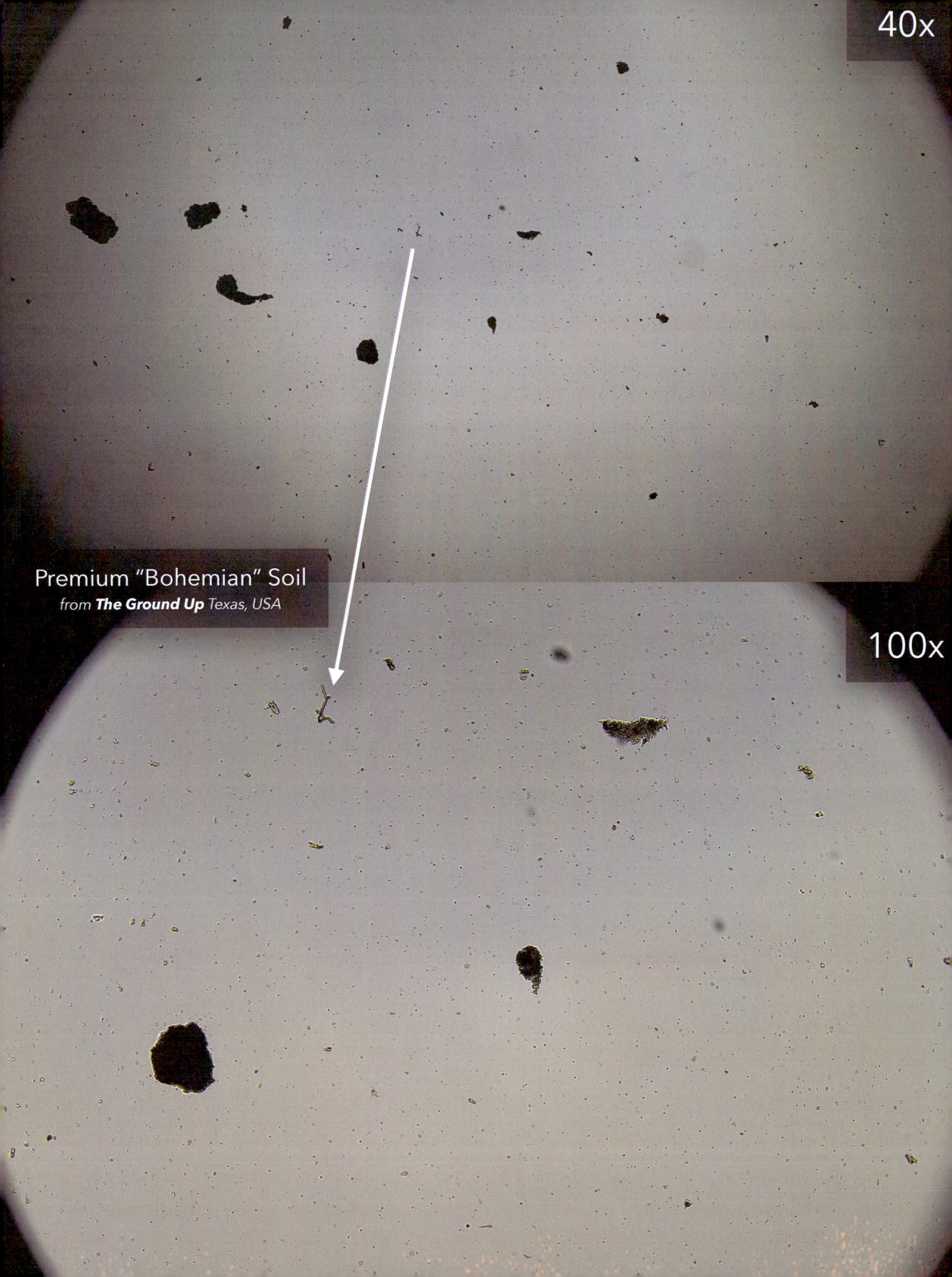

40x

100x

Premium "Bohemian" Soil
from *The Ground Up* Texas, USA

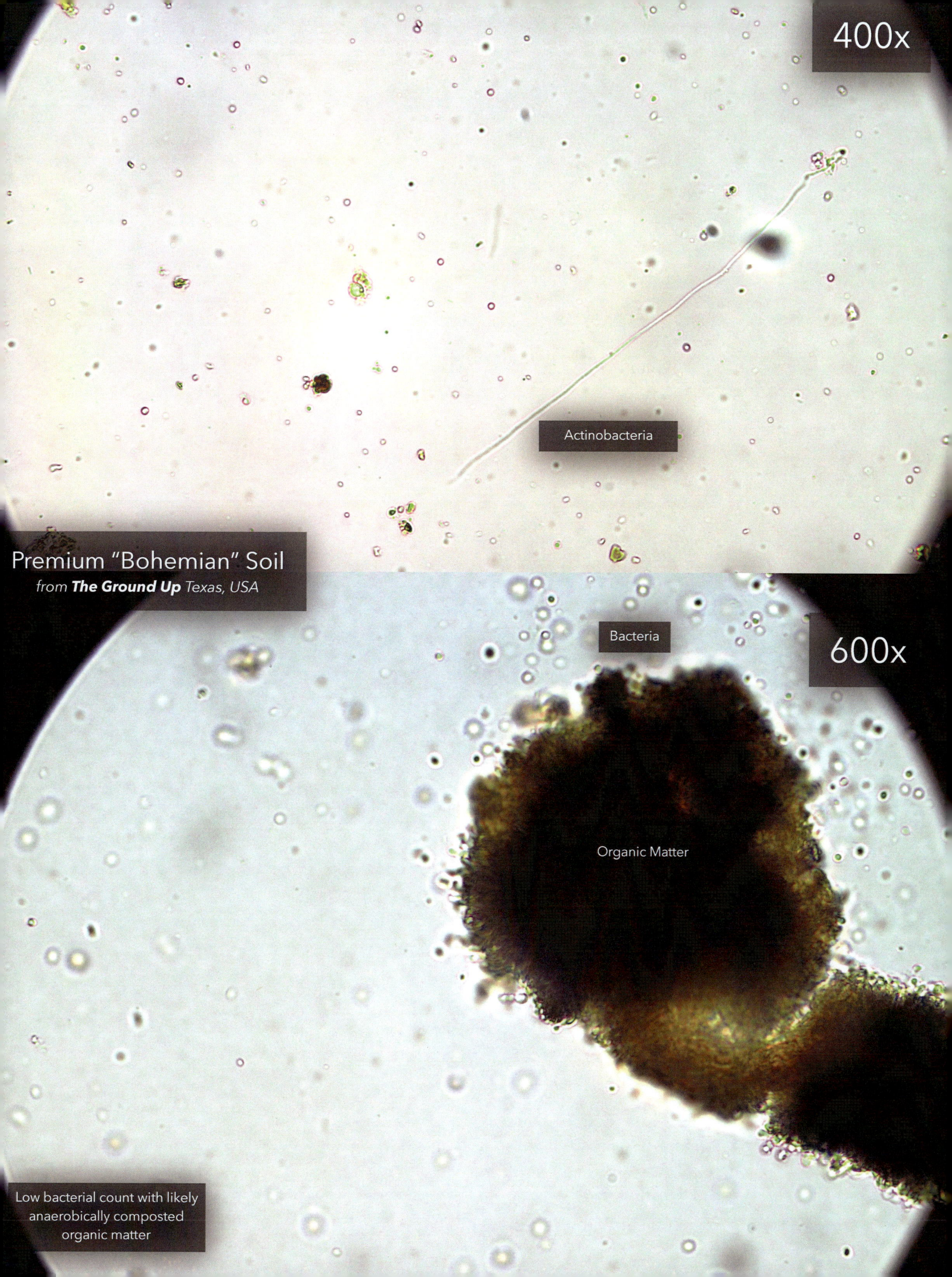

400x

Actinobacteria

Premium "Bohemian" Soil
from **The Ground Up** Texas, USA

Bacteria

600x

Organic Matter

Low bacterial count with likely
anaerobically composted
organic matter

management pattern, there are always changes in the microbial population of any given composting method over time. This is part of the reason why a mother compost or using EM makes so much sense: the biological inoculation from those sources will help guide and correct decomposition, guaranteeing a more uniform product.

In the hot composting world, the danger is ubiquitous: if a pile gets too hot, it will burn off (gas off) most if not all the major nutrients (mostly N and C). We can see the biologically rich composts are more likely to retain the nutrition through the lifecycles of the biology while the excessive heat in truly anaerobic composting generates a soot, a lack of nitrogen, and no higher level cycling of nutrients (no protozoa and no nematodes).

	ORGANIC MATTER	BACTERIA	FUNGI	PROTOZOA	NEMATODES	MICROARTHROPODS & WORMS	N	P	SALINITY	F:B	TOTAL MICROBIAL BIOMASS	PH
CATALYST BIOAMENDMENTS SAMPLE #1 CALIFORNIA, USA	Large, dark organic matter aggregates with many lighter smaller particles throughout	15 per 0.0025mm² ~6 billion bacteria per ml of soil Great Diversity of Bacteria, 6 – 10 actinobacteria per FOV @ 400x 1–2 Lactobacillus per FOV @ 400x	1 hyphal strand per FOV @ 400x but 6–12 spores per FOV @ 400x	2–6 testate amoebae per FOV @ 400x ~0.1 Flagellettes per FOV @ 400x No ciliates	1 – 2 per drop Omnivore and Fungal Feeders	Mites, Millipedes, Gnat Larvae, & earthworms (~1 earthworm per cup)	Very High	Med	930 ppm 0.09%	1.6:1 F: 61% B: 39%	775 ug C/g	8.0
WORMIES MICHIGAN, USA	Golden to dark black/ brown OM Highly Fungal but about equal in abundance to the mineral content	24 per 0.0025mm² ~9.6 billion bacteria per ml of soil 2 Actinobacteria per FOV @ 400x Lactobacilli 2 per FOV @ 400x	0.8 hyphae per FOV #@ 400x 3–4 spores per FOV @ 400x	0.2 testate amoebae per FOV @ 400x, No ciliates	0	Millipedes	Very High	High	1780 ppm 0.17%	4.8:1 F: 83% B: 17%	2013 ug C/g	7.5 (9.0 when dry)
NATURE'S WAY COMPOST TEXAS, USA	Sparse OM with 80-90% Sand and 5–10% Black soot (microbial soot = anaerobic decomposition)	25 per 0.0025mm² ~10 billion bacteria per ml of soil High in Actinobacteria, Bacterial Dominant (likely anaerobes)	0	2 ciliates per drop	0	0	Trace/ None	Med	602 ppm 0.06%	1.8:1 F: 64% B: 36%	883 ug C/g	8.0
HAPPY FROG POTTING SOIL BY FOXFARM CALIFORNIA, USA	Fungal digested organic matter, light to dark brown OM	19 per 0.0025mm² ~7.6 billion bacteria per ml of soil 2–4 actinobacteria per FOV @ 400x 3–4 oomycetes/ pathogens per FOV @ 400x	Black spore fungi with hyaline hyphae – looks like a mold	0	0	0	Trace/ None	Low	520 ppm 0.05%	1.6:1 F: 61% B: 39%	828 ug C/g	6.0
THE GROUND UP BOHEMIAN POTTING SOIL TEXAS, USA	Light brown to dark brown OM	14 per 0.0025mm2 ~5.6 billion bacteria per ml of soil 2 Lactobacilli per FOV @ 400x 4 Actinobacteria per FOV @ 400x 3 oomycetes/pathogens per FOV @ 400x	0	0	0	0	Trace/ None	Low	666 ppm 0.06%	3.8:1 F: 79% B: 21%	1808 ug C/g	8.0
THE GROUND UP LEAF MOLD COMPOST TEXAS, USA	Sand and OM - mostly lighter brown and either large anaerobic chunks or smaller particles with minor fungal activity under epifluorescence	Highly Bacterial with Very ACTIVE Bacteria Visible HGT 22 per 0.0025mm² ~8.8 billion bacteria per ml of soil Highly Actinobacterial 5 actinobacteria per FOV @ 400x	0.25 spores per FOV @ 400x No Fungal Hyphae	0	0	0	Trace/ None	Med	514 ppm 0.05%	4.0:1 F: 80% B: 20%	2311 ug C/g	8.0

VERMICOMPOST STUDY

Vermicomposting is simply worm composting – you may have them in a pit, in a pile, a worm tower, or a bin, but in all instances, the worms are doing the work for us. Their micro-tillage and digestion of the organic matter into worm castings slowly overturns and mixes soil over the seasons at a rate of 4 – 10% annually. Vermicompost can take approximately 3 months time to turn kitchen scraps and grass clippings into beautiful castings – the time can be shorter or longer depending on your climate, worms, setup, and inputs. It's also now seen as essential to most commercial composting since the worms dramatically lower the pathogen numbers within weeks. They are biological regulators – their digestion directs the bacterial and fungal direction and range of expression in the soil (keep in mind though: there are limits; we can easily kill our worms if we push them with too many challenging foods). Earthworm digestion contains a diversity of bacteria, typically: Bacillus, Aeromonas, Flavobacterium, Nocardia, Gordonia, Vibrio, Clostridium, Proteus, Serratia, Mycobacterium, Pseudomonas, Klebsiella, Azotobacter, and Enterobacter.

Windrow turned hot compost operations like Catalyst Bioamendments are letting worms find their piles or adding them themselves, Johnson-Su composting includes adding worms after the pile has gone through the thermophilic stage, and static piles and pits nearly always get found by earthworms common to the local bioregion. Darwin was fascinated by them, and few realize how vital and powerful their work is. What's amazing is how the worms tie together a broad range of composting methods, acting as a guide to those processes. Their diets influence the casting pH and biological expression of the soil life: fungal foods will lead to fungally inoculated castings. Just look at the fungal numbers on the Johnson-Su vermicomposting sample from my student: D.B. His compost is 85% fungal, and it was just neighborhood leaves. The high lignin foods lead to high fungal numbers because that's what is required to breakdown that type of organic matter. It's also interesting to note the correlation between total microbial biomass and high fungal %'s. They seem to be linked.

	ORGANIC MATTER	BACTERIA	FUNGI	PROTOZOA	NEMATODES	MICROARTHROPODS & WORMS	N	P	SALINITY	F:B	TOTAL MICROBIAL BIOMASS:	PH
STANGL JOHNSON-SU VERMICOMPOST #2 (ONTARIO CANADA (AGED - NOT FRESH)	Dark to golden brown, large to small particles throughout	1.5 per 0.0025mm² ~600 million bacteria per ml of soil Cocci and Bacilli are sparse, but actinobacteria 1–3 per FOV @ 400x	Small fragments every few FOV @ 400x	Testate Amoebae but sparse 1 every few FOV	0	Over 6 earthworms – red wigglers per cup	Very High	Med	3320 ppm 0.33%	**2.7:1** F: 73% B: 27%	1342 ug C/g	5.5

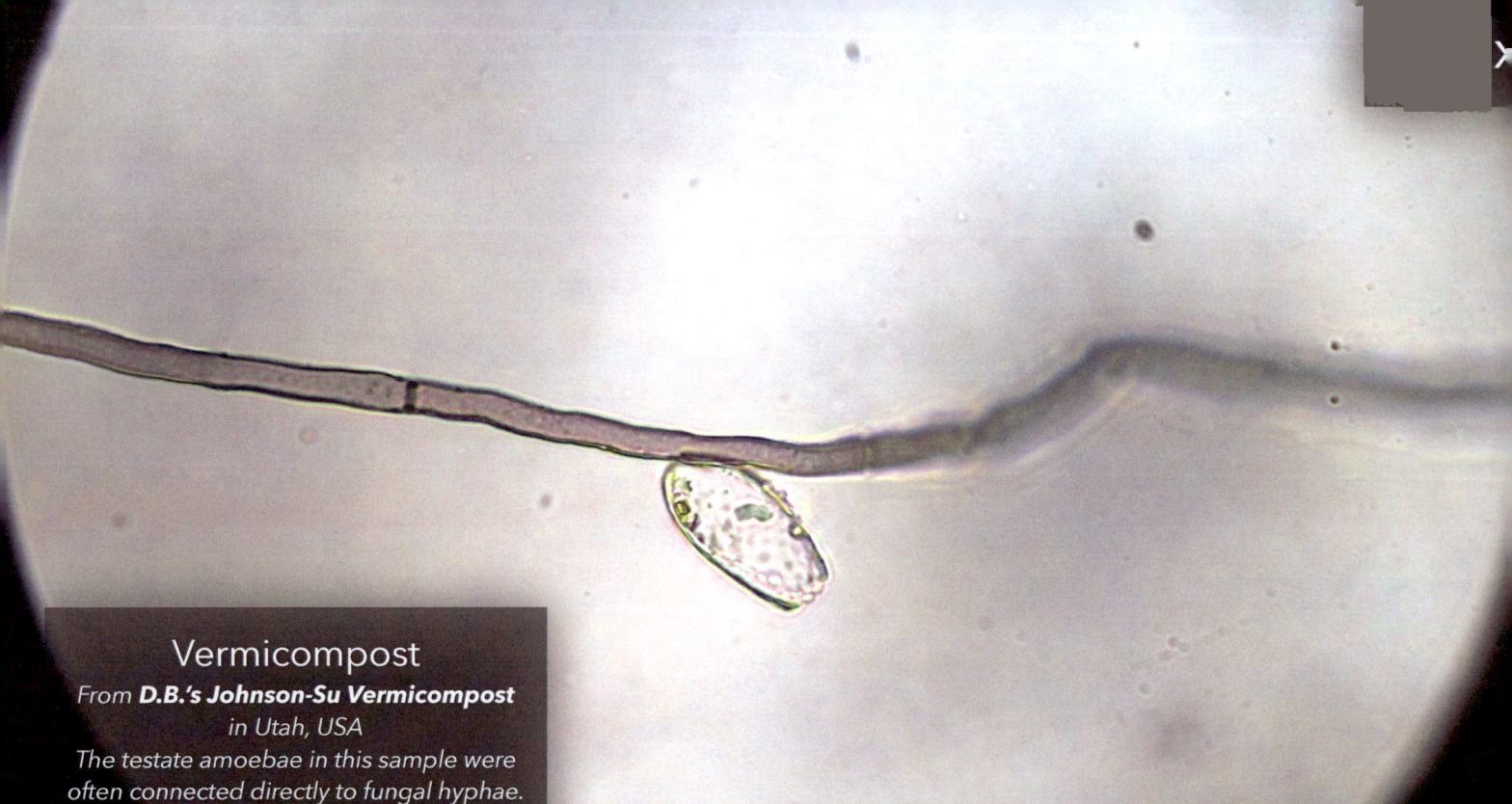

Vermicompost

From **D.B.'s Johnson-Su Vermicompost**
in Utah, USA
The testate amoebae in this sample were
often connected directly to fungal hyphae.
This was unique among the samples I've
tested and demonstrated new levels of
cooperation and structuring in the soil
between fungi and protozoa.

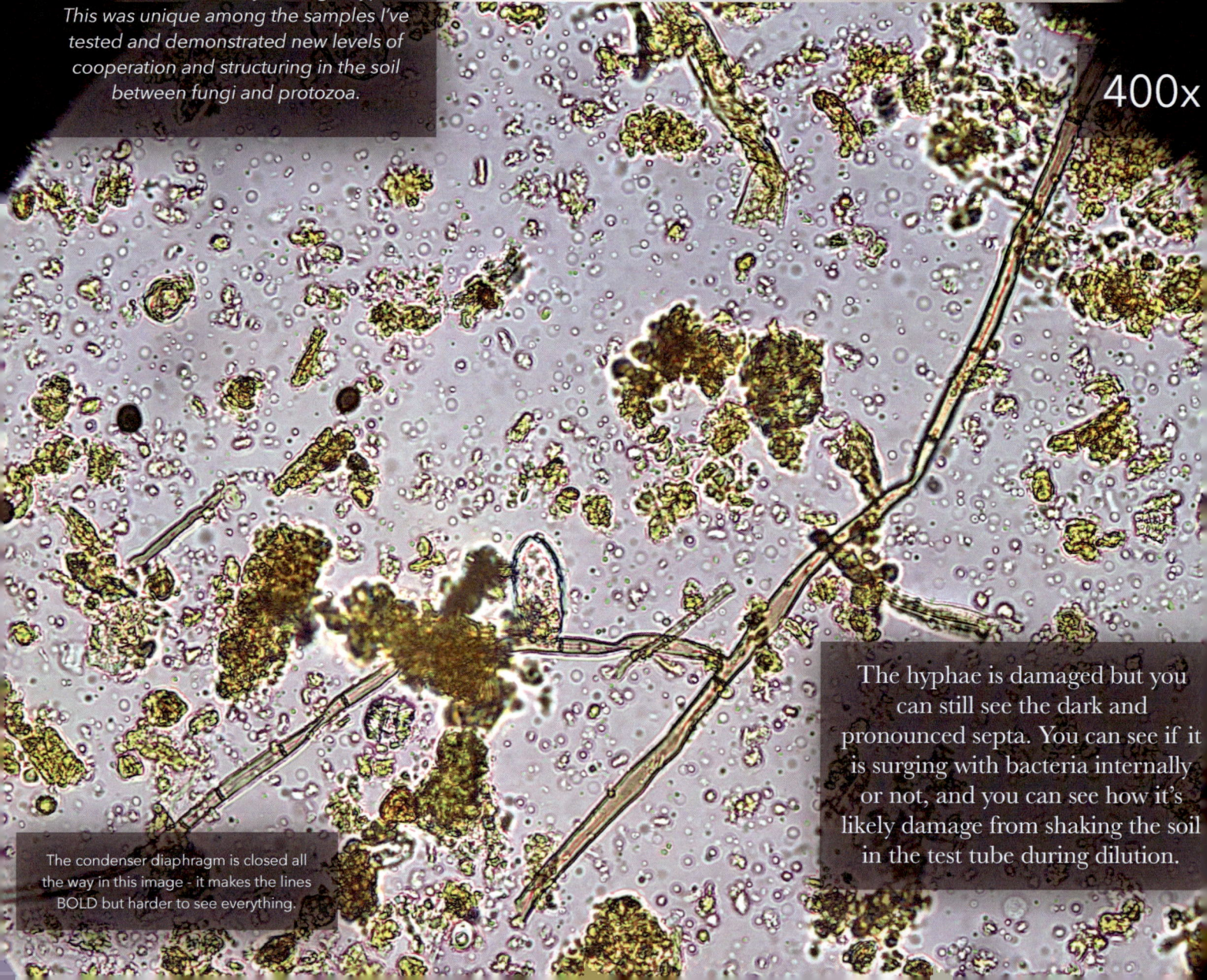

400x

The hyphae is damaged but you
can still see the dark and
pronounced septa. You can see if it
is surging with bacteria internally
or not, and you can see how it's
likely damage from shaking the soil
in the test tube during dilution.

The condenser diaphragm is closed all
the way in this image - it makes the lines
BOLD but harder to see everything.

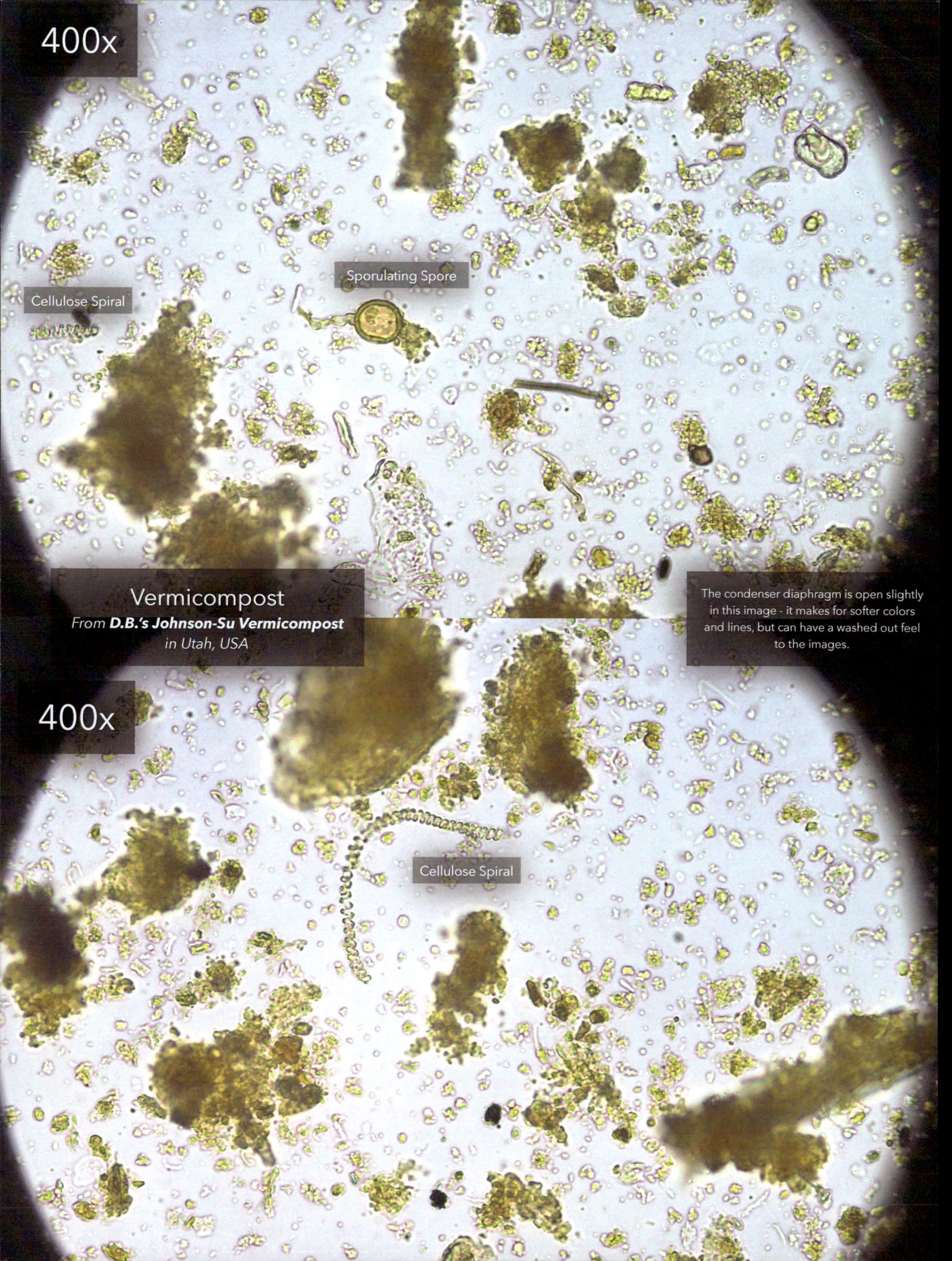

400x

Sporulating Spore

Cellulose Spiral

Vermicompost
From **D.B.'s Johnson-Su Vermicompost**
in Utah, USA

The condenser diaphragm is open slightly in this image - it makes for softer colors and lines, but can have a washed out feel to the images.

400x

Cellulose Spiral

400x

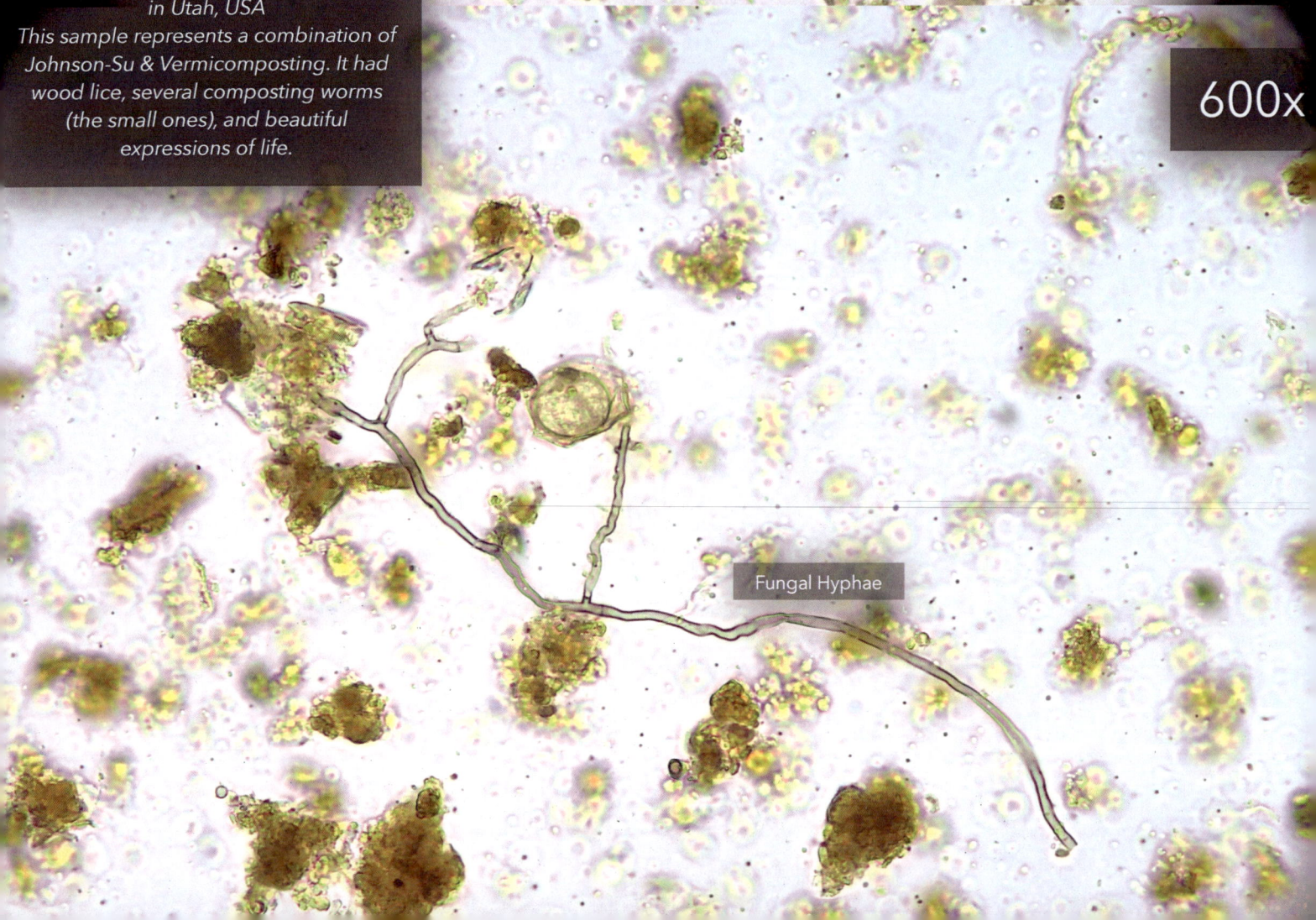

Fungal Hyphae

Fungal Hyphae

A Very Large Testate Amoeba
Likely a *Centropyxis Amoeba*

Vermicompost
From **D.B.'s Johnson-Su Vermicompost**
in Utah, USA
*This sample represents a combination of
Johnson-Su & Vermicomposting. It had
wood lice, several composting worms
(the small ones), and beautiful
expressions of life.*

600x

Fungal Hyphae

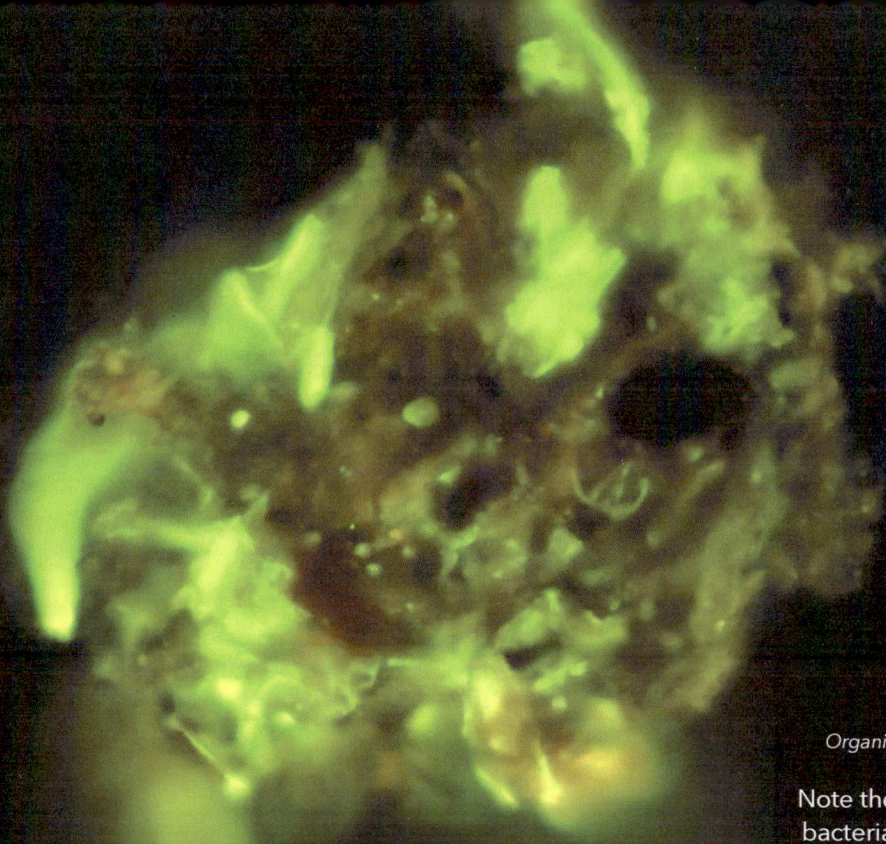

Vermicompost

Organic matter from Stangls commercial Johnson-Su compost finished with red wigglers.

Note the tiny glowing dots and rods are living bacteria and the glowing sections are fungal. This sample has been stained with acridine orange, so all living cells will autofluoresce under epifluorescence lighting at the right wavelengths.

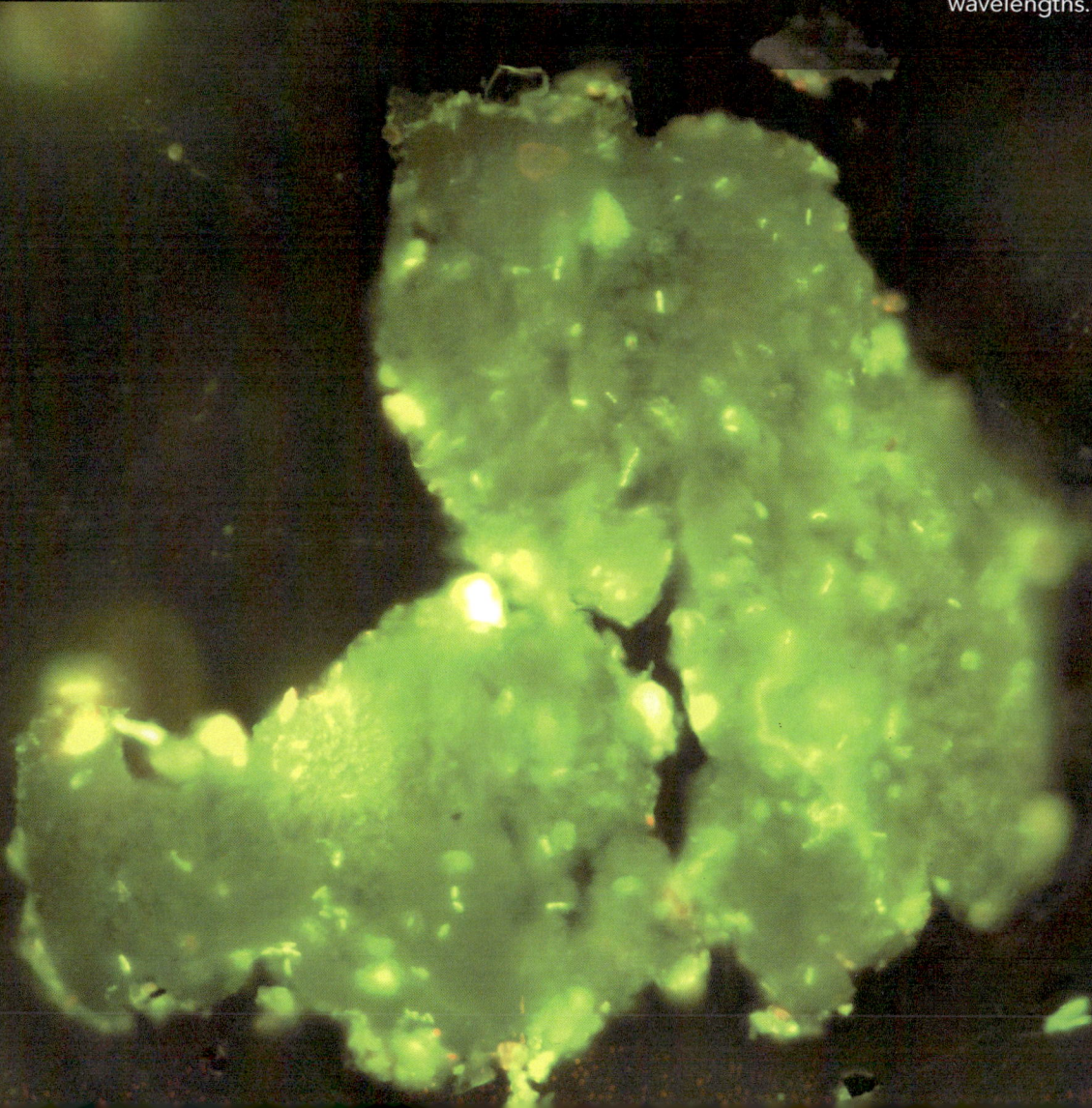

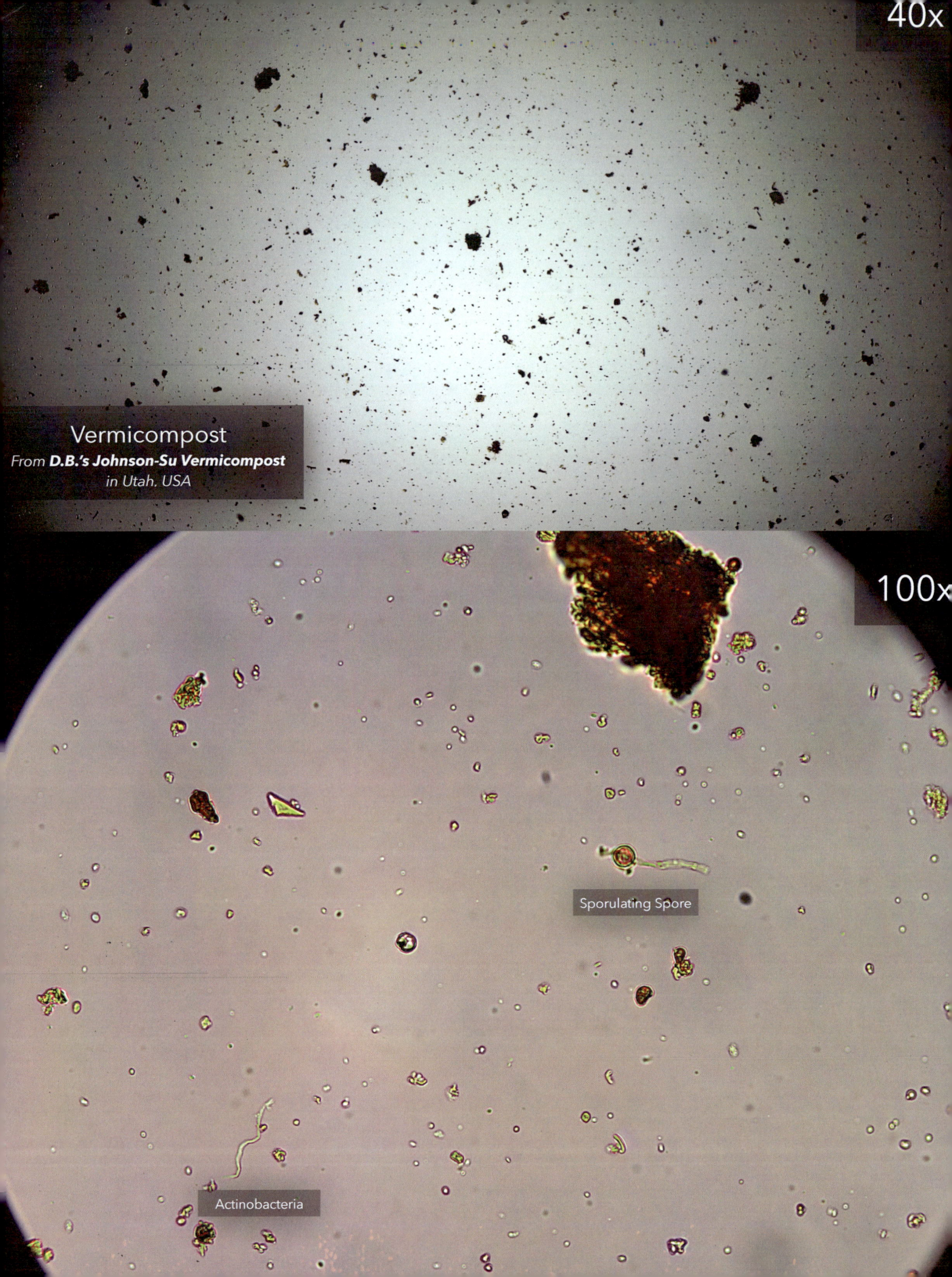

40x

Vermicompost
From **D.B.'s Johnson-Su Vermicompost**
in Utah, USA

100x

Sporulating Spore

Actinobacteria

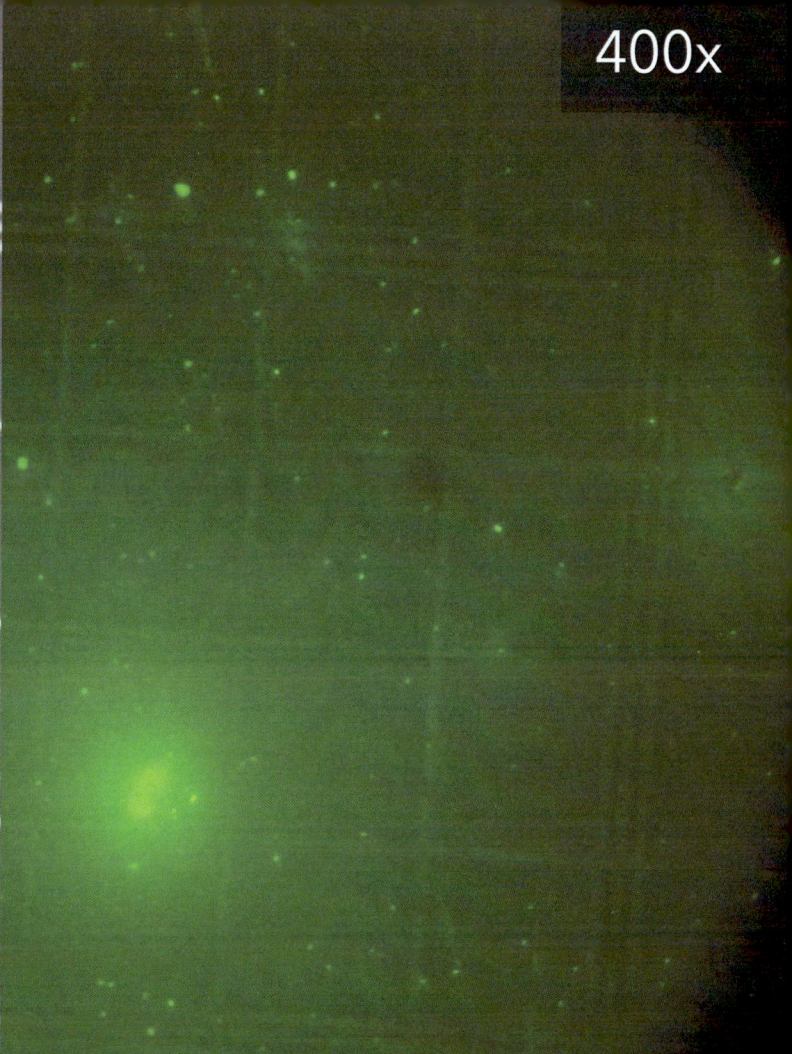

Vermicompost
Organic matter from Stangls commercial Johnson-Su compost finished with red wigglers.

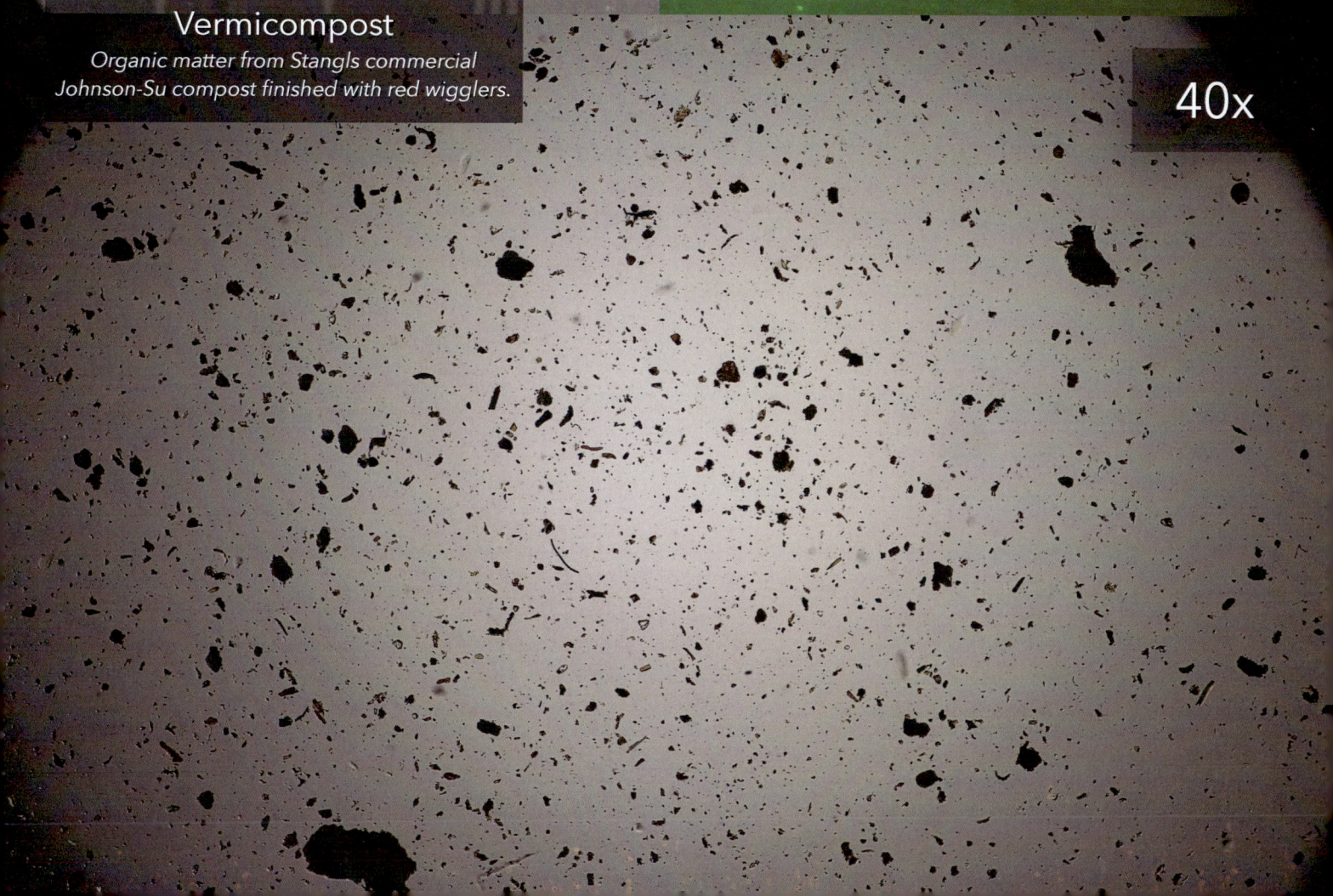

400x

100x

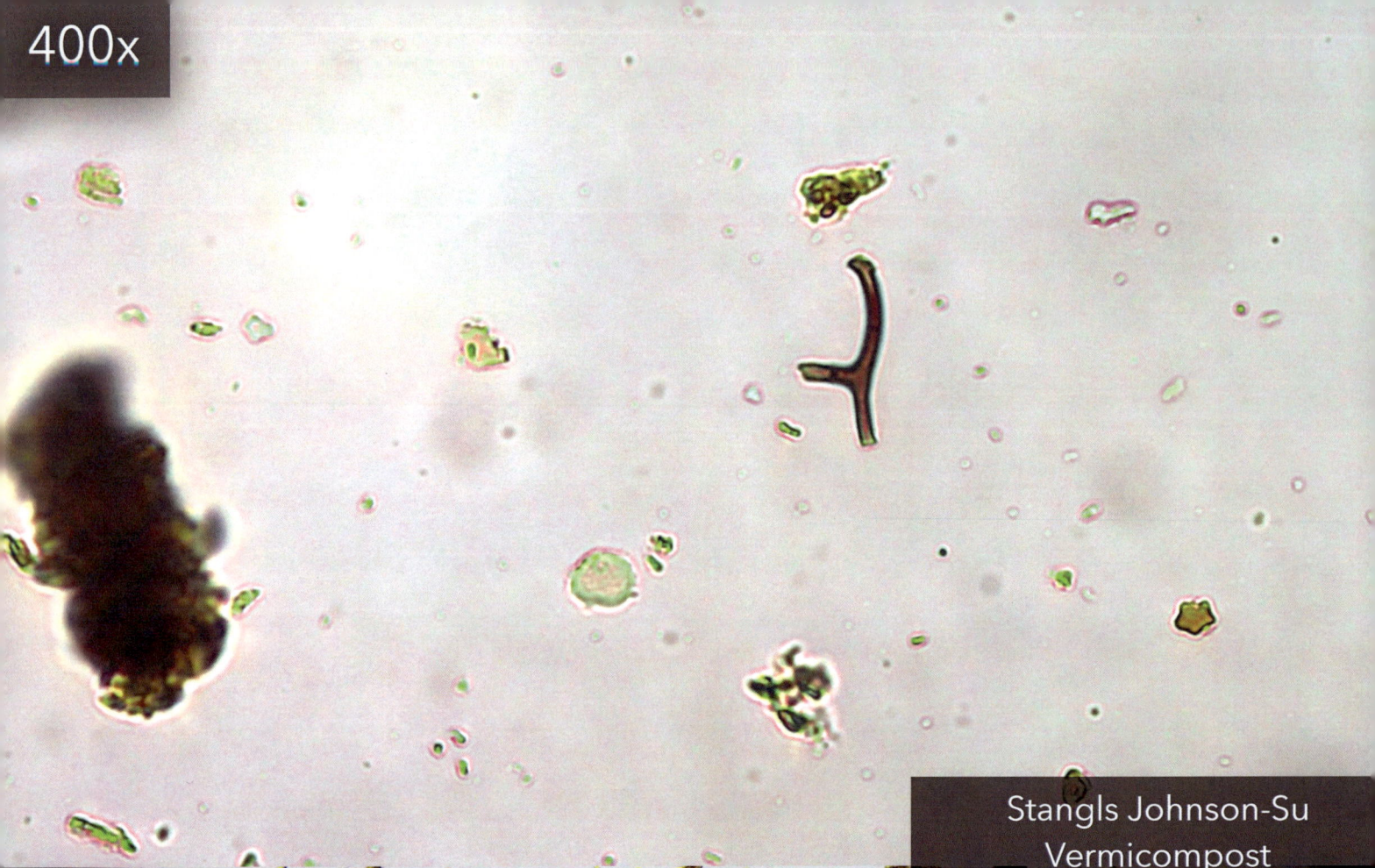

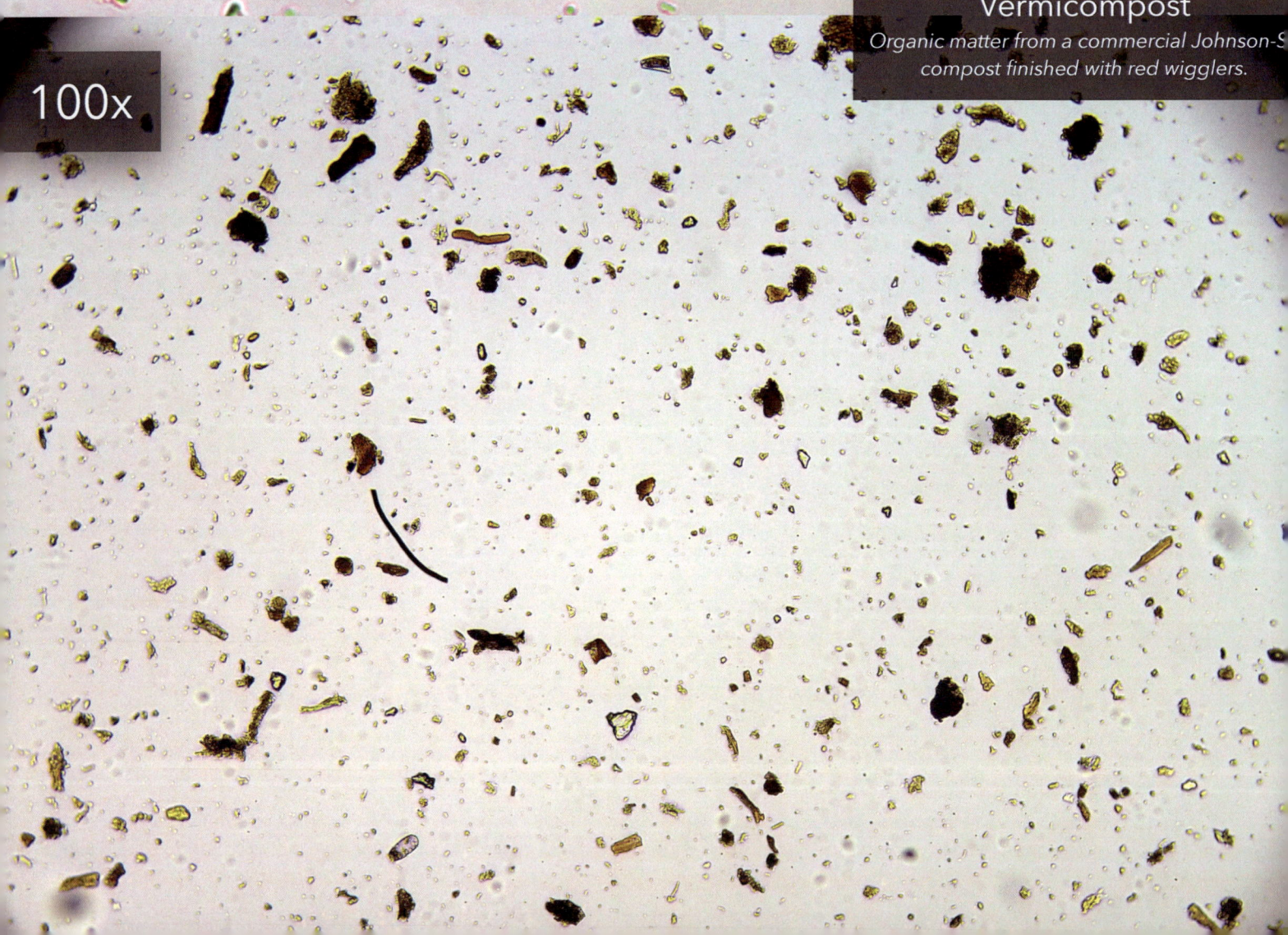

Stangls Johnson-Su
Vermicompost
Organic matter from a commercial Johnson-S
compost finished with red wigglers.

	ORGANIC MATTER	BACTERIA	FUNGI	PROTOZOA	NEMATODES	MICROARTHROPODS & WORMS	N	P	SALINITY	F:B	TOTAL MICROBIAL BIOMASS:	PH
D. B.'S JOHNSON-SU VERMICOMPOST #2 UTAH, USA	Fibers (lignin), Clayey, very little mineral Mostly golden brown with some darker OM	3 per 0.0025mm² ~1.2 billion bacteria per ml of soil High in lactobacilli: 3–5 per FOV @ 400x Wide diversity of Cocci and Bacilli 3 Actinobacteria per FOV @ 400x	Extensive, LONG Fungal Hyphae every FOV @ 400x and a diversity of spores: 5 spores per FOV @ 400x	3–7 testate amoebae per FOV @ 400x 4 ciliates per drop (large and small, slow & fast)	2 – 4 per drop Mostly Bacterial but also 1 Predator	Tiny red wigglers in almost every gram of soil	High	Trace	1752 ppm 0.17%	5.6:1 F: 85% B: 15%	2494 ug C/g	8.0

SUBSOIL, FIELD SOIL, COMPACTED DIRT, & BIOCHAR

It's important to have something to compare to create meaningful observations. If we don't know what to look at and what to ignore, we can be caught staring at artifacts of the lens or light and missing the entire scene playing out in the same field of vision. Most of us are going to be looking at our site's soils and degraded or damaged soils that we are seeking to fix. Once we begin to juxtapose the healthy soils or compost against dead dirt or soil with just bacteria in it, we begin to develop our eye for what is biologically active, what is diverse, what types of microbes are present, and which imbalances are holding back the full expression of nutrient cycling.

Looking at our soil from our site here in Texas under the microscope, just outside Austin, we can see that it is fine sand and silt, prone to compaction, and teeming with bacteria exclusively. I've seen no fungi, no nematodes, no protozoa, and low diversity of bacteria. It's important to remember that all the samples displayed in charts are 1:10 soil:water dilution, or if it's a hemocytometer, it's always 1:100 dilution, and every dilution always starts with only 1 ml of soil or compost – there are a few 1:5 dilution images but those are noted and not part of the chart data. That means that the compaction we see in some of these samples is dramatically increased in the actual field or garden soil. This is why my soil here turns into mud so easily. There's just nothing much there – it lacks nutrients (the clear silt and sand aren't carrying much), it lacks organic matter (that's why it's colorless and has few organic matter chunks), and it lacks fungi (so there's no structure being made). In this situation, compost, biofertilizers, rock dust, biochar, and cover crops are going to have a dramatic effect because there's no foods, no habitat, hardly any organic matter, no predators, no cycling, and no structure, so we only can go up from here!

100x

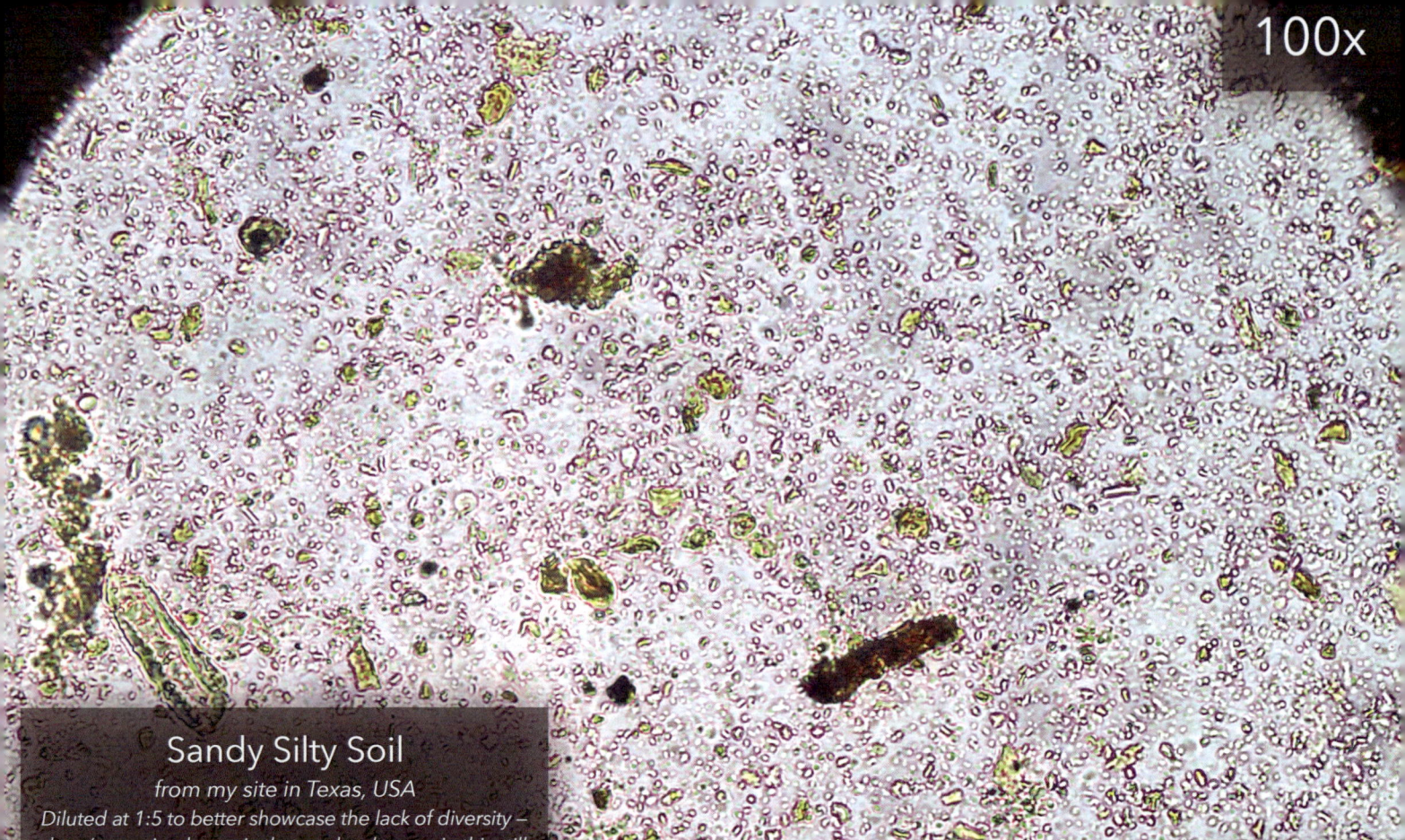

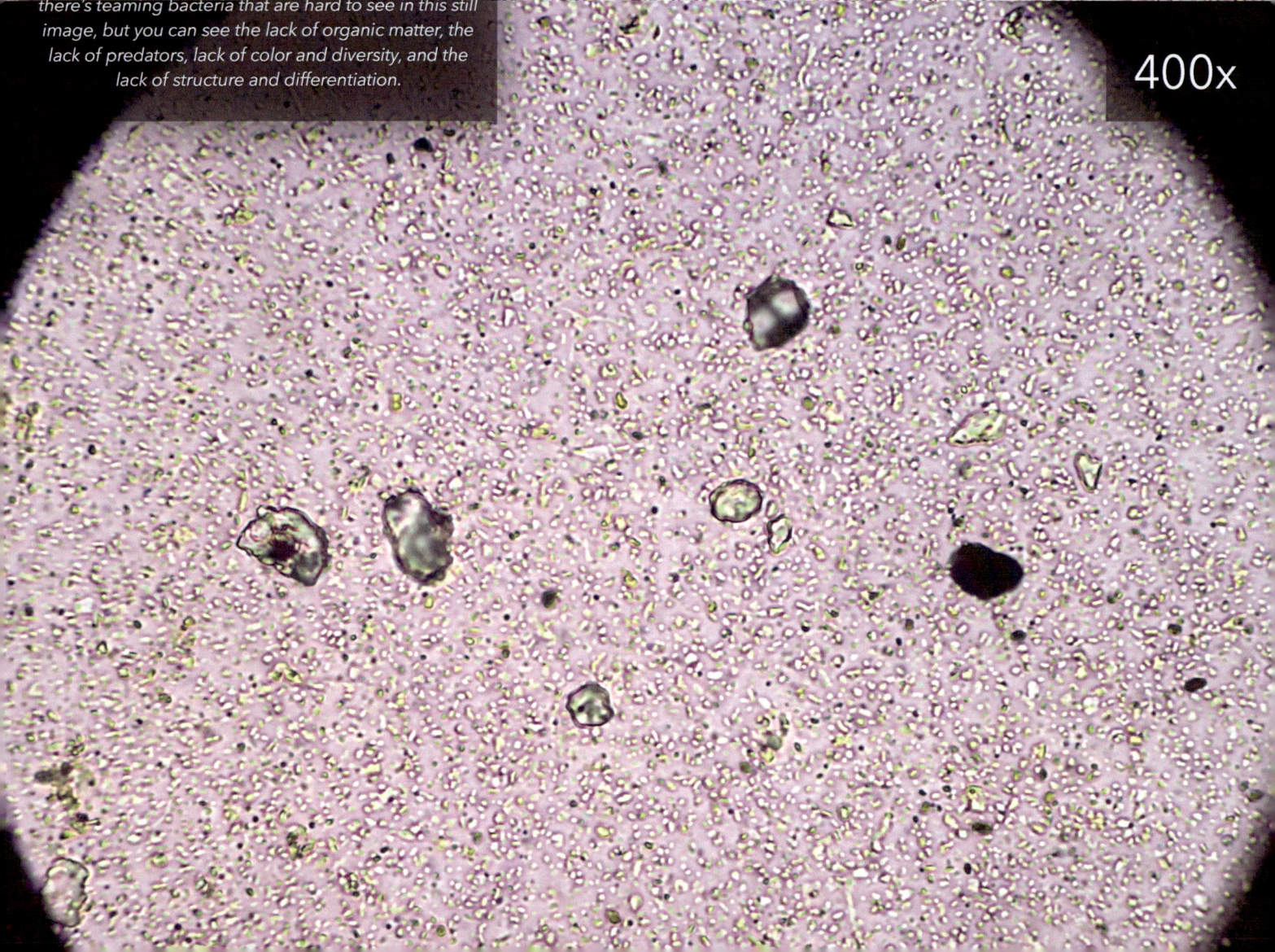

Sandy Silty Soil
from my site in Texas, USA
Diluted at 1:5 to better showcase the lack of diversity –
there's teaming bacteria that are hard to see in this still
image, but you can see the lack of organic matter, the
lack of predators, lack of color and diversity, and the
lack of structure and differentiation.

400x

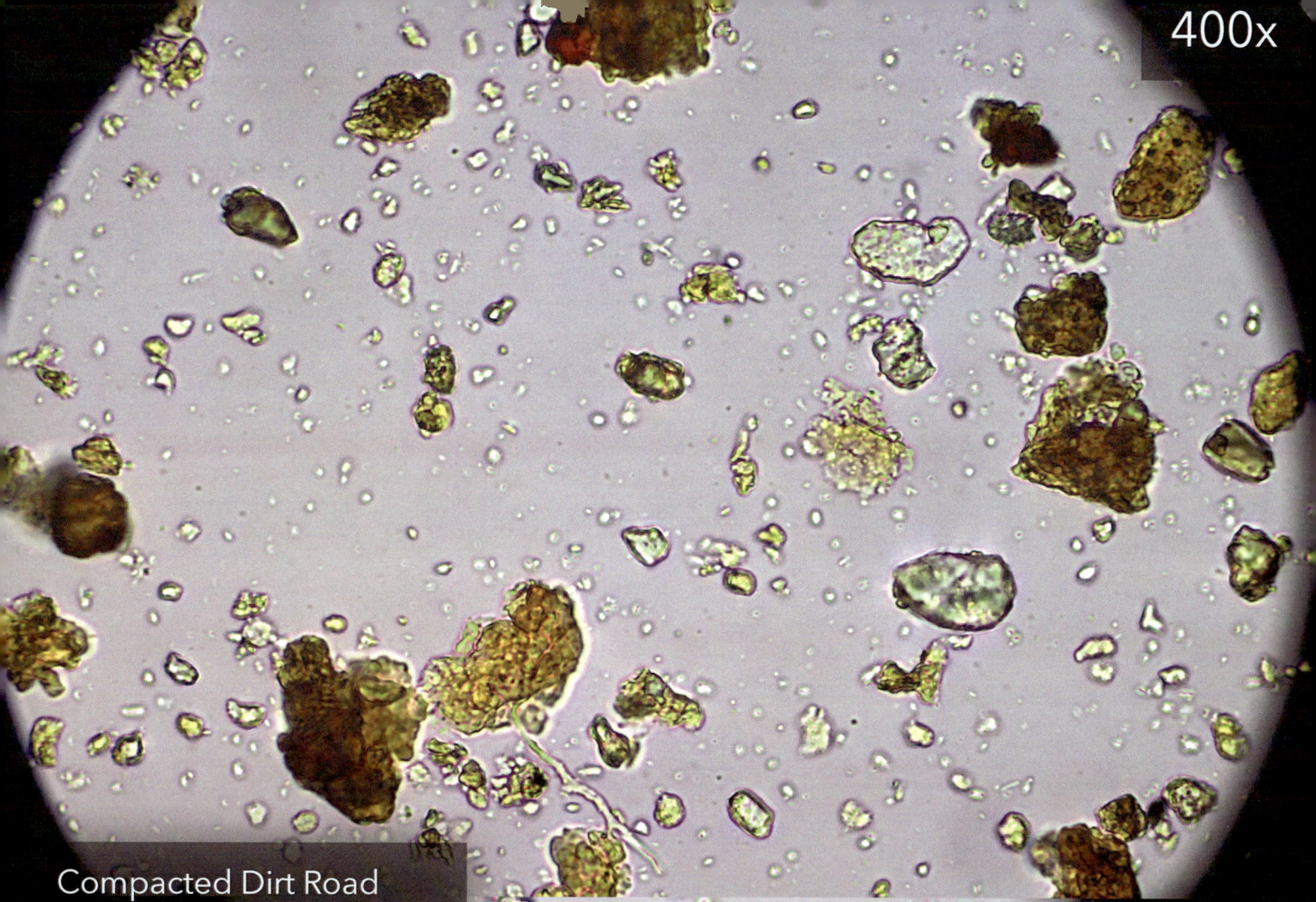

400x

Compacted Dirt Road
from my site in Texas, USA

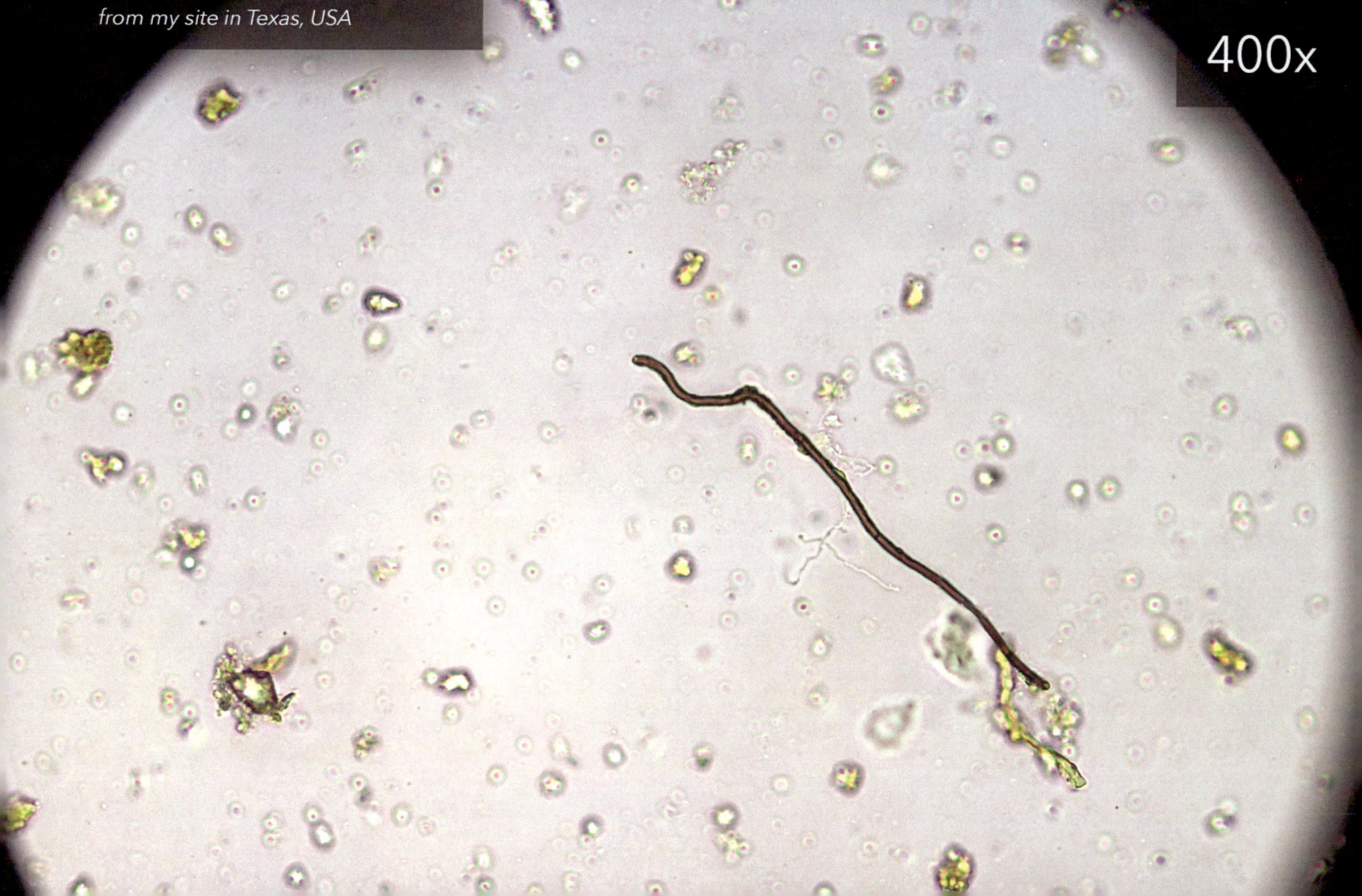

400x

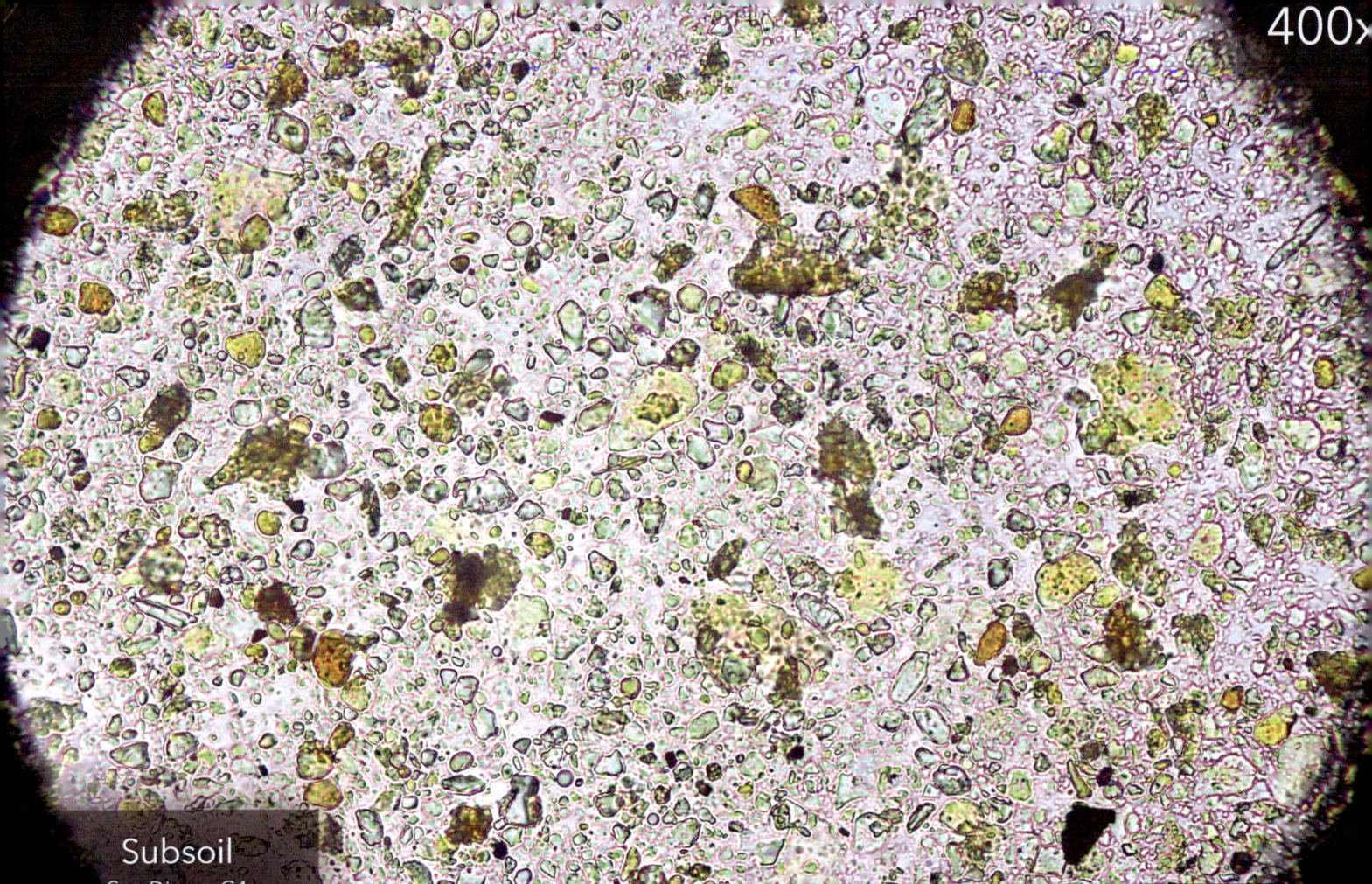

400x

Subsoil
San Diego, CA
Notice how compact
it is with minerals.

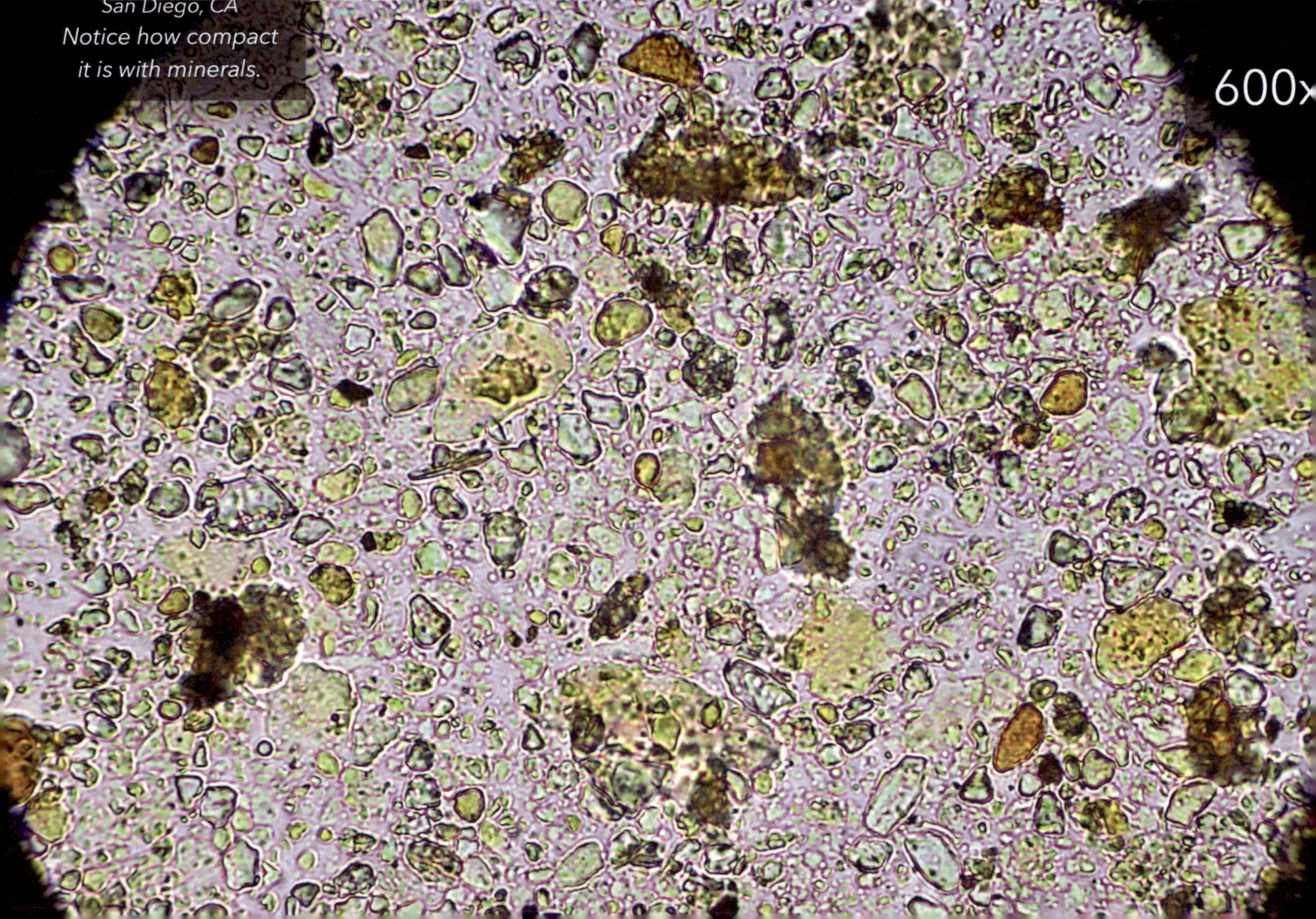

600x

40x

Subsoil
San Diego, CA
Notice how compact
it is with minerals.

40x

Compacted Dirt Road
from my site in Texas, USA

40x

Biochar
Pacific Biochar, CA
Notice the size.

40x

Sandy Silty Soil
Central Texas

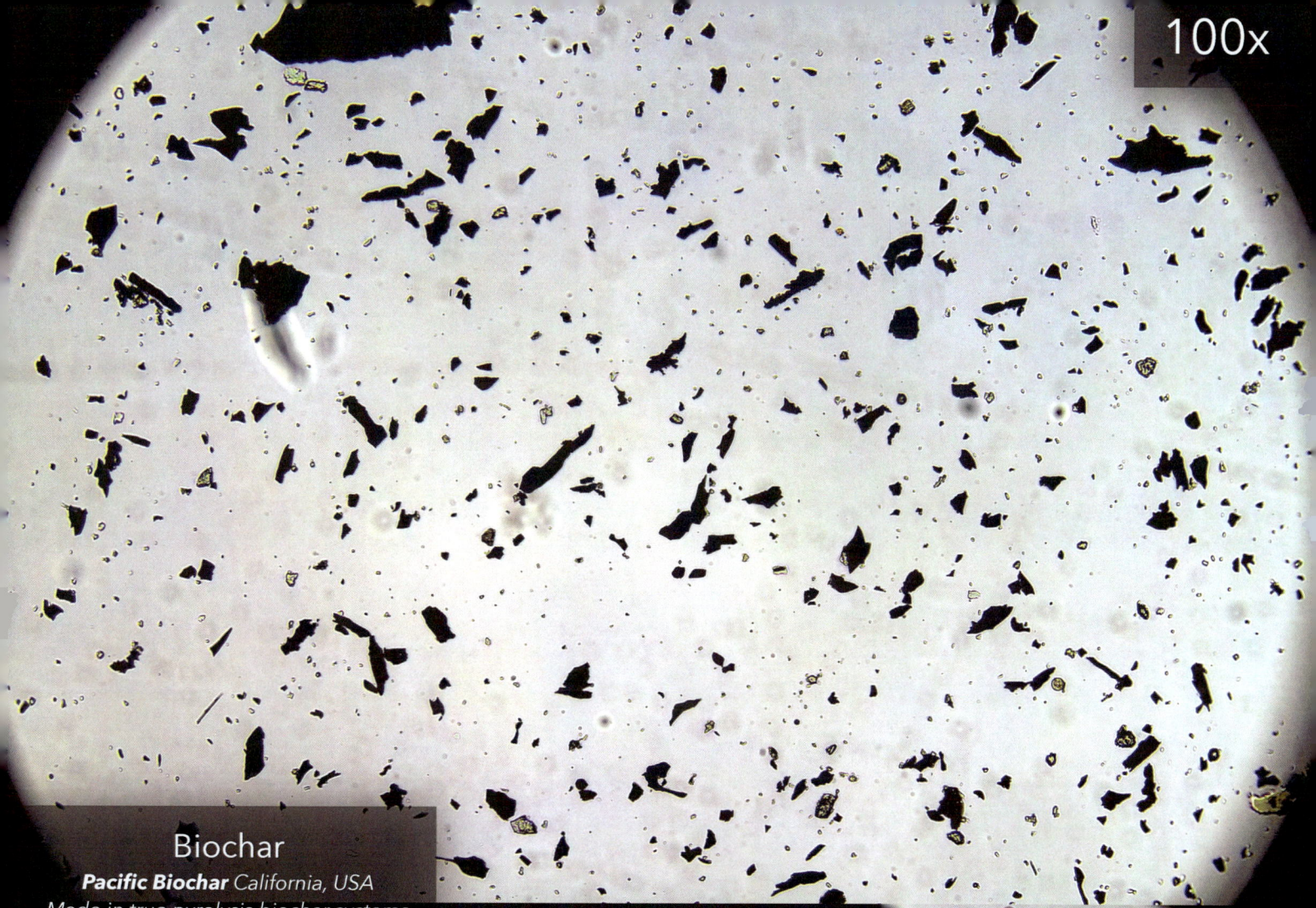

Biochar

Pacific Biochar *California, USA*
Made in true pyrolysis biochar systems
from woody biomass from California
native wilderness.

100x

Biochar in Dark Field
Pacific Biochar *California, USA*

Another thing to keep in mind is that highly mineral-based soils are not going to suspend in solutions like organic matter and microbes will. While we do see sand and biochar under the microscope, these are small samples. None of us are taking the very bottom of the test tube because it's large pieces of grit that don't work well with cover slides and always sink when we let the test tube rest after mixing.

	ORGANIC MATTER	BACTERIA	FUNGI	PROTOZOA	NEMATODES	N	P	SALINITY	F:B	TOTAL MICROBIAL BIOMASS	PH
SANDY SILTY SOIL TEXAS, USA	Sparse and light colored – nearly all mineral	3 per 0.0025mm² ~1.2 billion bacteria per ml of soil Bacilli & cocci are dominant w/ some actinobacteria	0	0	0	Trace/ None	Med	72 ppm 0.007%	**0.7:1** F: 40% B: 60%	354 ug C/g	8.0
SUBSOIL CALIFORNIA, USA	Extremely low in organic matter	20 per 0.025mm² ~8 billion bacteria per ml of soil Teeming with Bacteria and 2-3 actinobacteria per FOV @ 400x	Sparse spores, no hyphae	0	0	Trace/ None	Low	206 ppm 0.02%	**0.6:1** F:36% B:64%	291 ug C/g	7.0
COMPACTED SOIL - DIRT ROAD TEXAS, USA	Sparse, ranging in color (next to a regenerative orchard), reddish minerals (iron oxides), lots of silicates	23 per 0.025mm² ~9.2 billion bacteria per ml of soil Bacilli & cocci are dominant w/ actinobacteria in almost every FOV @ 400x	Sparse 0.3 per FOV @ 400x	0	0	Trace/ None	Low	28 ppm 0.002%	**1.3:1** F: 57% B: 43%	667 ug C/g	7.0

227

EM® STUDY

I've been studying, using, and teaching about Effective Microbes® (EM®, EM-1®, EM-1 Pro®, EM-3) for years now, and this amazing consortium of primarily yeast, lactobacillus spp., streptomyces, rhodopseudomonas, and e.coli can be replicated by anyone anywhere. These brews lack predators, so the nutrients we add as foods get incorporated into the biology: making them perfect to deliver that nutrition through rhizophagy directly to plants inside their roots. These microbes are also common to all types of compost, so despite sometimes being labeled as anaerobic, these are actually facultative anaerobes that thrive in compost, no matter the method utilized. Thus,

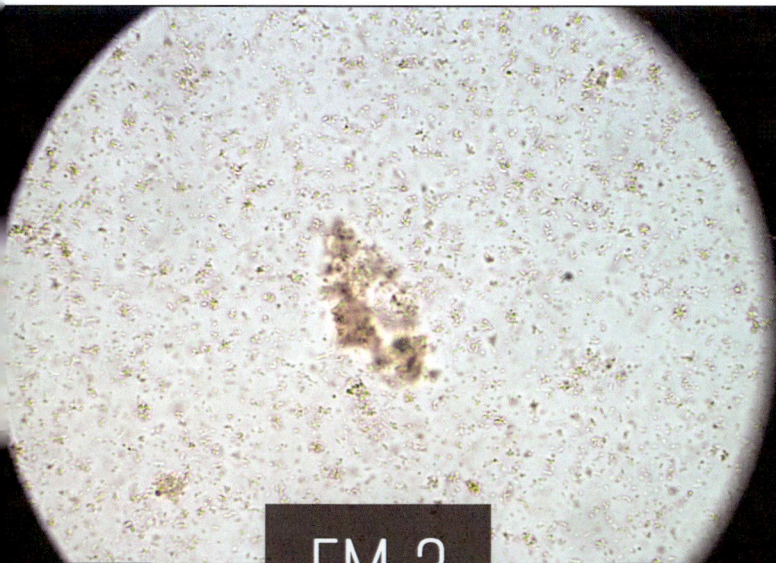

EM-3

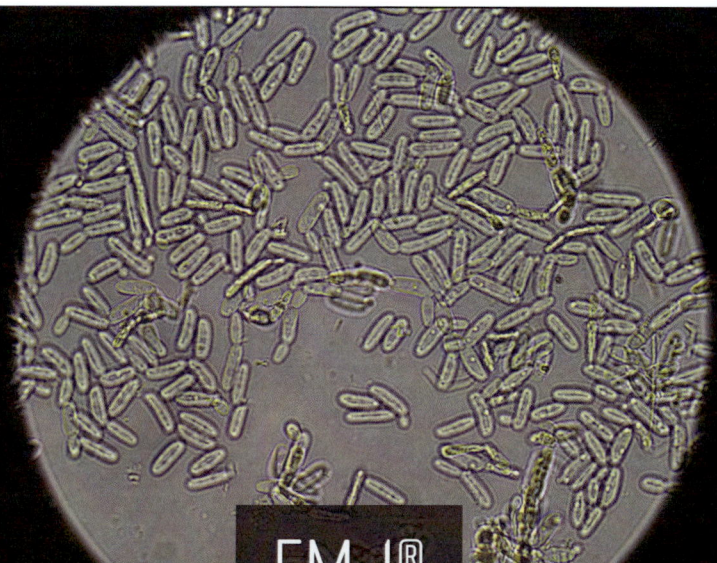

EM-1®

THIS ONE IS FOR **PNSB**.

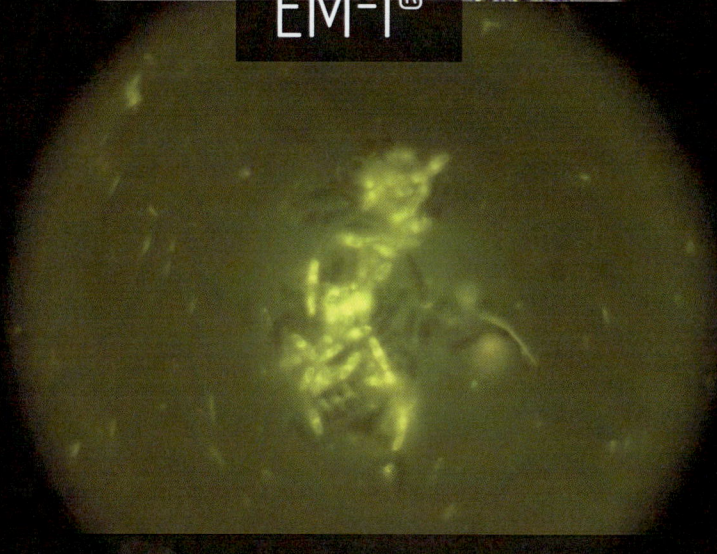

THIS ONE IS FOR **LAB & YEAST**.

1000X

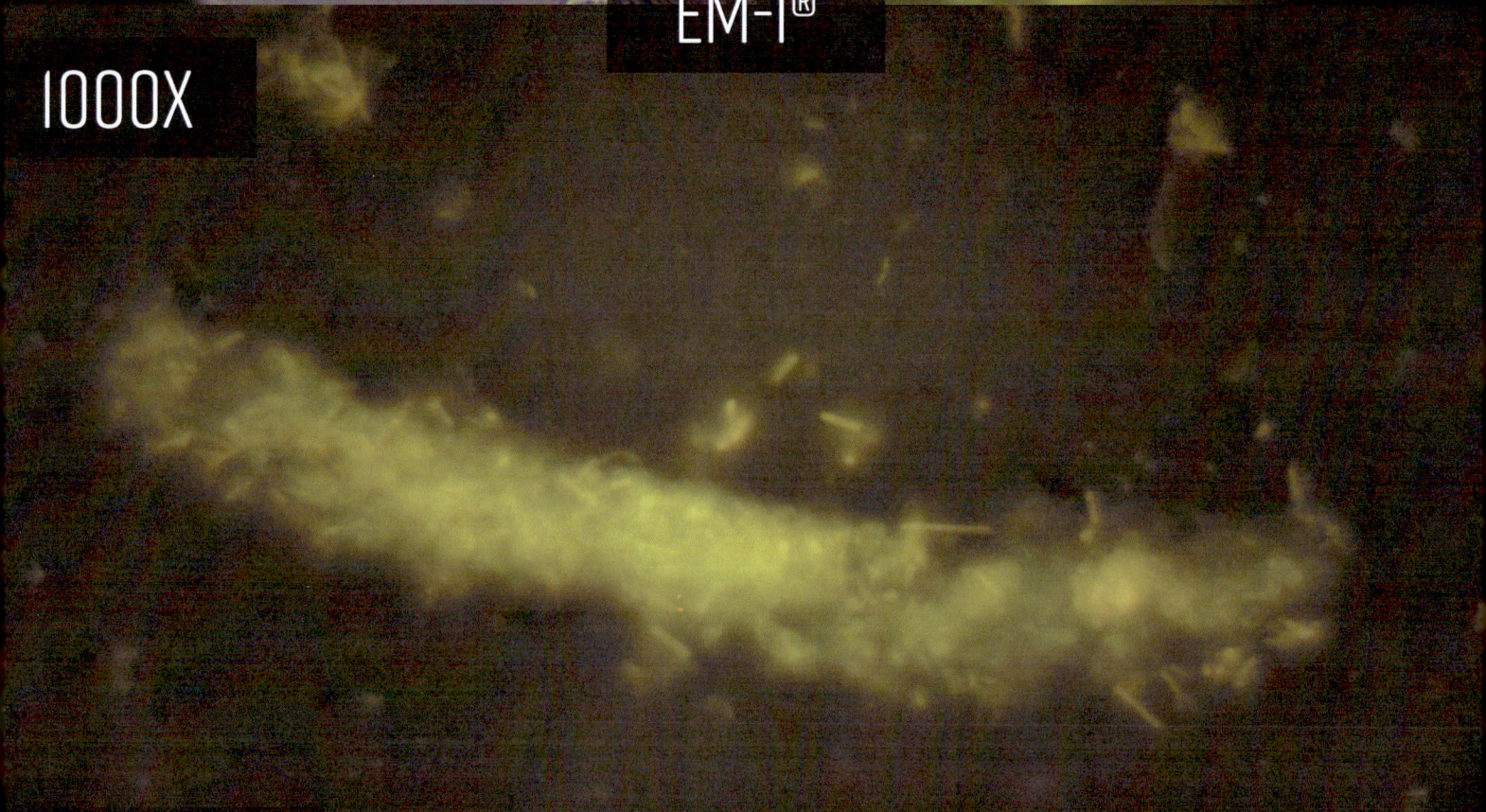

EPIFLUORESCENCE

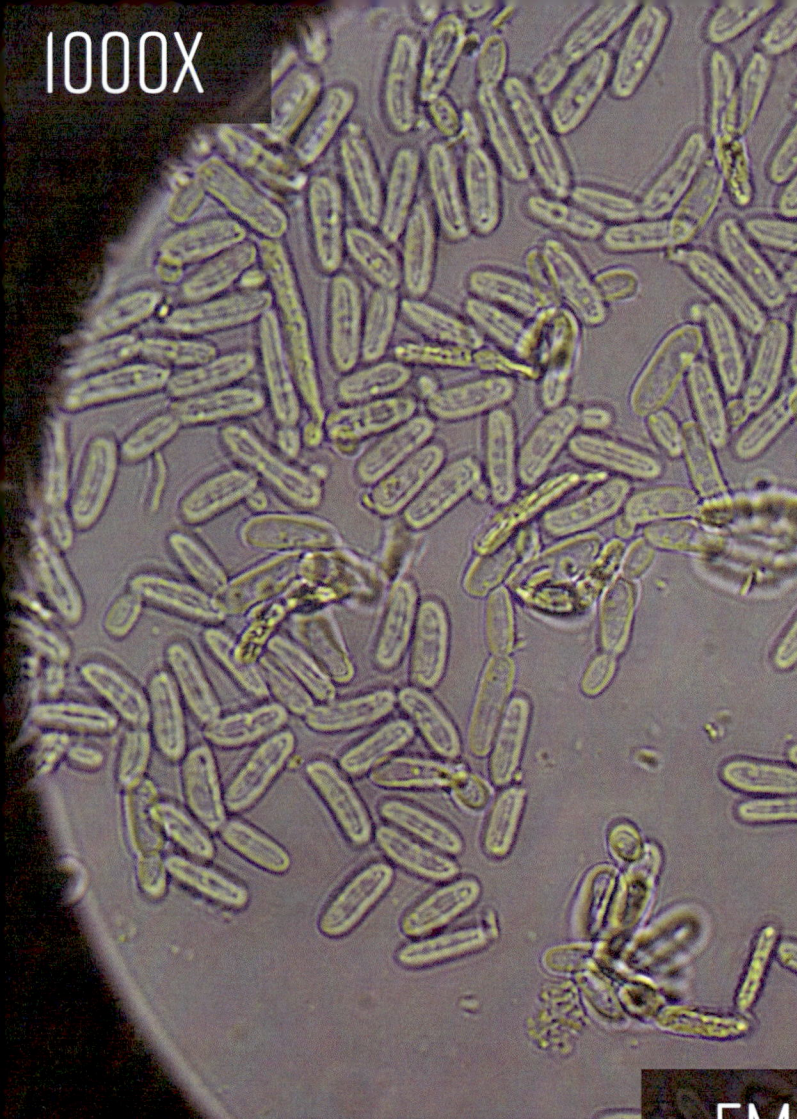

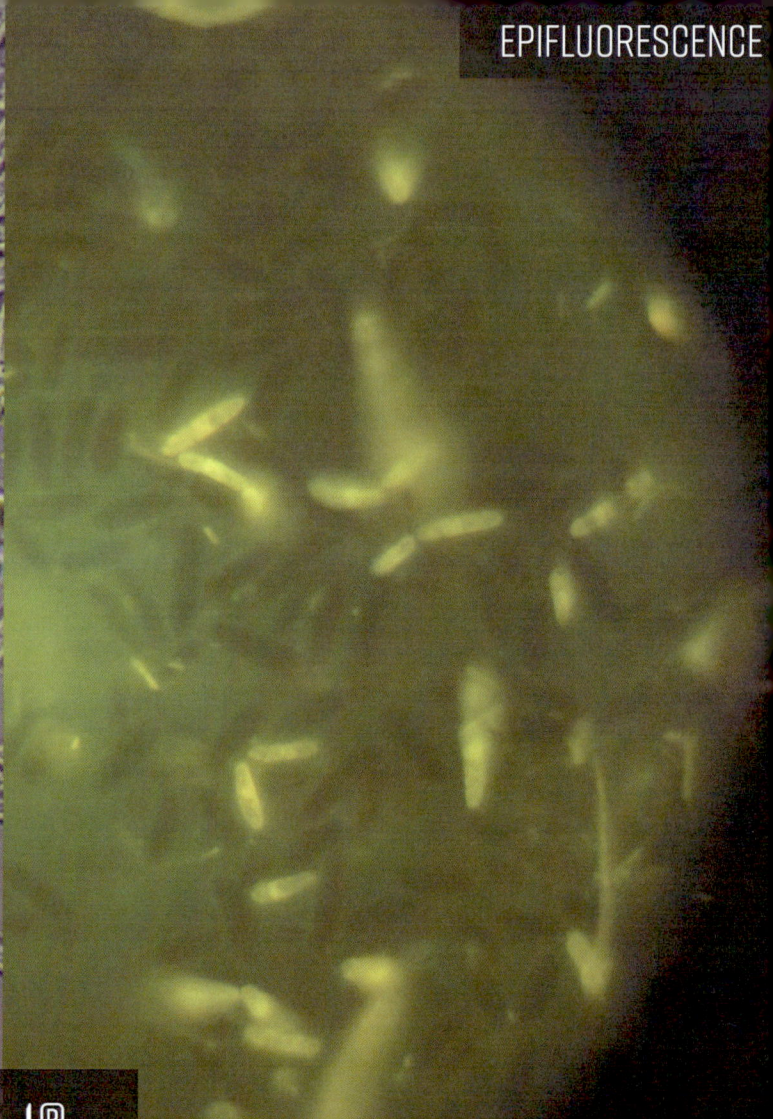

EM-1®

1000X

EPIFLUORESCENCE

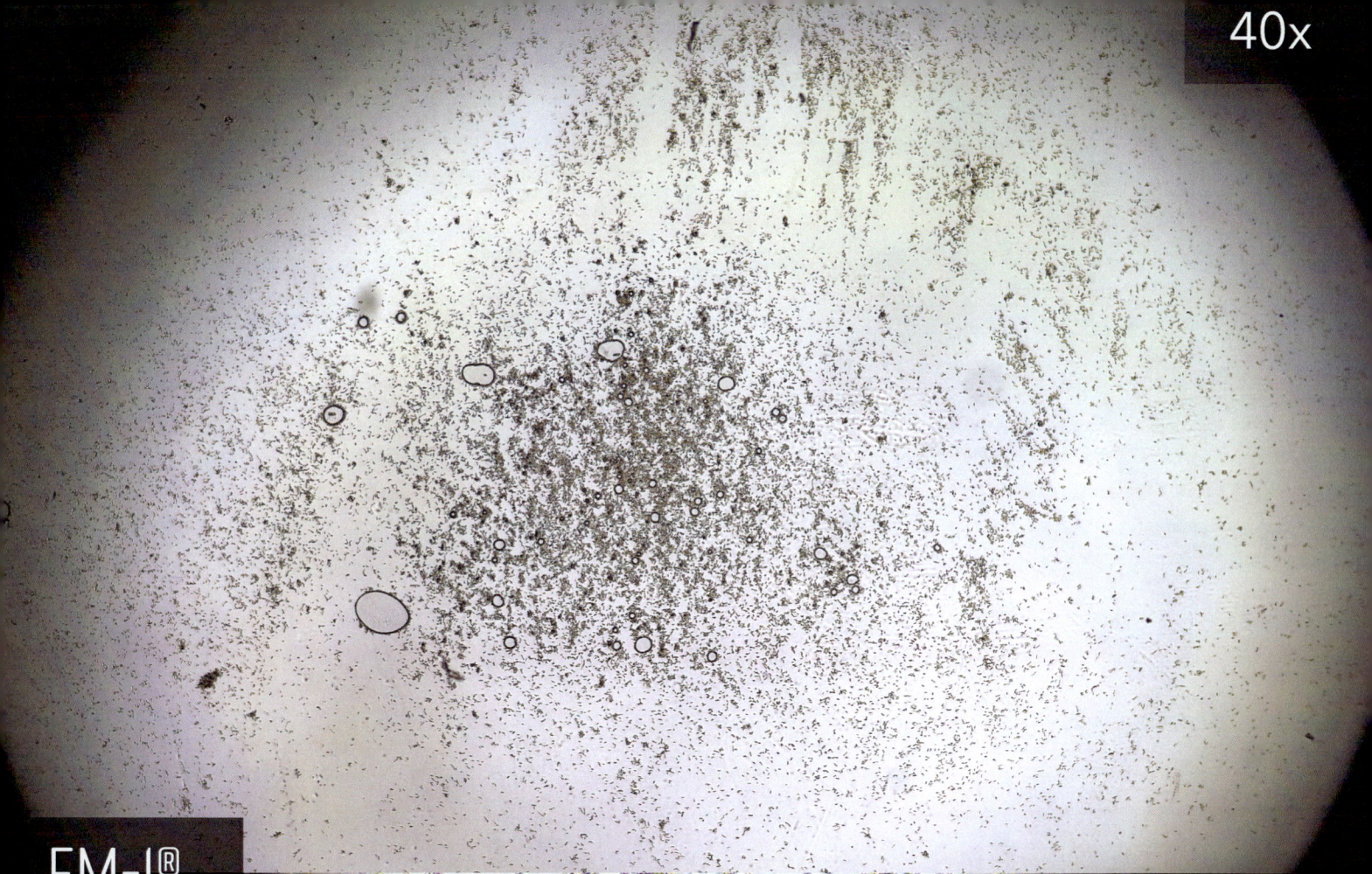
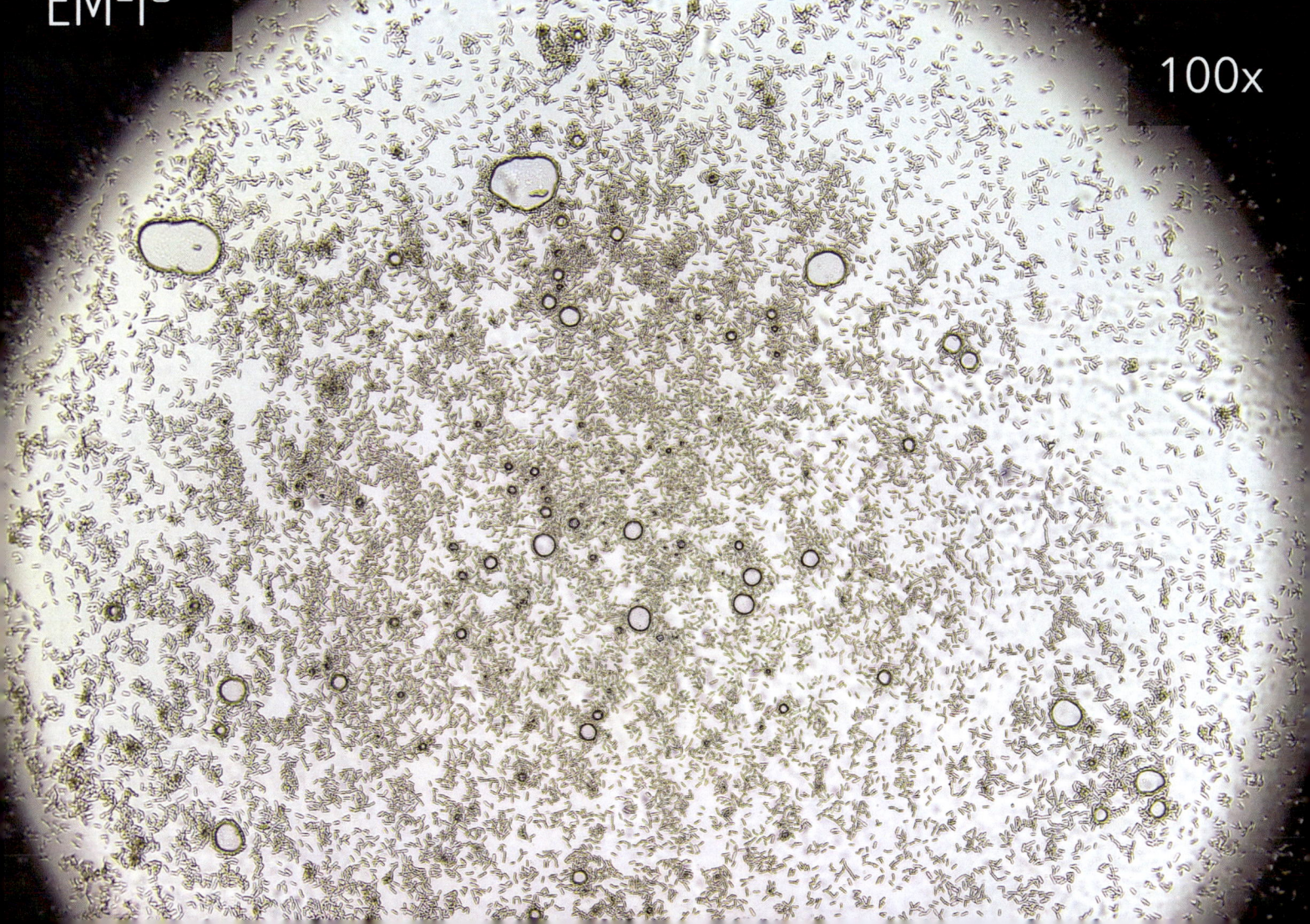

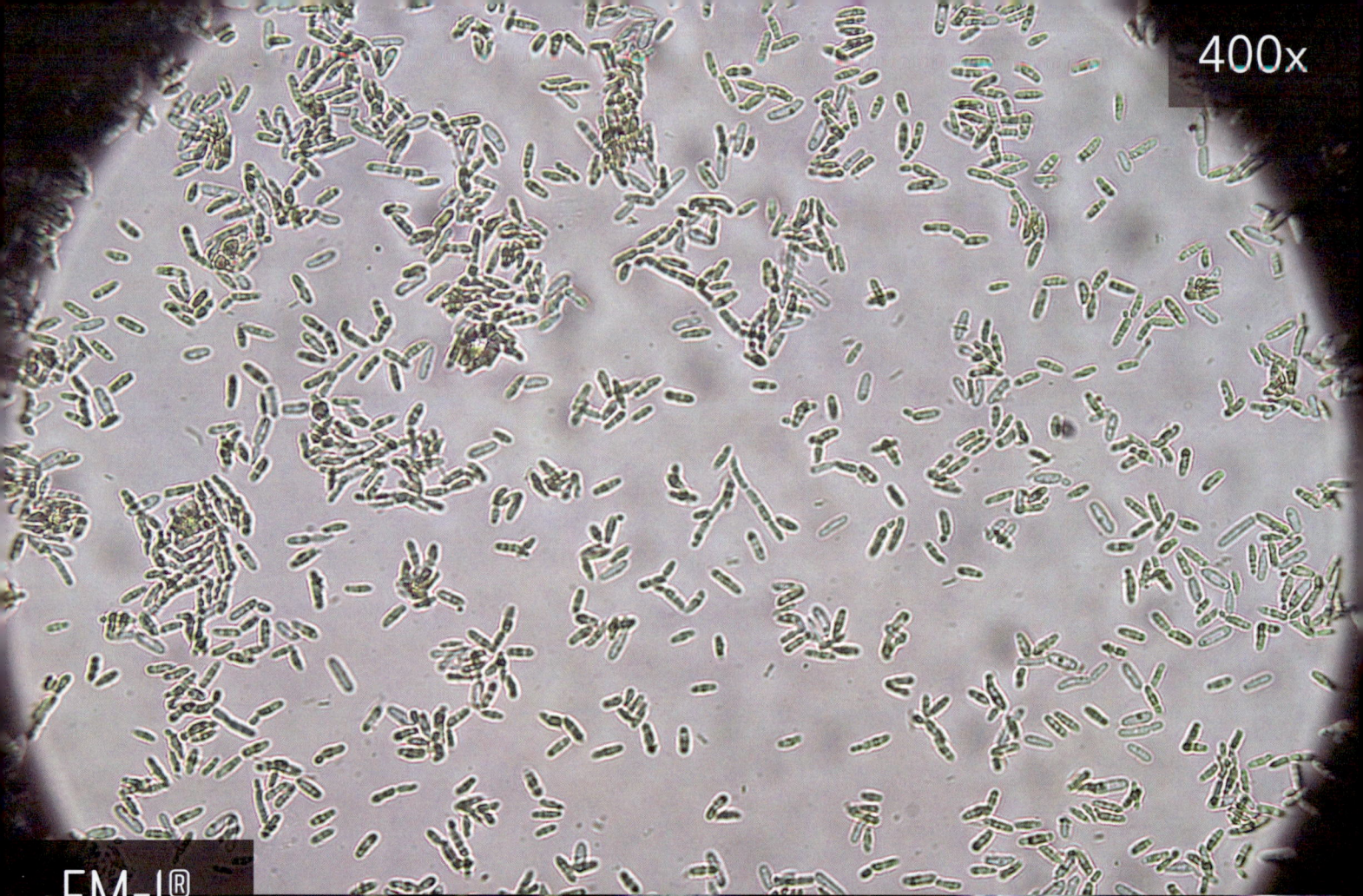
400x

EM-1®

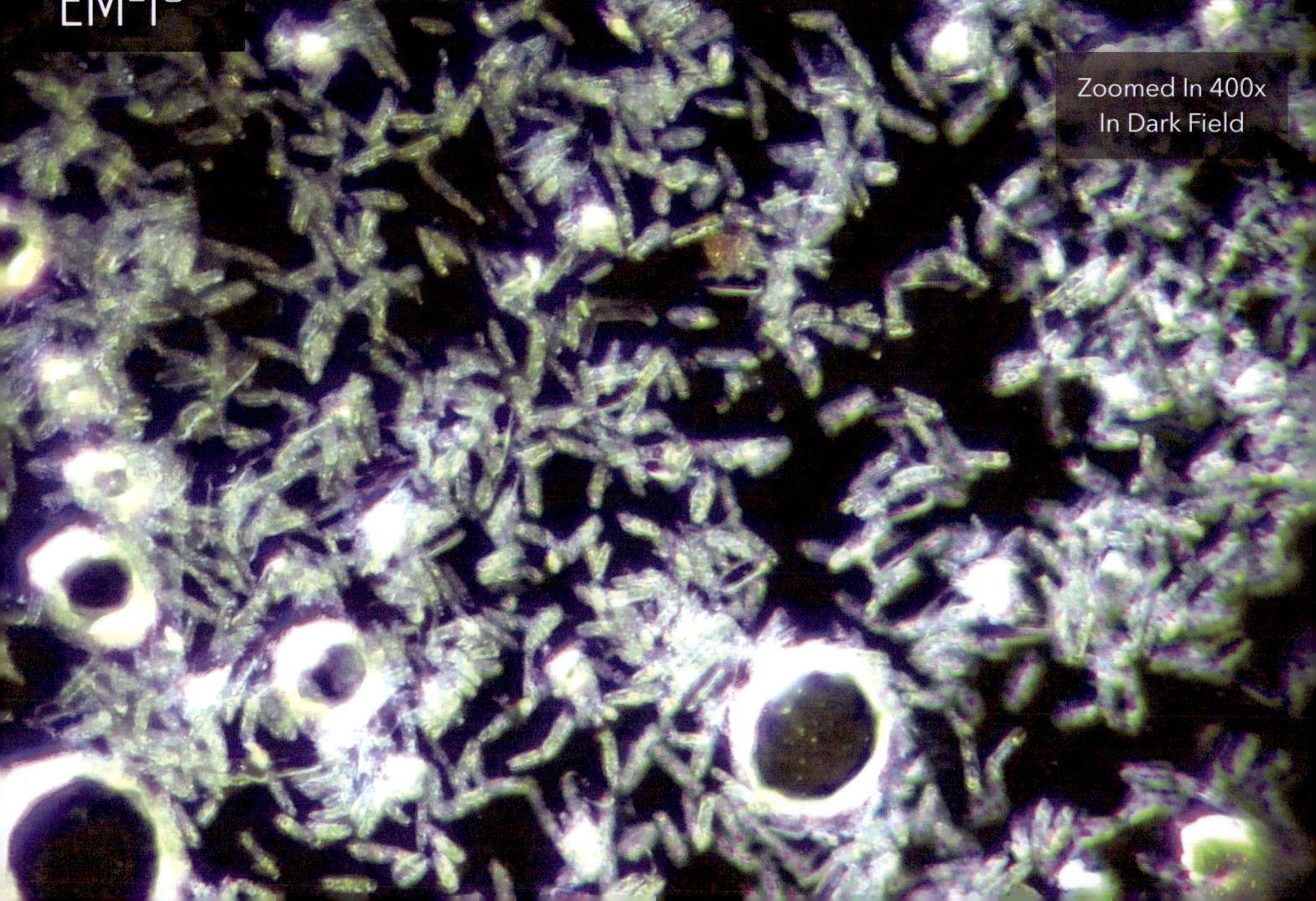
Zoomed In 400x
In Dark Field

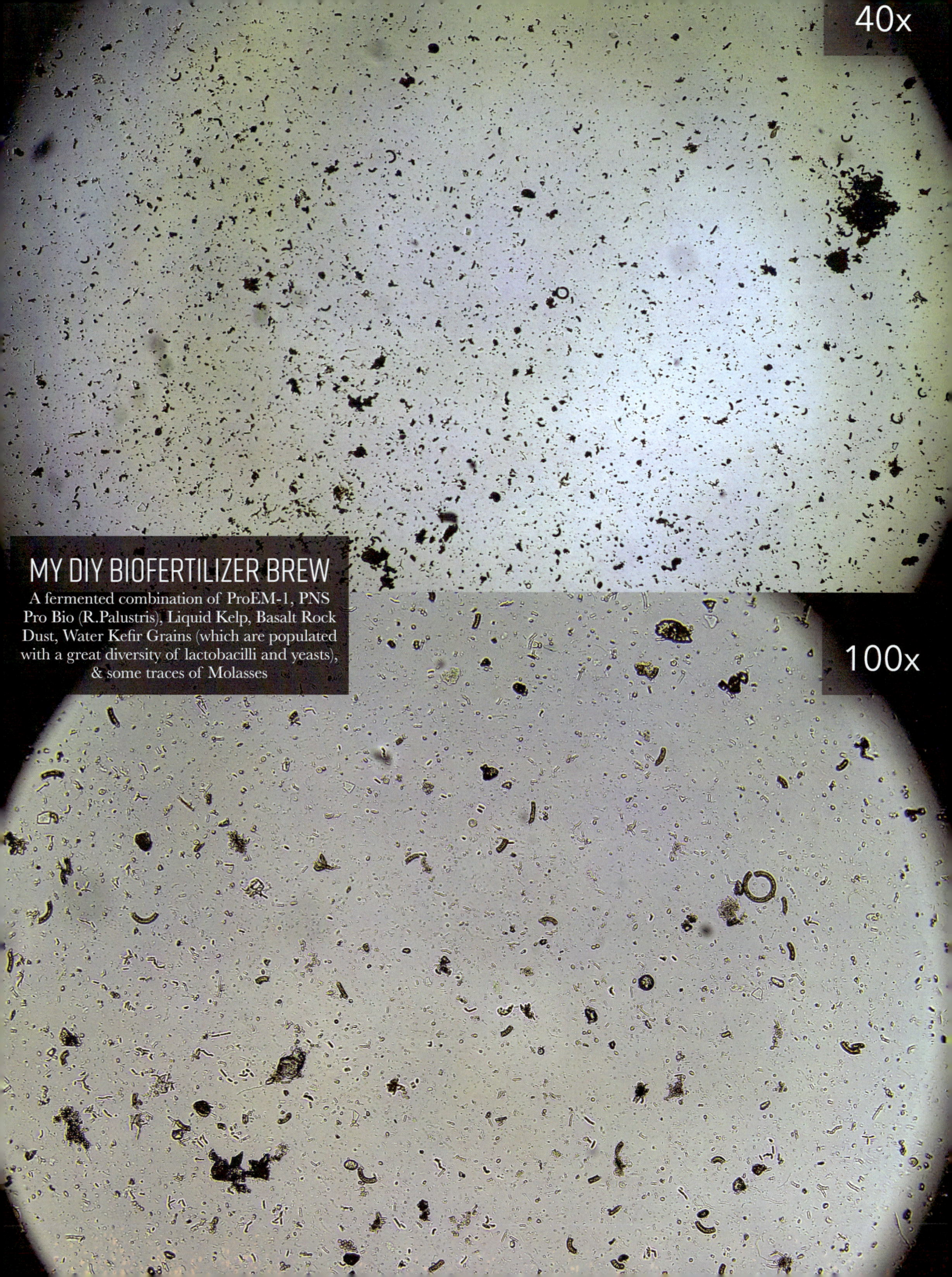

40x

100x

MY DIY BIOFERTILIZER BREW

A fermented combination of ProEM-1, PNS Pro Bio (R.Palustris), Liquid Kelp, Basalt Rock Dust, Water Kefir Grains (which are populated with a great diversity of lactobacilli and yeasts), & some traces of Molasses

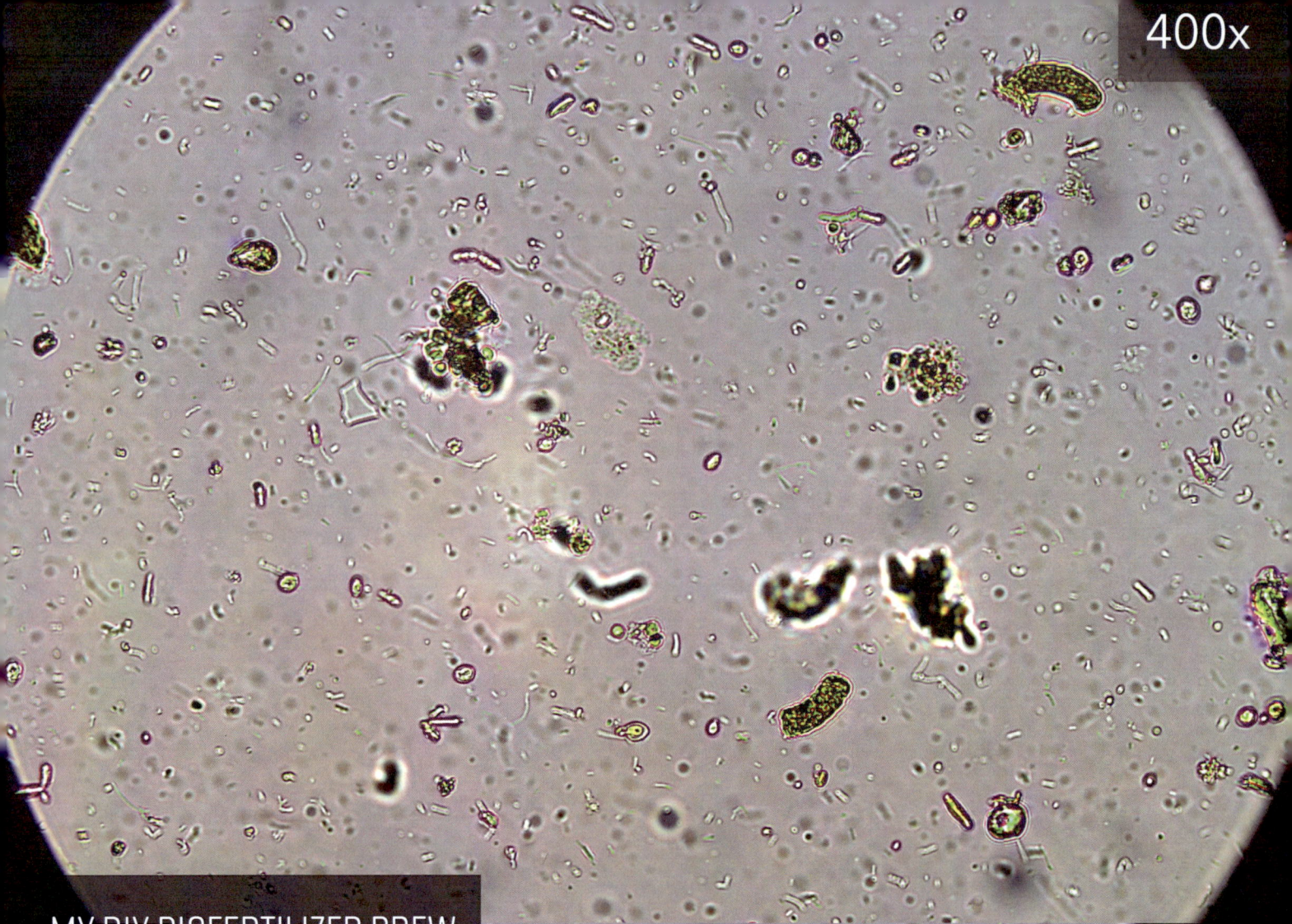

400x

MY DIY BIOFERTILIZER BREW

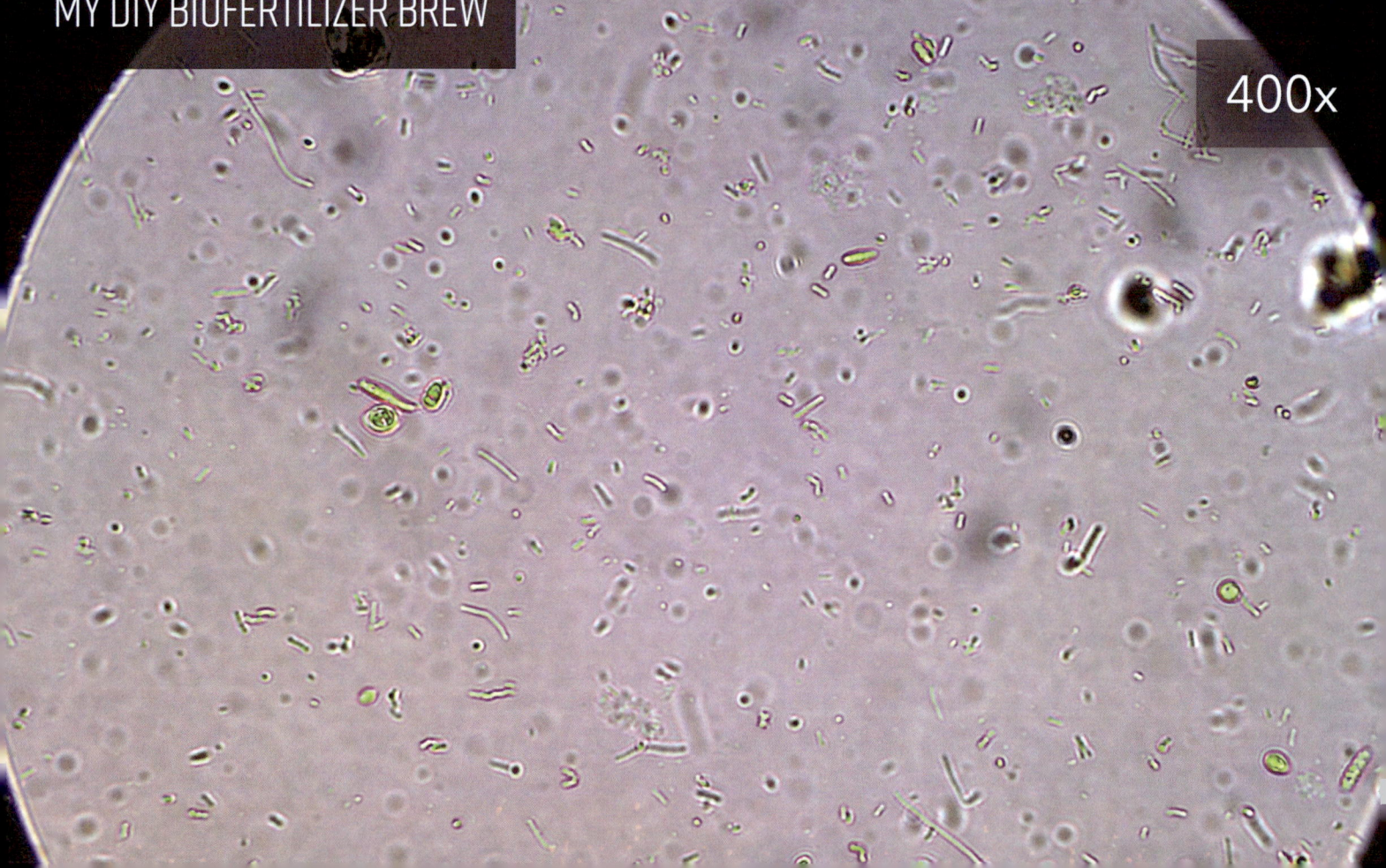

400x

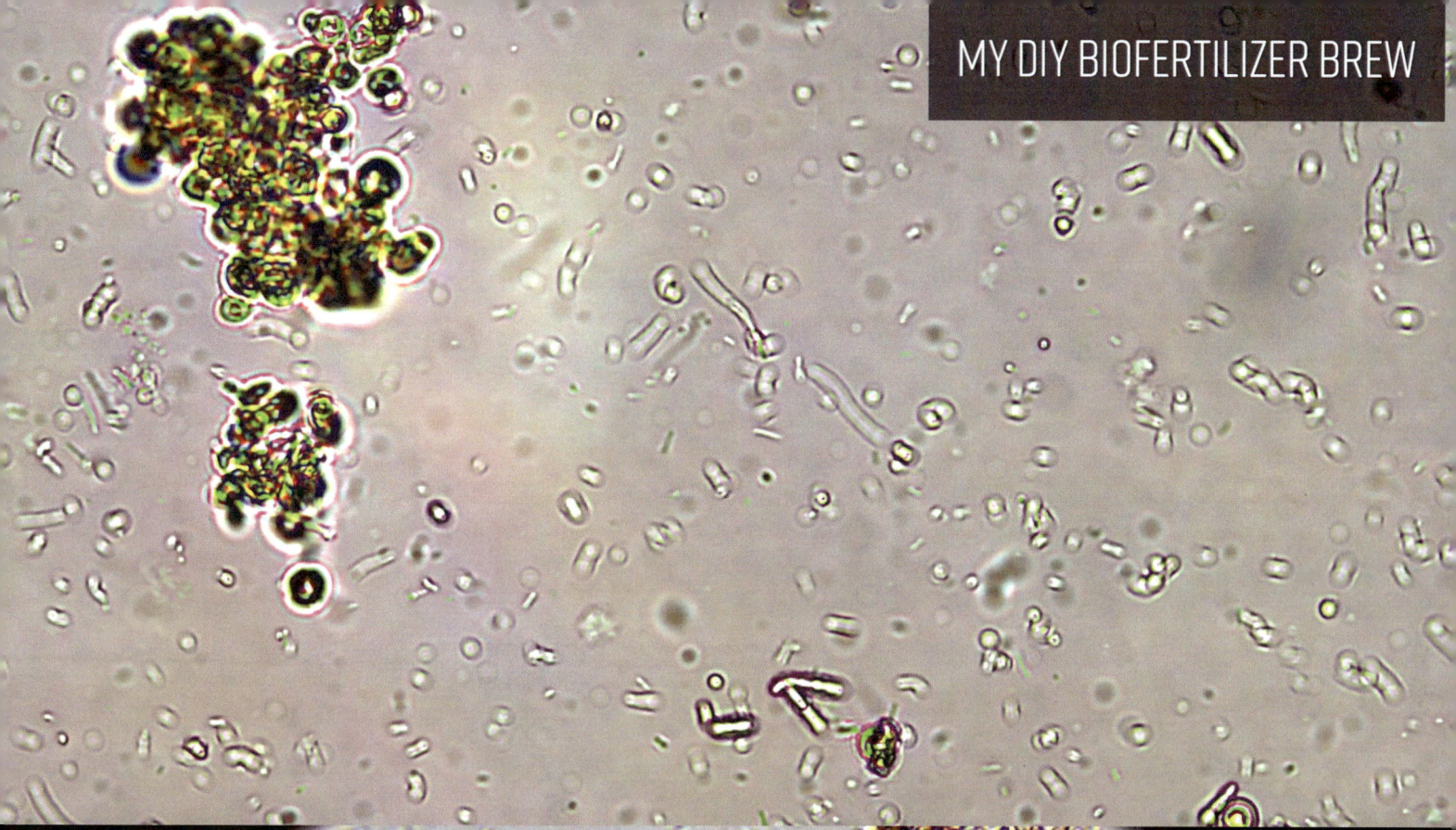

MY DIY BIOFERTILIZER BREW

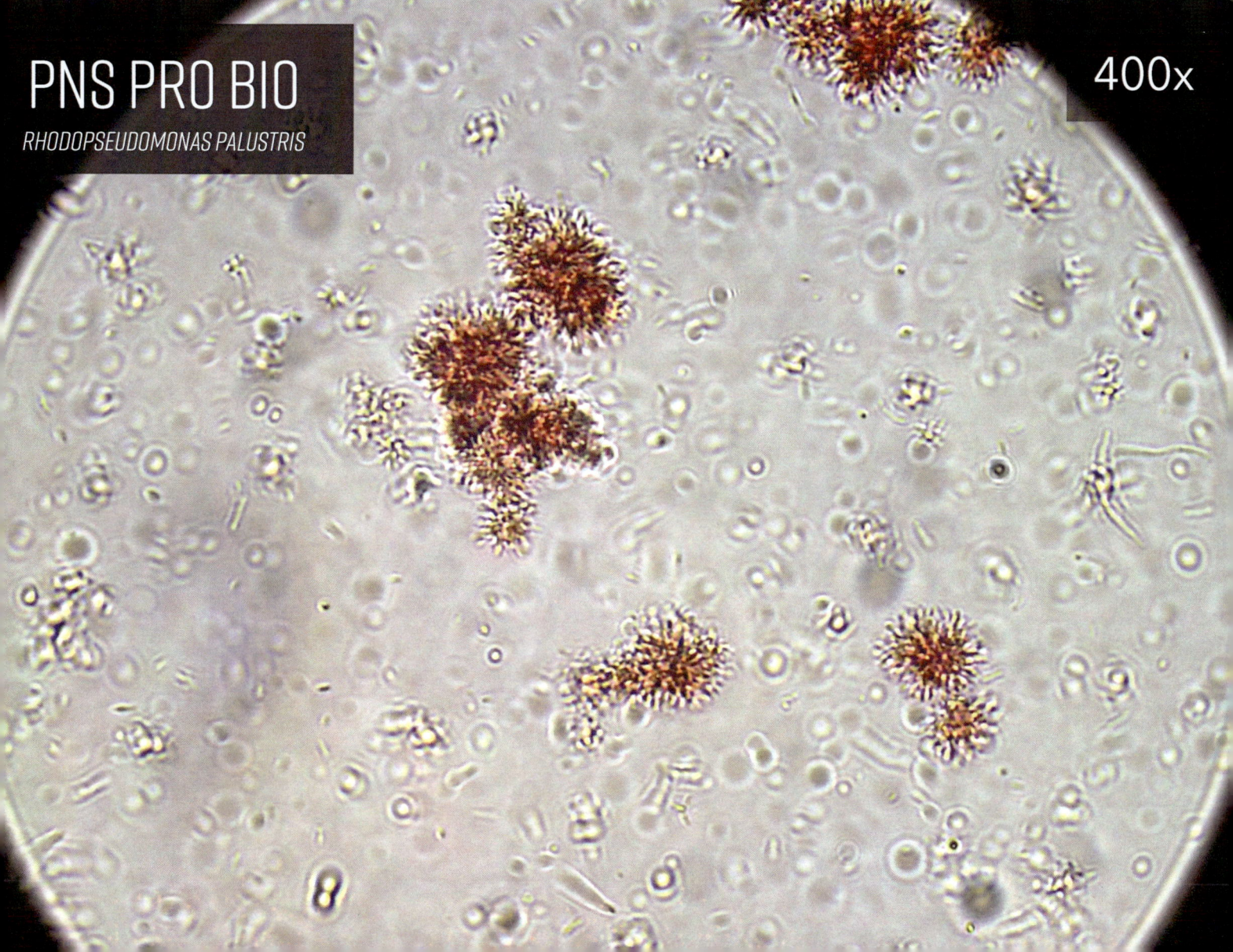

PNS PRO BIO
RHODOPSEUDOMONAS PALUSTRIS

400x

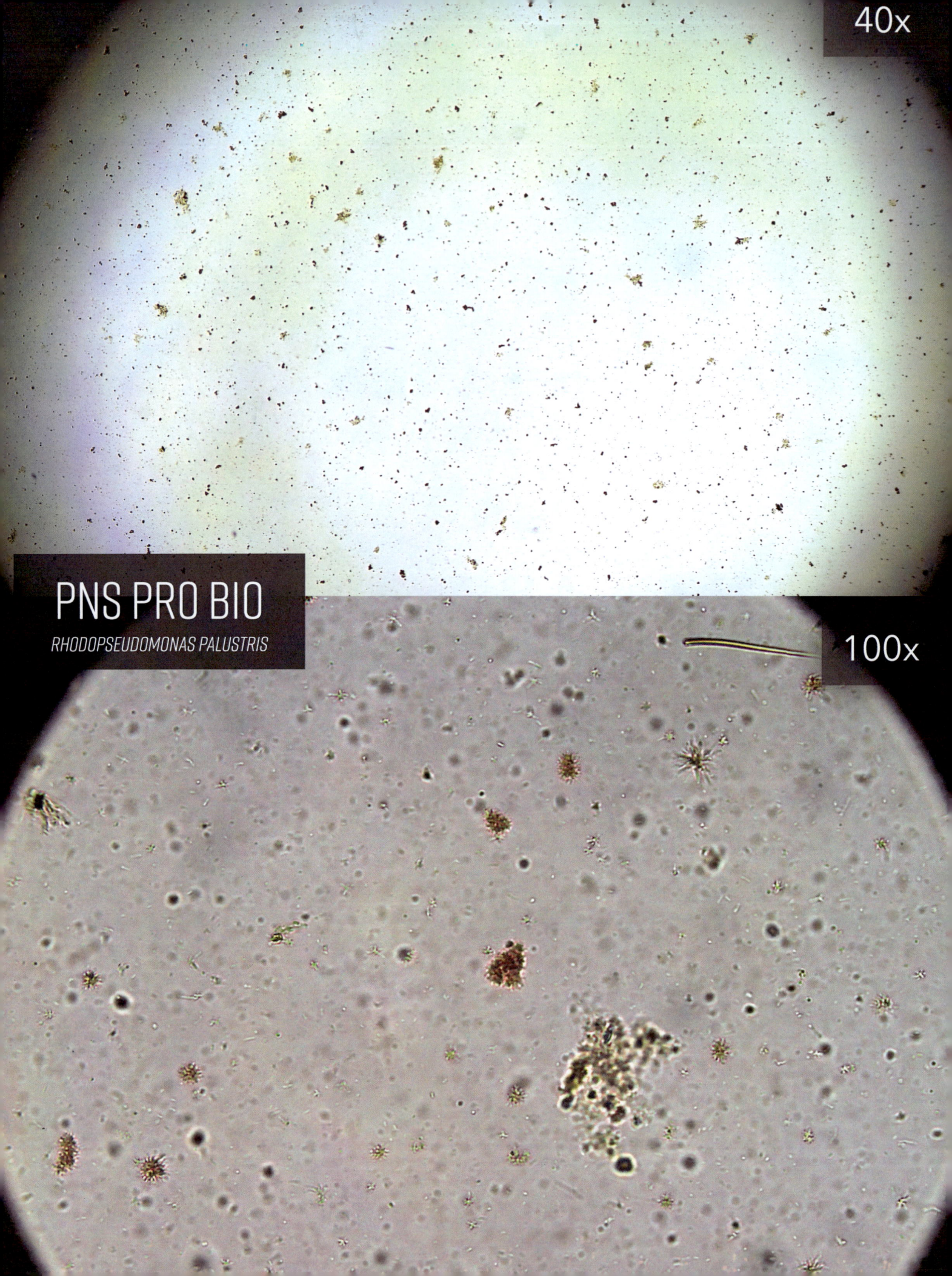

40x

PNS PRO BIO

RHODOPSEUDOMONAS PALUSTRIS

100x

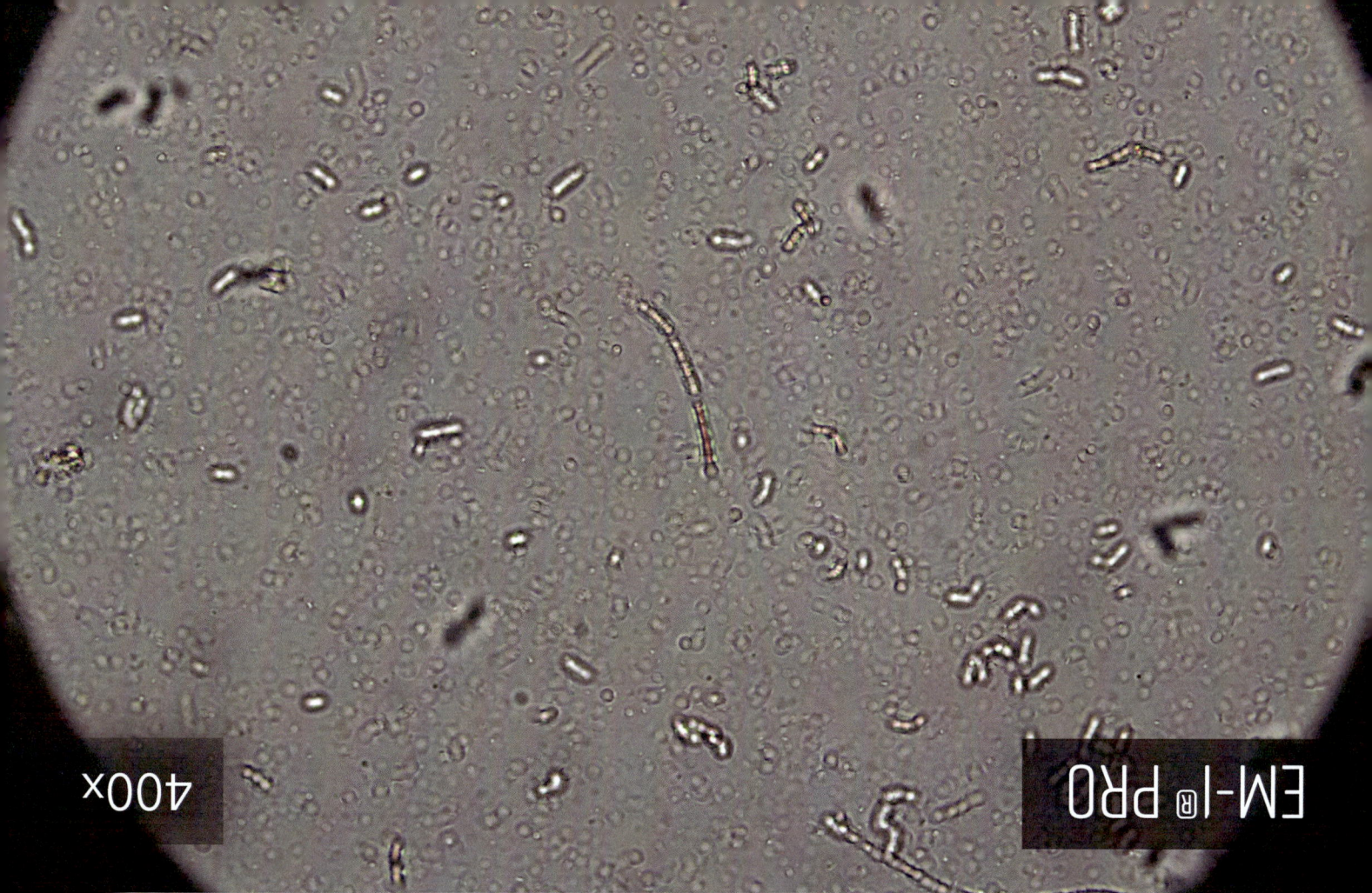

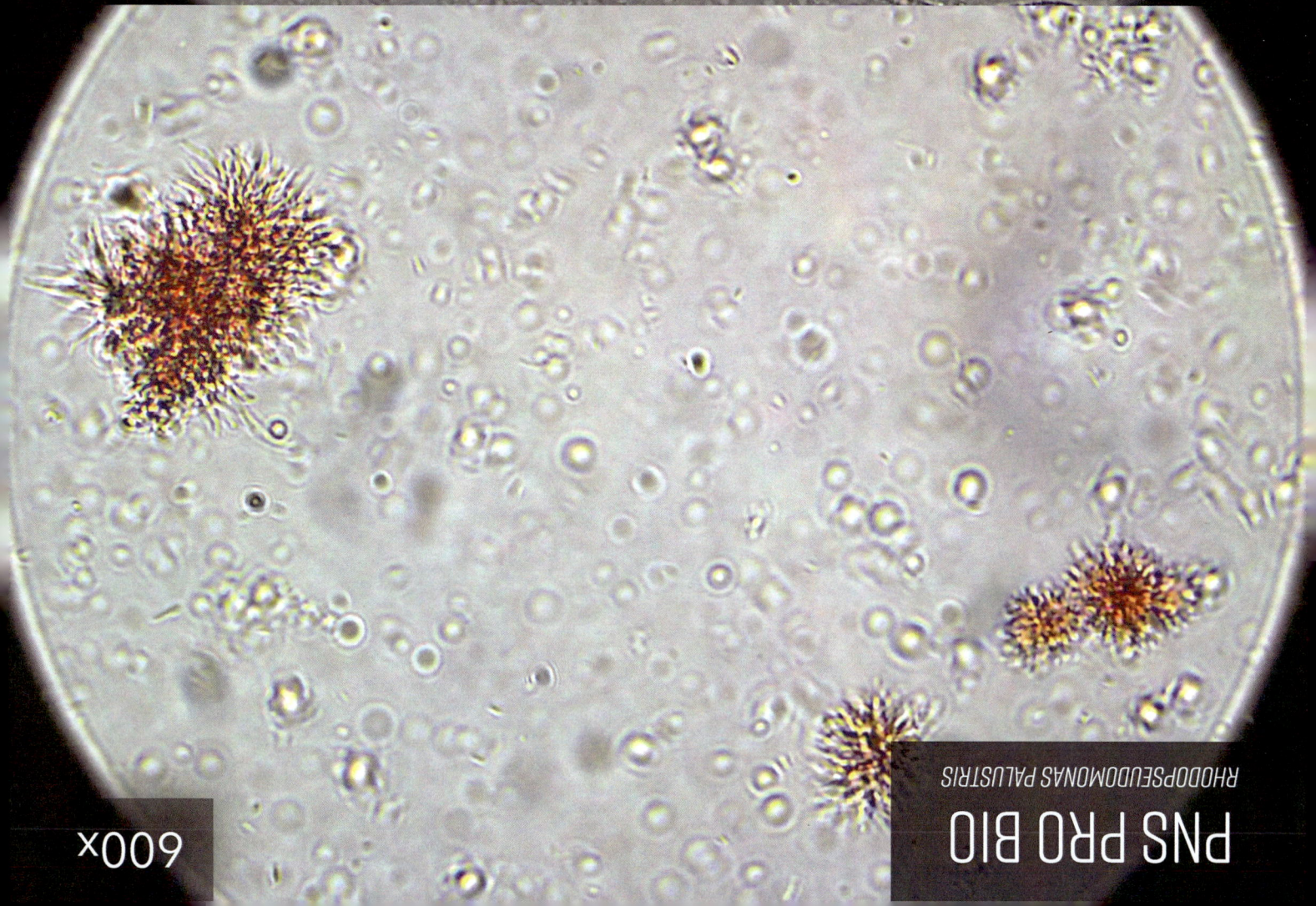

400x

EM-I® PRO

600x

PNS PRO BIO

RHODOPSEUDOMONAS PALUSTRIS

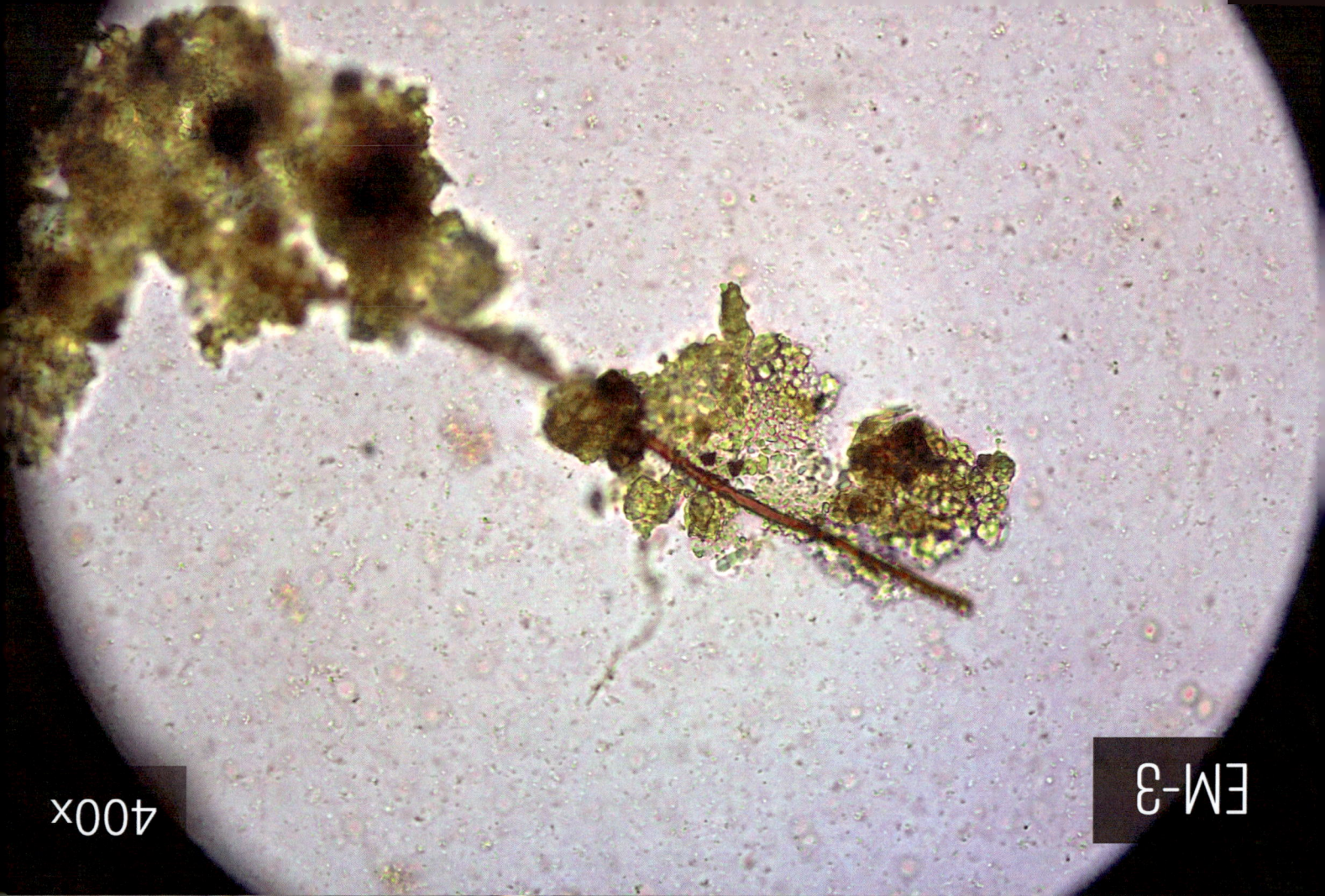

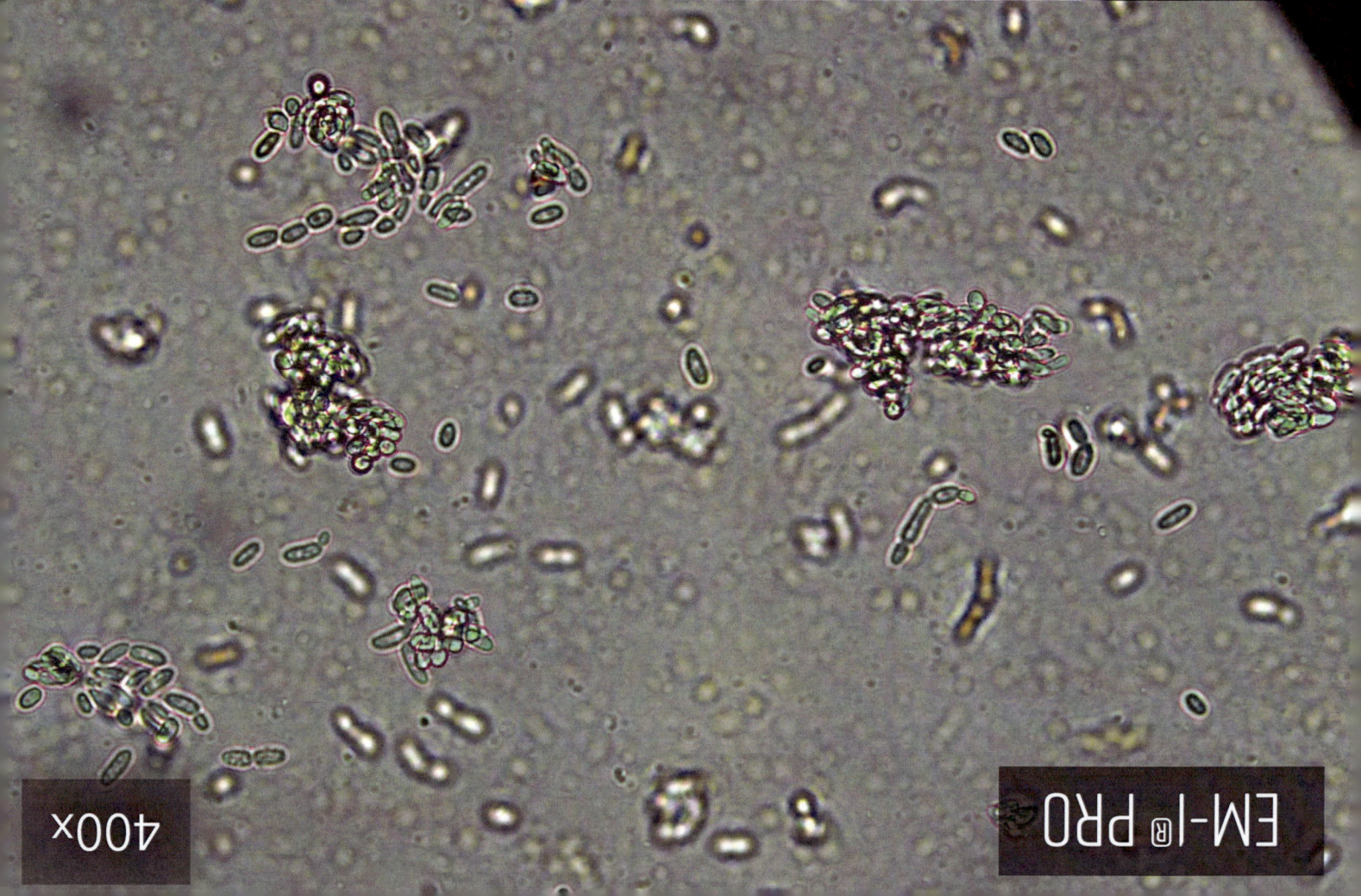

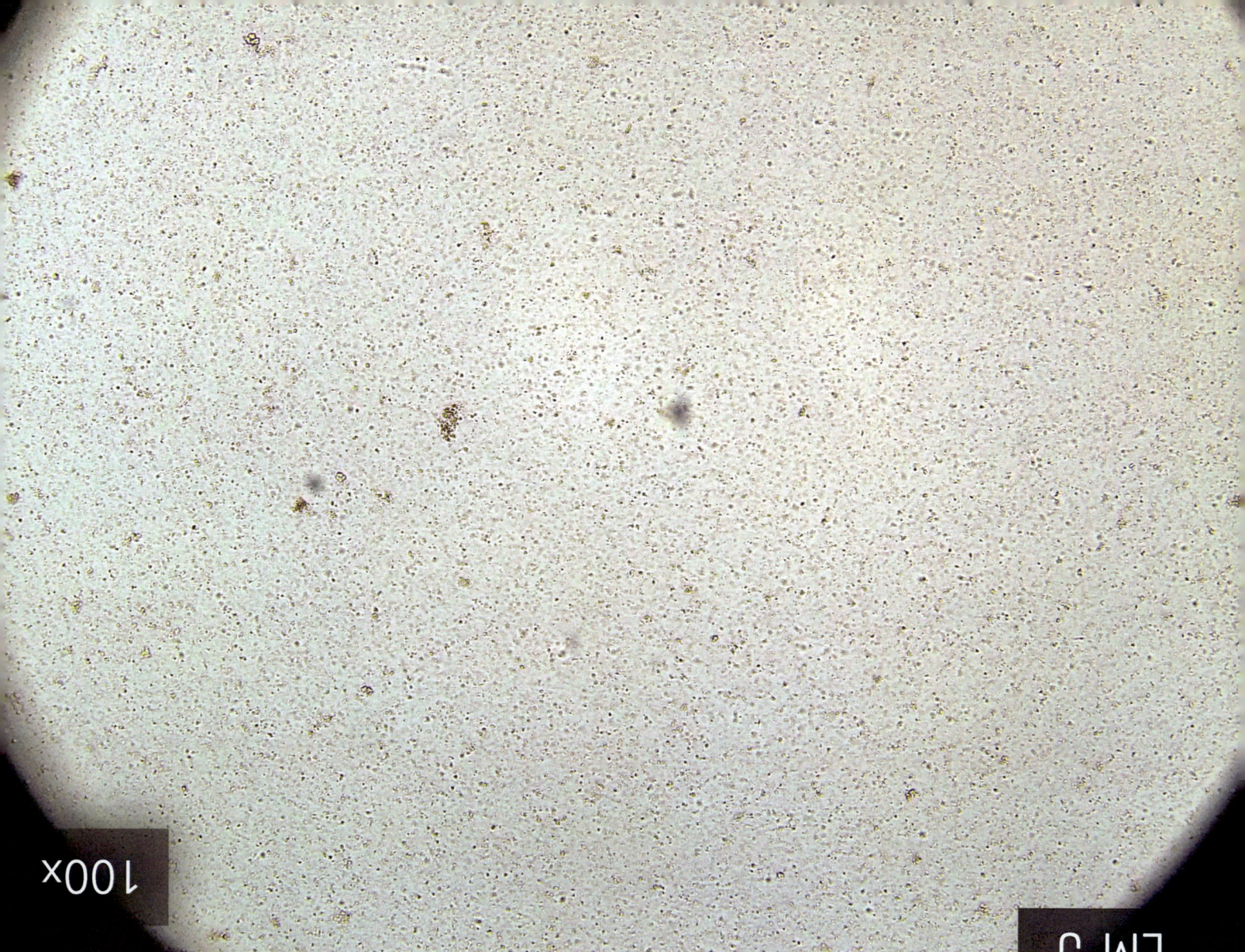

100x

EM-3

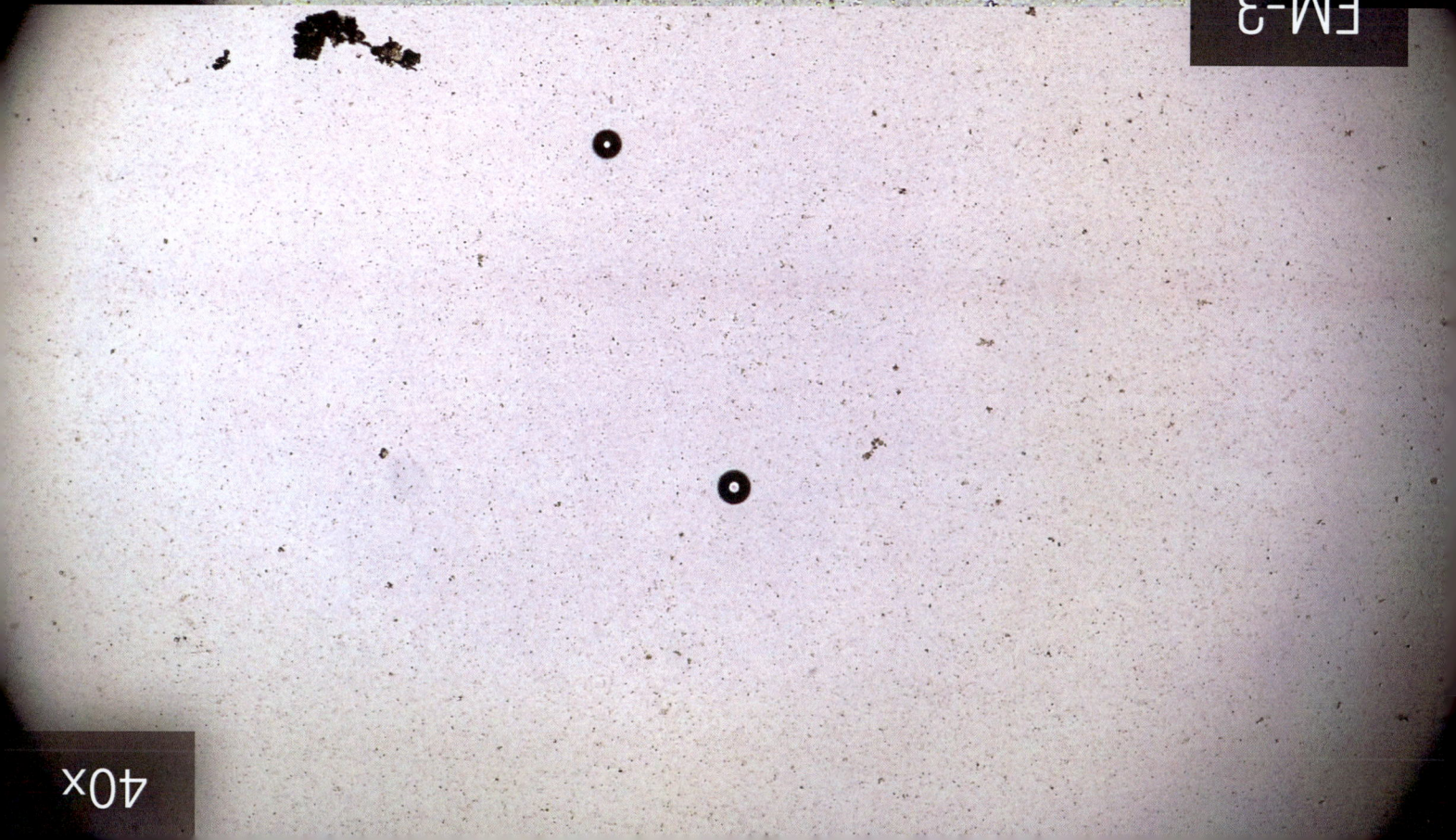

40x

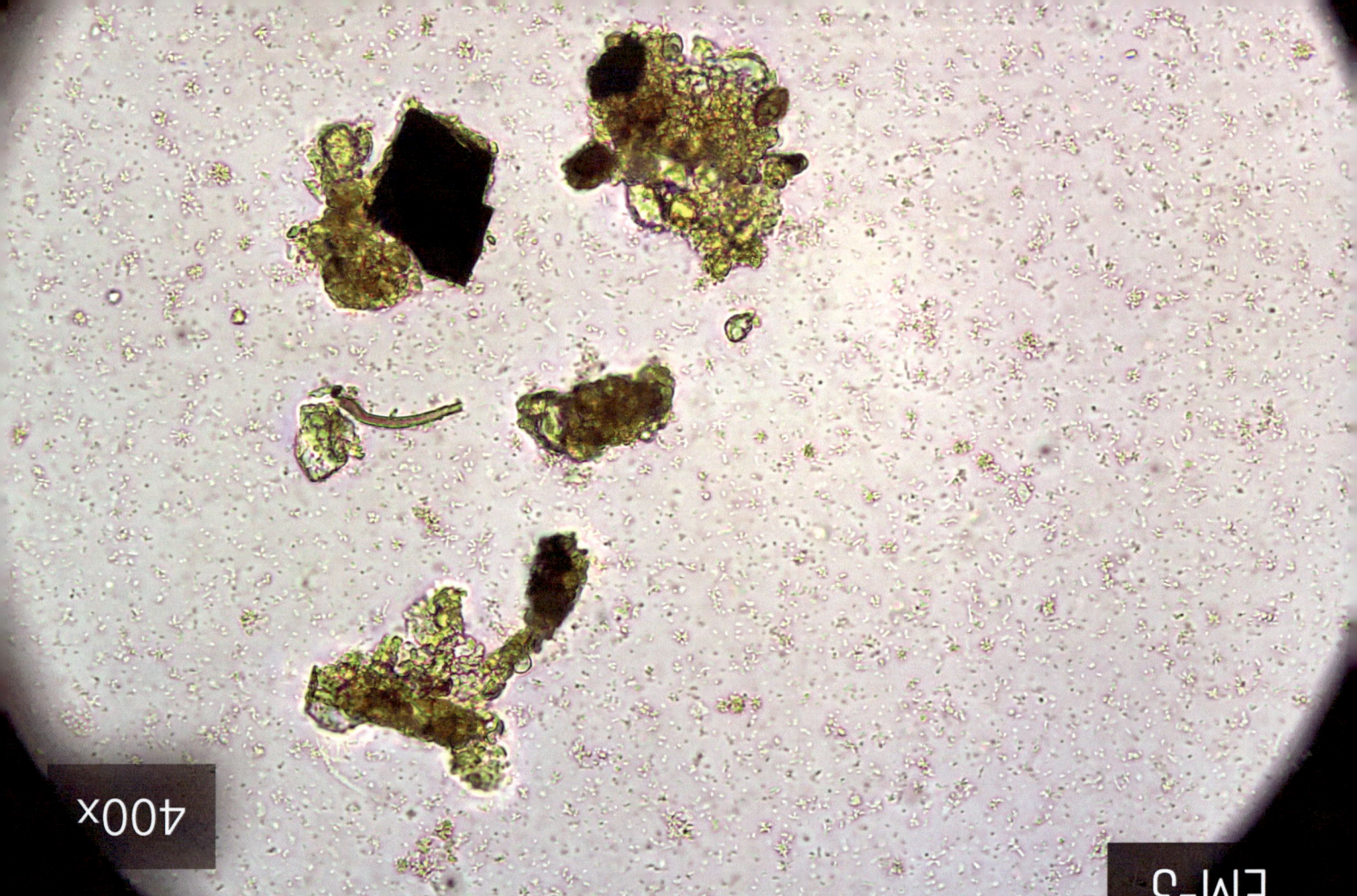

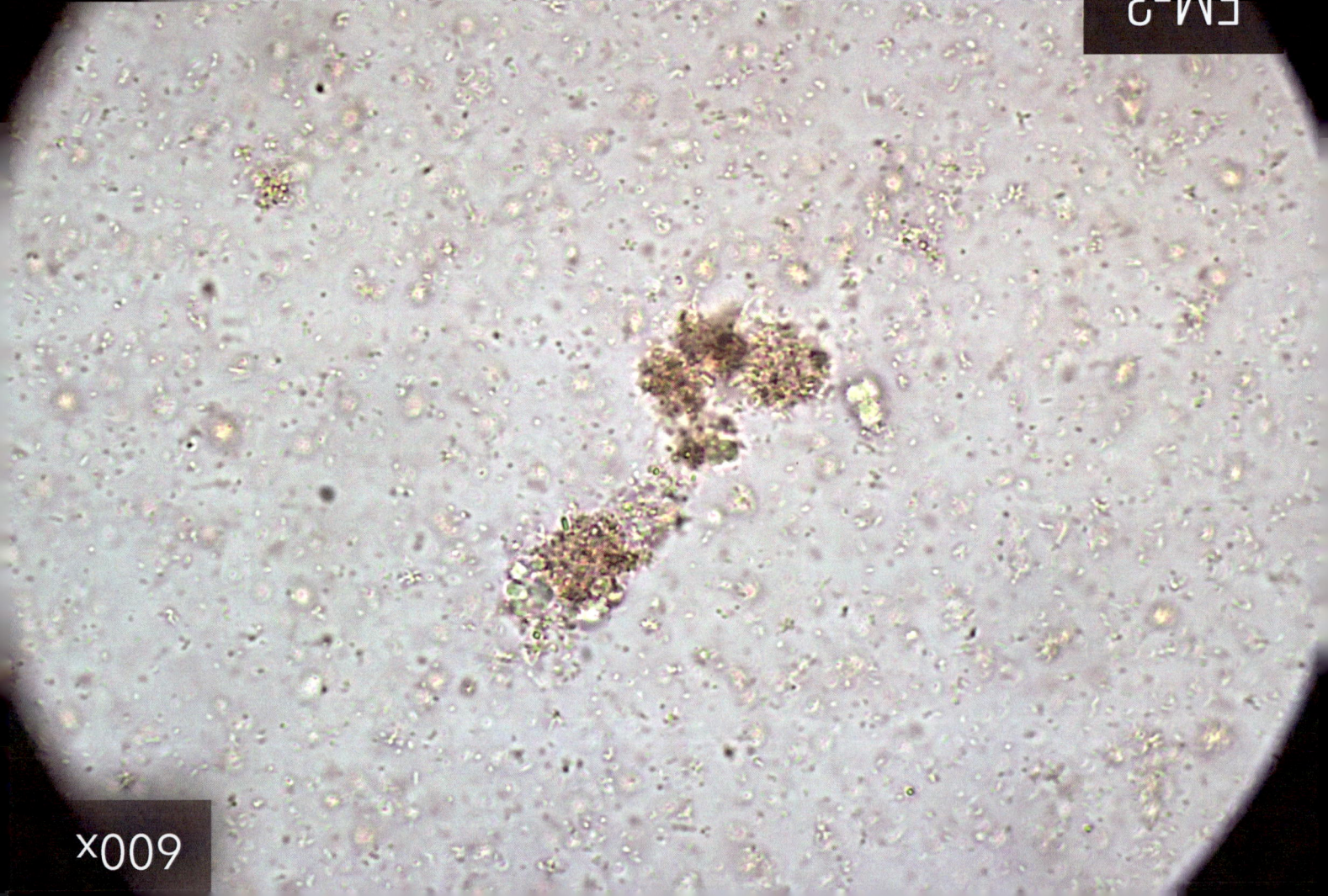

ORGANIC MATTER	BACTERIA	FUNGI	PROTOZOA	NEMATODES	N	P	SALINITY	REDOX	pH
EM-1 PRO	0	3 per 0.0025mm² ~1.2 billion bacteria per ml of soil Sparse bacilli	Yeasts in every FOV in small clusters @ 400x at 1:100 dilution	0	0	Trace/ Zero	Zero	1260 ppm 0.12%	407 RmV	3.5
DIY BIOFERTILIZER	0	9 per 0.0025mm² ~3.6 billion bacteria per ml of soil Diverse Cocci, Bacilli, & Lactobacilli A LOT of actinobacteria - the most out of any sample High motility & HGT activity	Yeasts in every FOV in multiple clusters @ 400x at 1:100 dilution	0	0	Trace	Zero	2360 ppm 0.23%	350 RmV	3.5

we can add these microbes alone or in groups to our composts and soils to boost the foundational layer of the microbial food web.

One of the most astonishing things about EM® is how little is known and understood about these microbes yet they've been studied in science for hundreds of years (yeast in particular). Most folks stop at the name "EM®" which is trademarked only – the microbes themselves are free to be used by anyone commercially. When we test these brews and other biofertilizer brews, we can verify if we are growing the right microbes – thus, we can verify the claims made by products as well as the methods we are using. I DNA sequenced EM-1® and found there was only 1 count of rhodopseudomonas palustris in over 10,000 reads – not very promising, but then again if we read the bottle, it only guarantees lactobacillus and yeast being present. That's why I've recommended and have used Pro EM-1® which is designed for human consumption (USDA organic approved) and has a guarantee that rhodopseudomonas palustris is also included. Under the microscope, we can see that Pro EM-1® has very little yeast and EM-3 has no yeasts, so a microscope is very handy when we want to verify what we are actually working with.

A friend, the former VP of Teraganix – the US distributor of EMRO's Effective Microbes® line of products, recently sent me a bottle of EM-3 which is not sold in the US (it's all in Japanese). EM-3 is used to create EM-1® according to my friend from Teraganix. This makes sense since it is rich in purple nonsulfur bacteria (rhodopseudomonas palustris) and has the telltale high pH 8 associated with brewing those microbes. This is likely the main reason PNSB disappears with extensions of EM-1® or EM-1 Pro® over time as they are acidic brews that reach 3.4 – 3.5 pH (quite the opposite of what EM-3 is). This is to be expected as all the PNSB cultures I've collected and tested were pH 8–9 range, smelled absolutely horrendous, and had the same telltale red color (they aren't purple but red). It's nontoxic and nonhazardous but the smell is terrible – just be prepared for that!

	ORGANIC MATTER	BACTERIA	FUNGI	PROTOZOA	NEMATODES	N	P	SALINITY	REDOX	PH
EM-3	Mostly mineral but combined with organic matter too	7 per 0.0025mm² ~2.8 billion bacteria per ml of soil Mostly Bacilli	Sparse fungal Hyphae connected to the organic matter and minerals 0.1 fungal hyphae per FOV @ 400x	0	0	Trace	High	3120 ppm 0.31%	-5 RmV	8
PNS PRO BIO	0	3 per 0.0025mm² ~1.2 billion bacteria per ml of soil Appears to be just bacilli and mostly if not entirely R.Palustris	0	0	0	Zero	Trace/ Zero	Med 896 ppm 0.08%	-90 RmV	9.0

How is this all useful?

If we can develop our eye to recognize when our piles are ready to use and to what purpose we can put them because we can recognize the biology and diagnose the health of the nutrient cycles, we are miles ahead of ourselves without those skills, and the same applies for biofertilizers like EM® or our own DIY brews: you want to know that you're getting a consistent and predictable product, that you are applying it strategically so it has maximum efficacy and benefit, and that you aren't wasting time, effort, or money. Our DIY brews or our EM products should have essentially no nitrogen or phosphorus in soluble forms, so when you do a mineral test it doesn't show up – that doesn't mean it's gone: we can see the P embodied in some of the yeasts in the EM-1 epifluorescent image. That implies that all the P was soaked up by the microbes and internalized (and that some yeasts didn't get any!!), and all the nitrogen is now either amino acids or incorporated into their bodies as well.

We can check our brews for key species to our biofertilizers blends. We can develop our own benchmarks for numbers and ranges of microbes per FOV or per drop or per 0.05mm² of a hemocytometer inner grid. We can monitor our work over time and maintain the standards we establish and catch things early when they get imbalanced, but we have to look often, we have to establish a protocol we repeat, and we have to catalog our findings in a way we can compare them across our own timeline of findings and against those of others.

IMO STUDY

Not only can we view our roots, compost, and biofertilizer brews, but we can look at our natural farming indigenous microbe (IMO) collections. These are commonly, but not always, collected on undercooked white rice which combines the endophytic and spoilage microbes in and on the rice like zygomycota (likely Rhizopus and Mucor) and lactobacillus spp. (which survives and is stimulated by high heat) with the indigenous bacteria present in the soil and air around the collection box. By verifying that the fungi is teeming internally with bacteria, we can assume that it is nonpathogenic (as it has its own food source) – I learned this from Dr. James Francis White as he gave me feedback on this IMO review. There are also different colors and expressions throughout the lifecycle that indicate differences in the fungi present – we can use keys to identify these to an extent, but being able to rule them out as pathogenic is the key. From that perspective, there's a timing to the collection process that is key: the fungi will always run through their food sources and then sporulate, so capturing your collection before that time is essential to harvest and promote the indigenous microbes – otherwise, you could just end up with the dominant fungi, their spores, and its surviving endosymbiotic bacteria (which may or may not be significant and may or may not be indigenous).

Comparing provides the most insight – we can easily see that these two collections are different. They are both likely zygomycota, a group of spoilage fungi that look similar to each other. Basically, the IMO technique traps or selects for spoilage fungi that is hyper efficient at breaking down organic matter swiftly. We can see the vigorous saprophytic hyphae (lacking septa and surging with bacteria), and we can see their conidia and conidiospores everywhere! What else do you see? Do you see places where the bacteria are escaping the fungi? Bacteria is mostly contained in these images – spores instead dominate the FOV. Zygomycota fungi are known primarily by their teleomorphs and only recently were their sexual stage forms discovered, and only for the most common at that – there is a huge majority of zygomycota that we have no idea what their sexual stage looks like. The classic green mold, Trichoderma, has a teleomorph that is a wood saprophyte, Hypocrea lixii, which can be infection-yellow, orange, brown, or black!

IMO-1 Collection #1
from forestland in Eastern Texas, USA

IMO-1 Collection #2
from forestland in Eastern Texas, USA

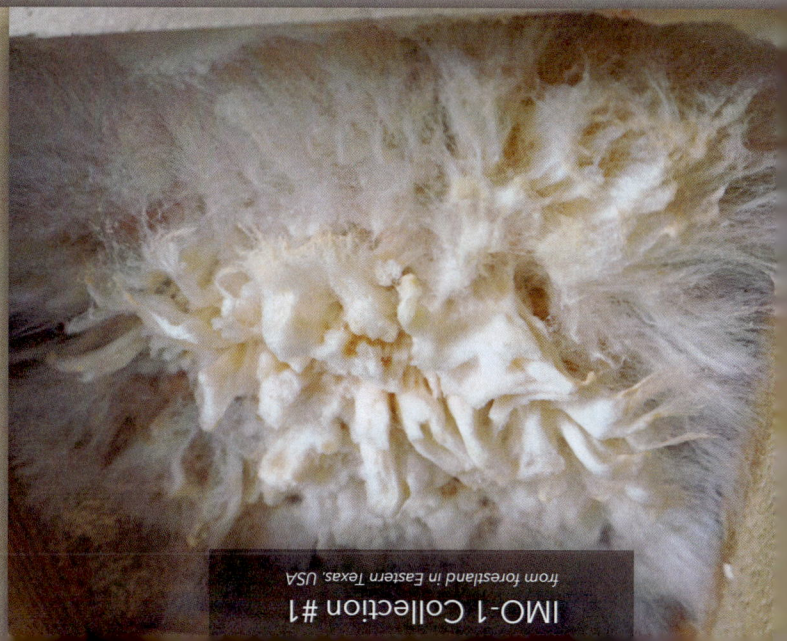

Matt Powers © 2023

100X

400X

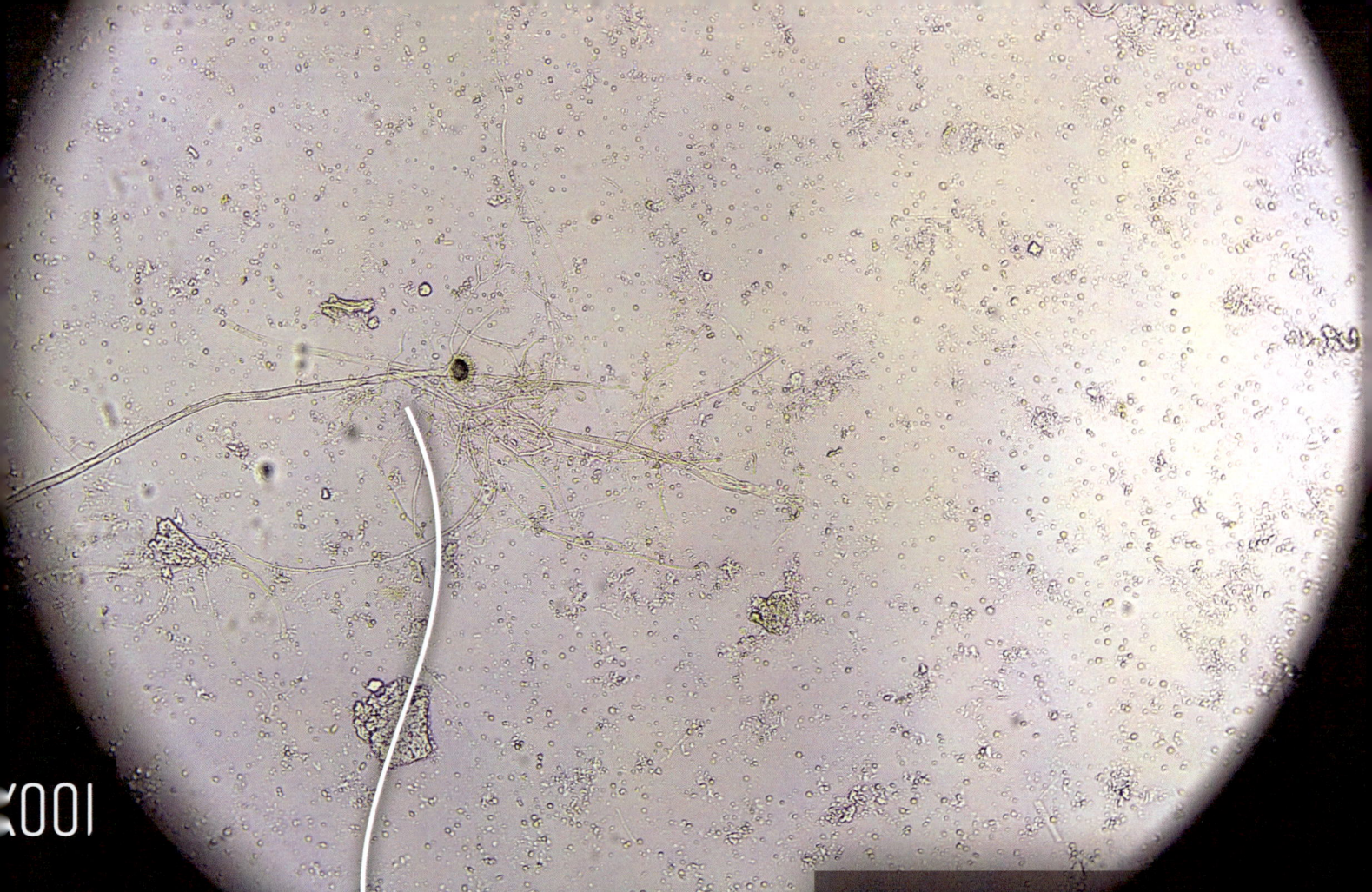

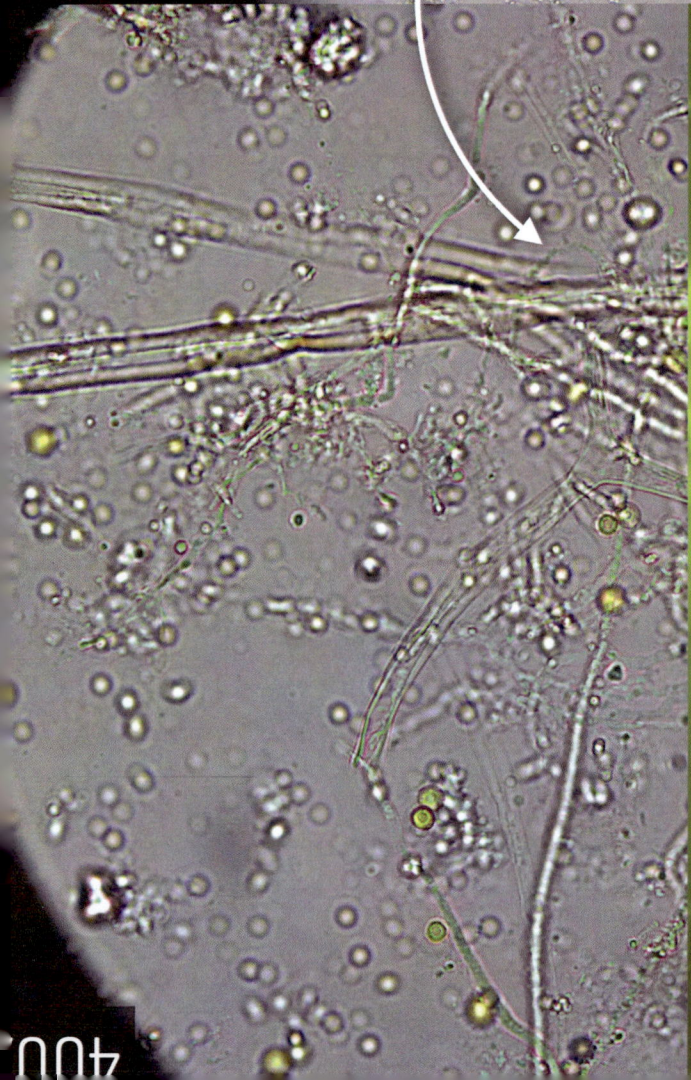

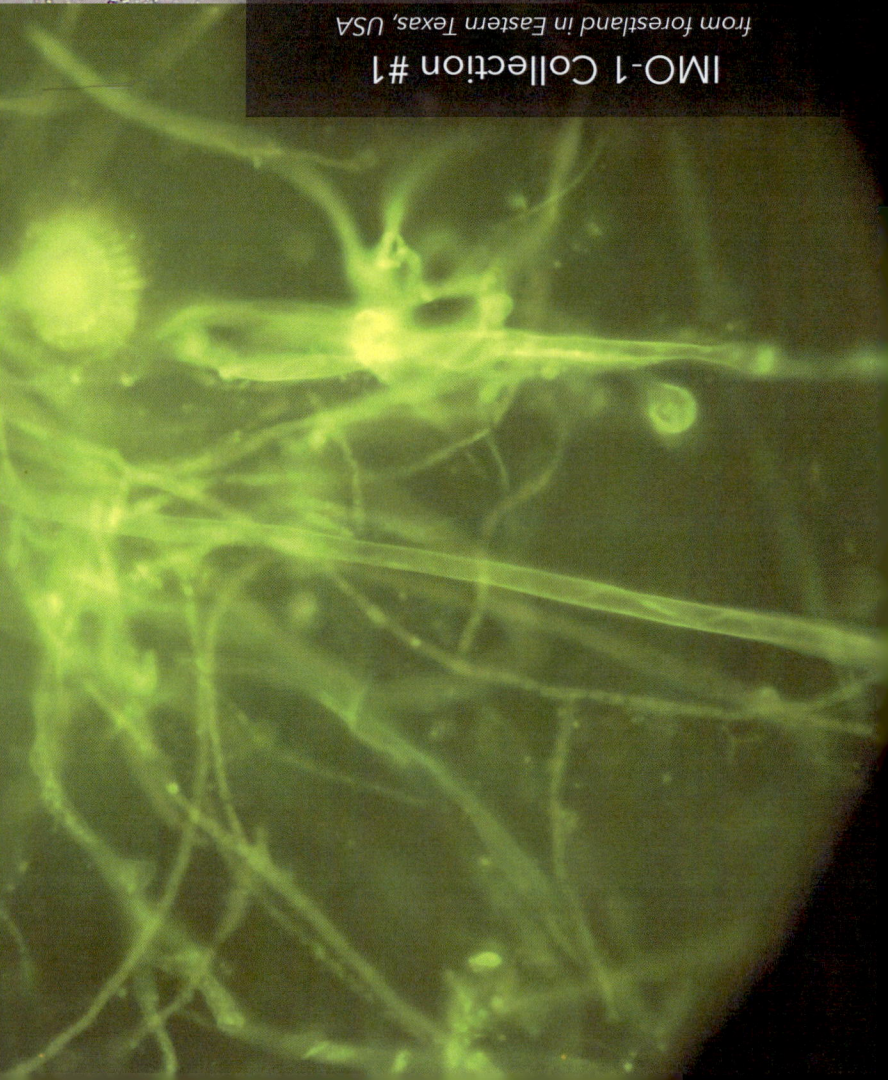

IMO-1 Collection #1
from forestland in Eastern Texas, USA

40X

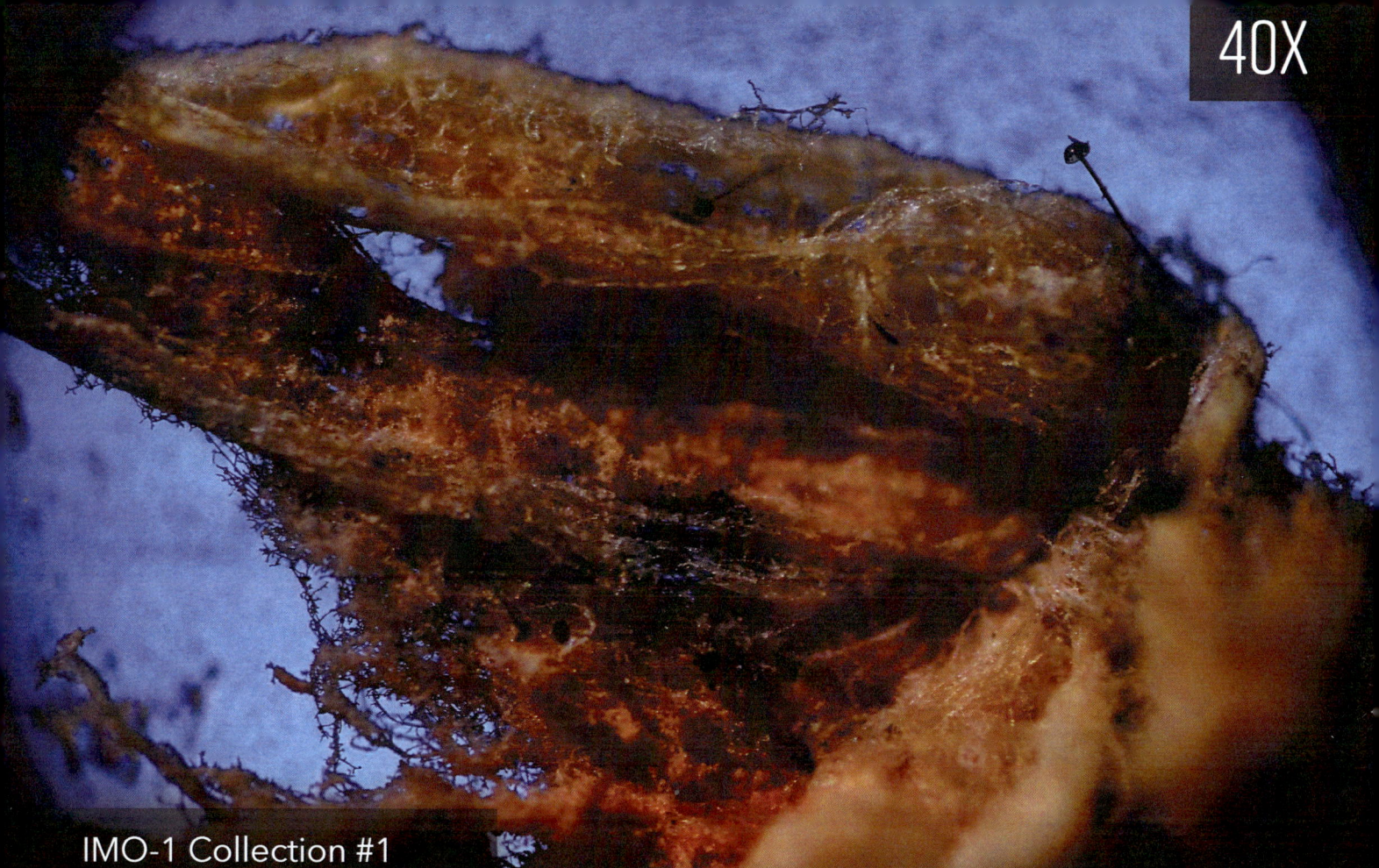

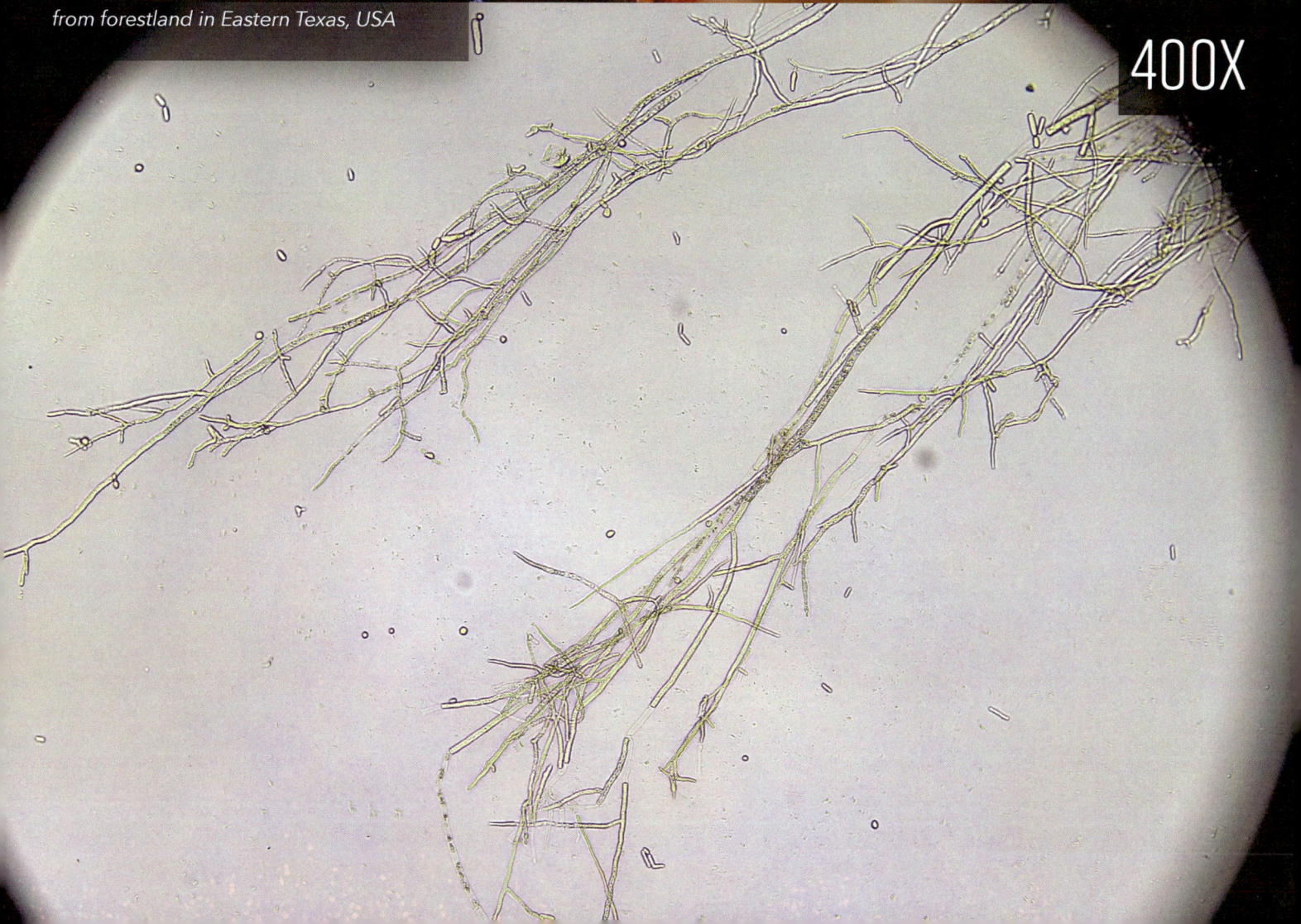

IMO-1 Collection #1
from forestland in Eastern Texas, USA

400X

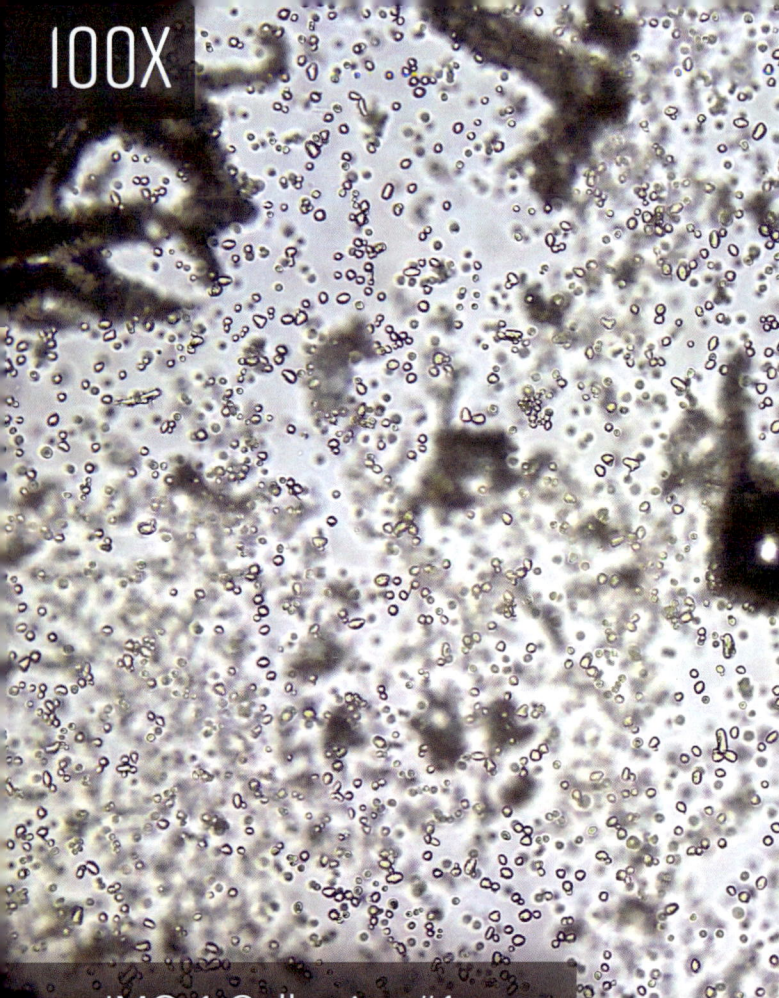

100X

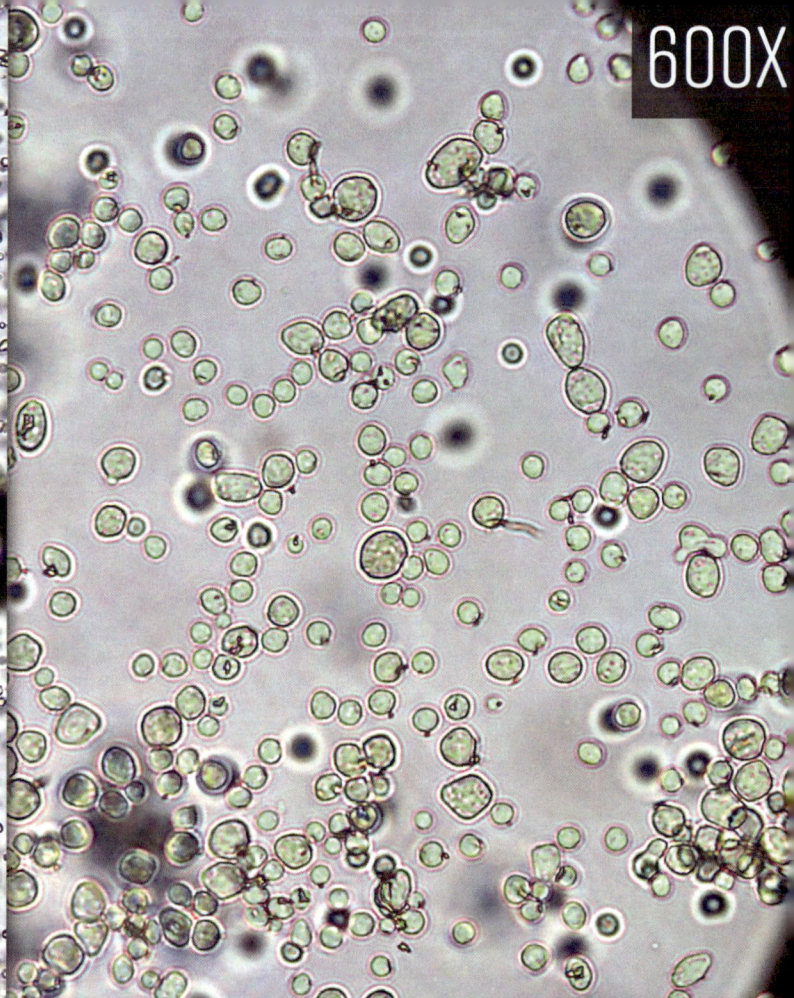

600X

IMO-1 Collection #1
from forestland in Eastern Texas, USA

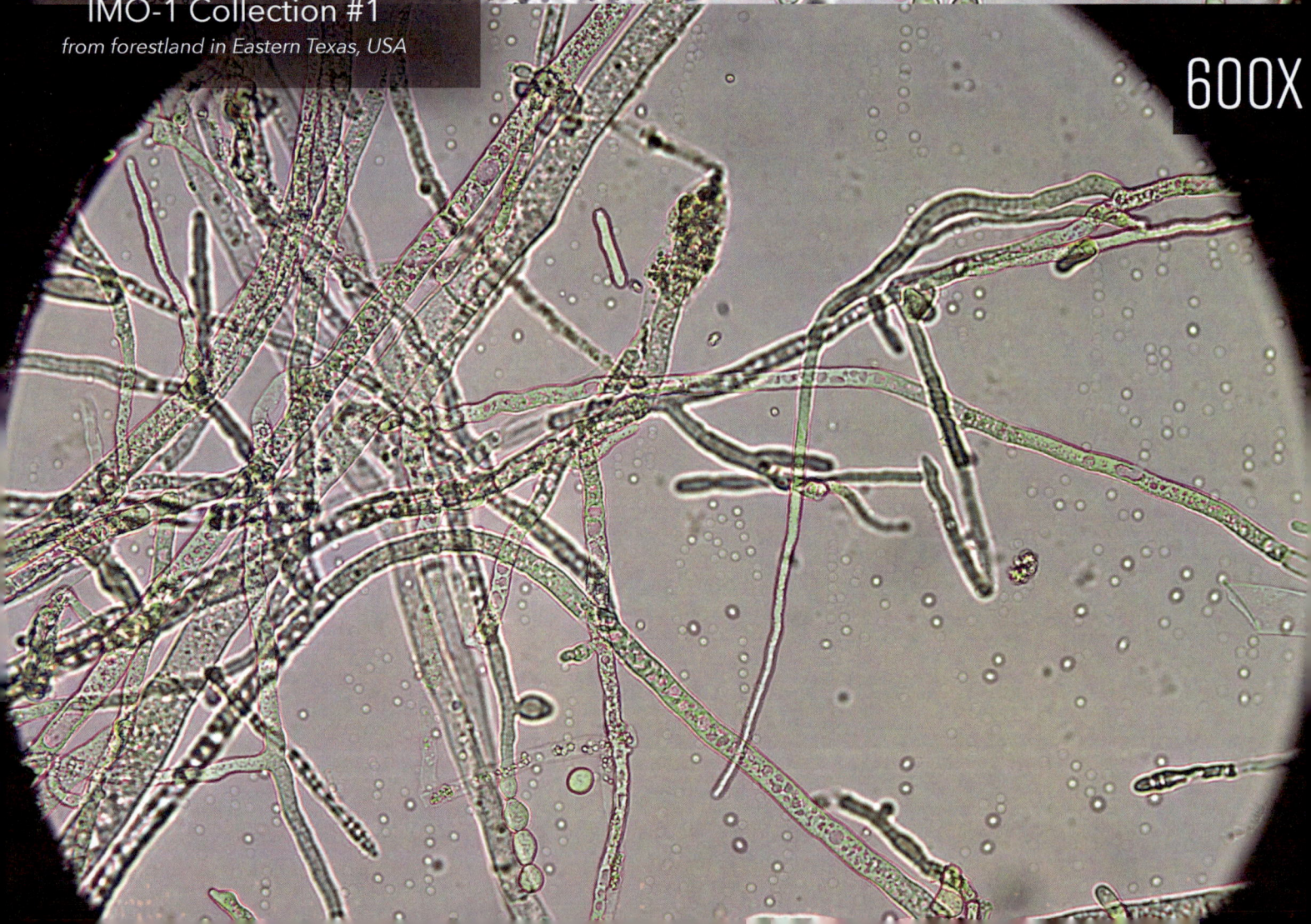

600X

600X

100X

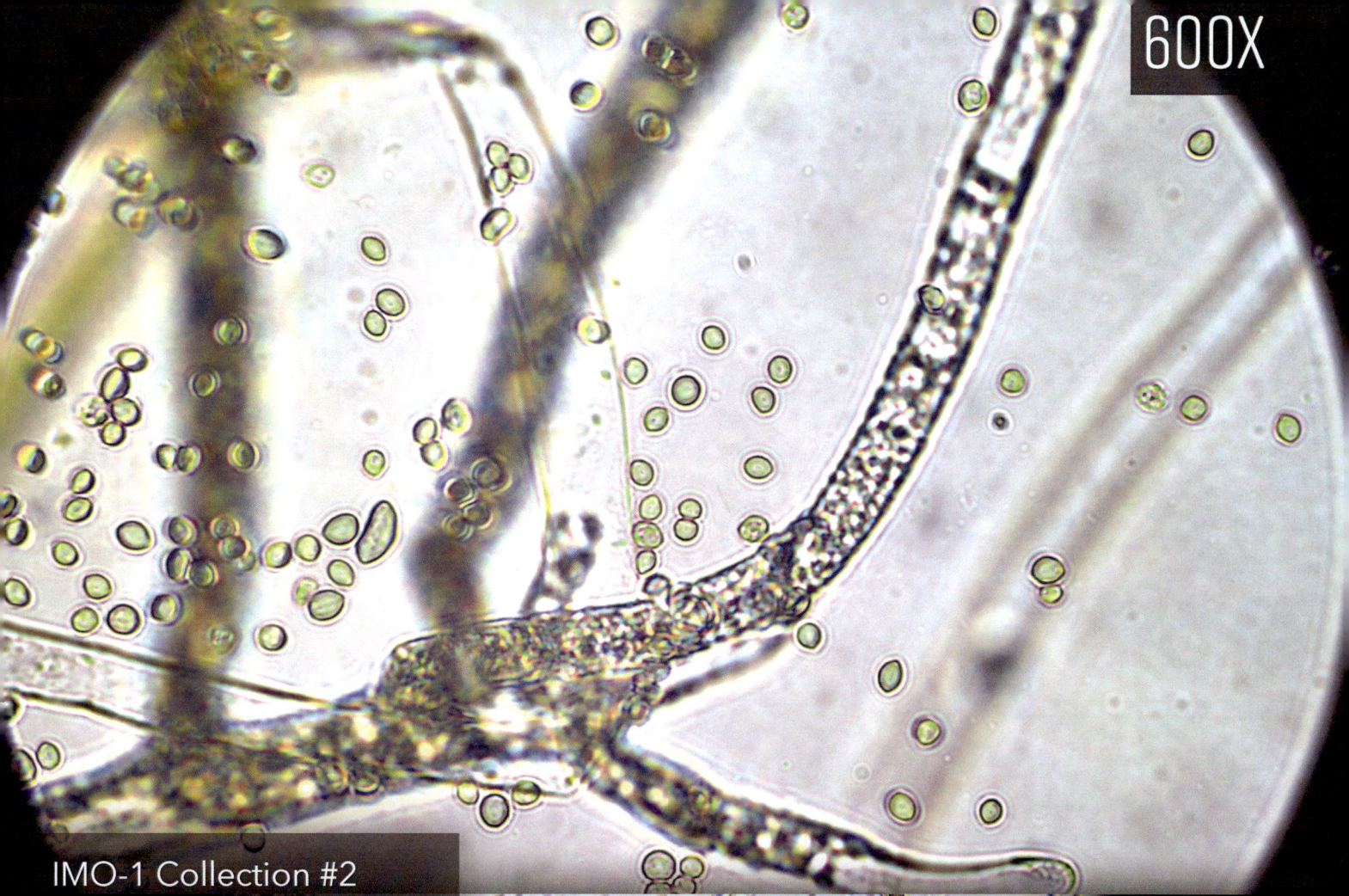

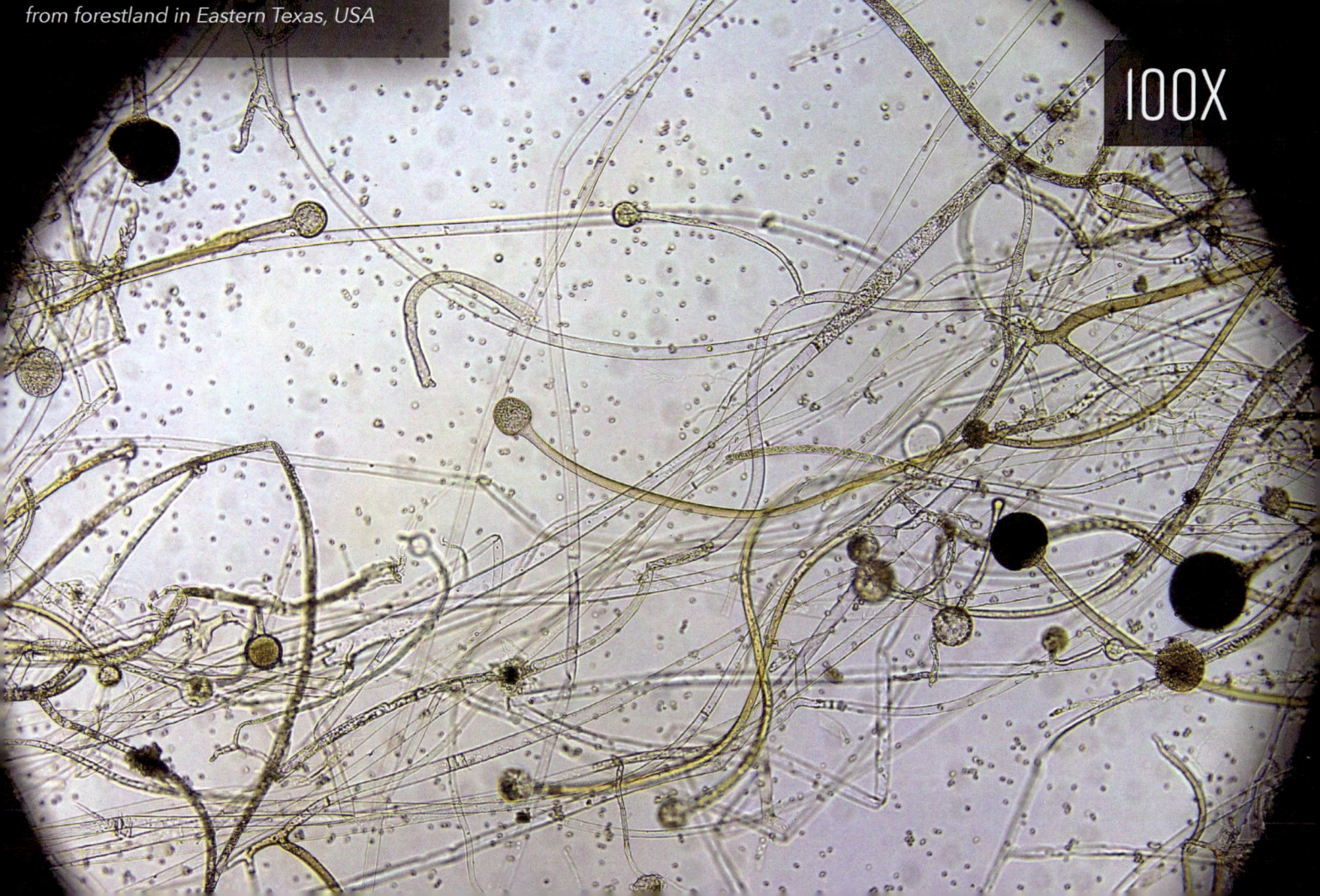

IMO-1 Collection #2
from forestland in Eastern Texas, USA

600X

600X

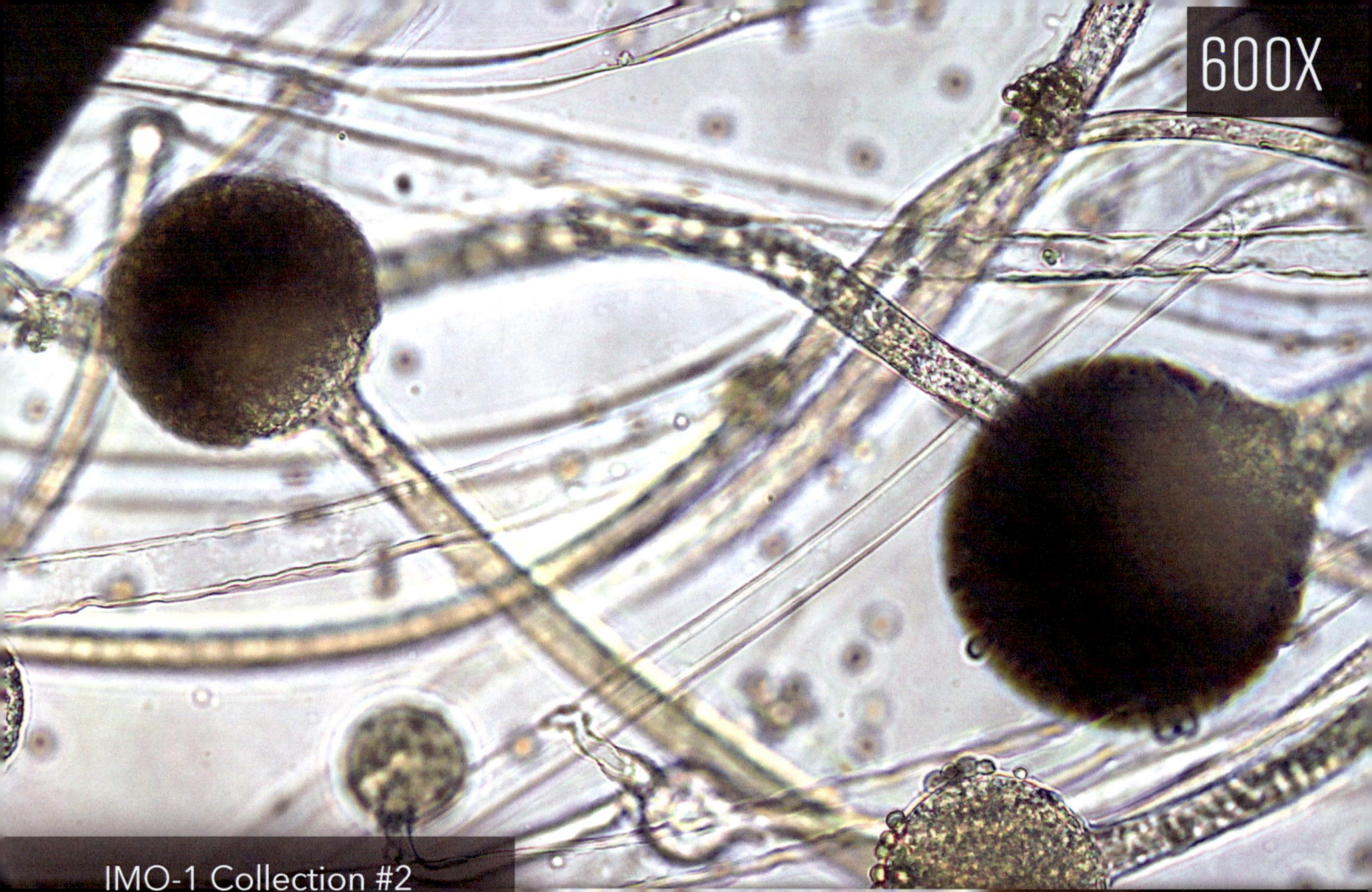

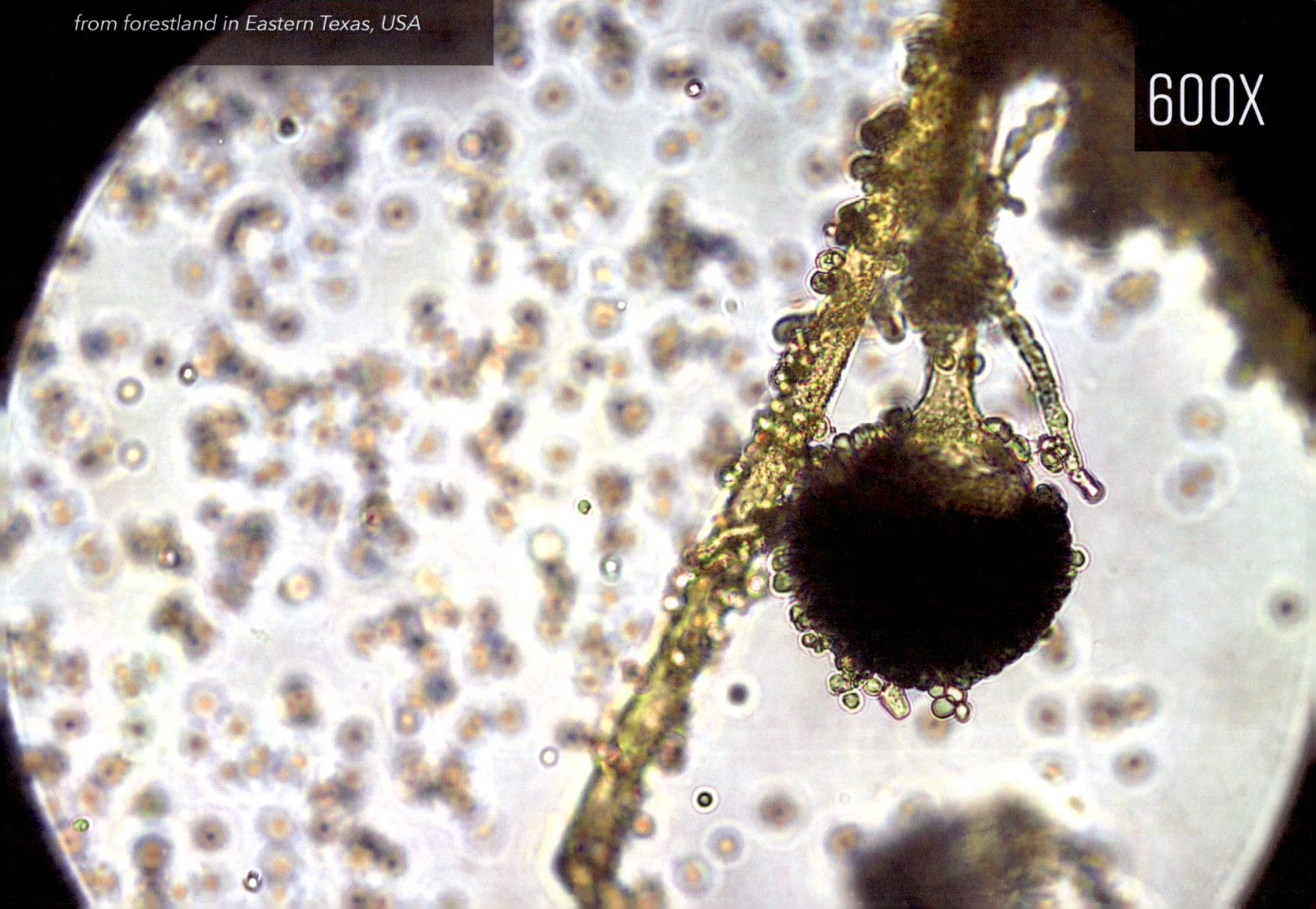

IMO-1 Collection #2
from forestland in Eastern Texas, USA

This is why we must keep an open mind and our humility close at hand: there's just so much we are still learning. The people at the cutting edge of soil science and microscopy are making new discoveries because they are asking questions based on the constant, rigorous study and experimentation they are doing: they are in the student mindset and nature is the teacher. That is the pattern to learn FAST and to incorporate what you learn DEEPLY.

THE FUTURE

THE WORLD OF MICROSCOPY HAS BEEN HOLDING ITS BREATH...

Few have tried to connect the phylogenic and non-phylogenic worlds – most just jump the chasm, leaving the old ways behind, or stay safely on the side they started on. This book is an attempt to bridge these divergent views on the same world though I know it is a first step in a direction that requires a journey to complete. I also believe it requires a community that is open: something that is not often found in the scientific or online communities at large, only in pockets around true educators still hanging on in a largely broken system. What we have already mapped should not be abandoned just because the names and organizational tags we've given them have changed. Yes, it does mean that we can't identify anything definitively from images and videos analyzed and collected using the old methods, but I believe that the morphological expressions range in relation to genetic expression and HGT – it's just more fluidic across boundaries than previously thought. The expressions have MEANING even if the organisms being examined have genetics that are divergent – they are responses to the environmental stressors that are causing similar expressions in morphology: *how can that not be important??* The morphological expression then is in relation to the environment and thus has more real-time utility as information for us as gardeners and farmers, and perhaps it has more importance than the genetic sequence as a whole (especially because as we read the DNA, we cannot tell how the organisms are reading and expressing the DNA themselves in response to their environments which they are, without question, doing constantly and effectively). That means that just because they have that gene, it doesn't mean it's expressed.

It doesn't make sense to throw the baby out with the bathwater especially when DNA can be divergent – it simply raises new questions. It will take more DNA research and microscopy work to know more, but we are drawing these two halves together into a whole steadily day by day, slide by slide, sequence by sequence.

START YOUR JOURNEY

With this book, you can participate in that process. You can explore, document, study, and compare your soils, your composts, and your samples' biology to the anyone else's. Comparisons are simple but yield complex insights, deeper clarity, and connections not recognized any other way. Always when possible find something to compare your sample to – ideally compare it to what you'd like it to be – though comparing it to what you don't want it to be can also be very useful.

WHAT TO DO NEXT

- **Develop Your Eye** - practice as much as possible, examine as wide array of samples as possible, and use this book as your guide to develop your own repertoire and pattern.

- **Develop Your Cultivation Skills** – learn how to build soil, make compost, brew biofertilizers, use the microscope, and test the soil in a variety of ways in your home lab. Take a course like **Regenerative Soil Microscopy** to take your skills further than the written word can. If you haven't already, take the first course and/or get the first book in the *Regenerative Soil Trilogy*.

- **Share & Participate in Community Feedback** – get a buddy, join a course with a community, and/or join the **R-Soil Database**. Share your work. Get feedback. Compare your work to others. Study others work. We haven't fully mapped the expressions of all microbes nor have we identified more than 1 – 10% of what is truly out there. We are just at the beginning. This is why **R-Soil Database** is so necessary.

THE FUTURE IS COLLABORATIVE

Not just in terms of people working together to further the science farther than has ever been possible before, but in terms of our testing: *we have to work with a variety of perspectives, tests, and paradigms.* It is the combination of and conversation had between perspectives as a community that gives us the greatest insights at the fastest rate. It's true democracy in action. It's positive people-power. It's citizen-science. It's community-derived solutions that come through high frequency cycles of iteration: this is exactly the way that nature adapts at a local or bioregional level. This is how the flock of birds fly as one – it's a collection of individuals making good choices because they are in feedback with their community. Any of the birds can at any time break away – they are not entranced or mindless. They are instead attentive, attuned, and in the symphonic dance and joy of the moment… of collaboration. This is my vision for the **R-Soil Database** and the annual **R-Soil** online conference.

Use this book as a guide to launch you into the world of soil microscopy, to better understand your soils, composts, inoculants, and other inputs, to make better decisions in managing your systems for more nutrient-dense and resilient plants and animals, and, over time, as a primer for your own adaptations, protocols, and experiments *since innovating new methods is the only way we get new answers and often the only way we begin to ask new questions.*

I challenge you to join me on the journey – we will not only unlock the full potential of our soils, but our food will fundamentally change in quality such that it will affect human and animal health. Food will become our medicine (again) – sadly, it has in recent decades become our poison; it will be redeemed through this process. I hope you will join me in the **Regenerative Soil Microscopy** online courses, the **R-Soil Database**, and **R-Soil**, the annual online conference. The world is waiting, perched on the edge of regenerative change – it's up to us to do the work to move it forward.

Join us and start exploring, evaluating, documenting,
and healing soil, compost, & more wherever you are in the world.

Use this book, the online database, the online courses,
& the free guides at
http://RegenerativeSoilScience.com,
and start looking at the world through different lenses!!

Grow Abundantly, Learn Daily, & Live Regeneratively,

Matt Powers

ABOUT THE AUTHOR

Matt Powers (M.Ed) is an author, educator, citizen scientist, entrepreneur, and family guy who teaches people all over the world how to live more regeneratively. Personally driven by a deep desire to have the best food possible for his wife and cancer-survivor, Adriana, and their two boys, Matt, a former public high school teacher with a Masters degree in Education, is creator of over a dozen online courses and author of over 20 books on permaculture and regenerative soil science like **The Permaculture Student series** and the **Regenerative Soil Trilogy**. Matt is also the host of **R-Future** & **R-Soil**, the annual online conferences, and **A Regenerative Future**, the podcast and Youtube show, where he interviews leaders in the regenerative space and shares his own work and insights.

MATT IS ON A MISSION TO EMPOWER PEOPLE EVERYWHERE TO LIVE MORE REGENERATIVELY.

Learn more about & from Matt Here:

www.ThePermacultureStudent.com

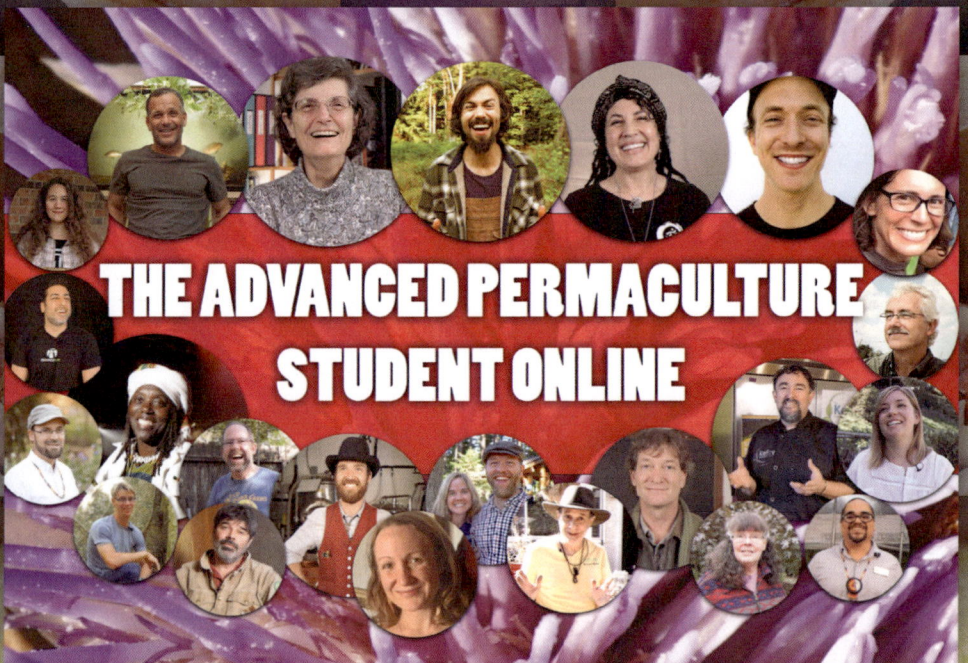

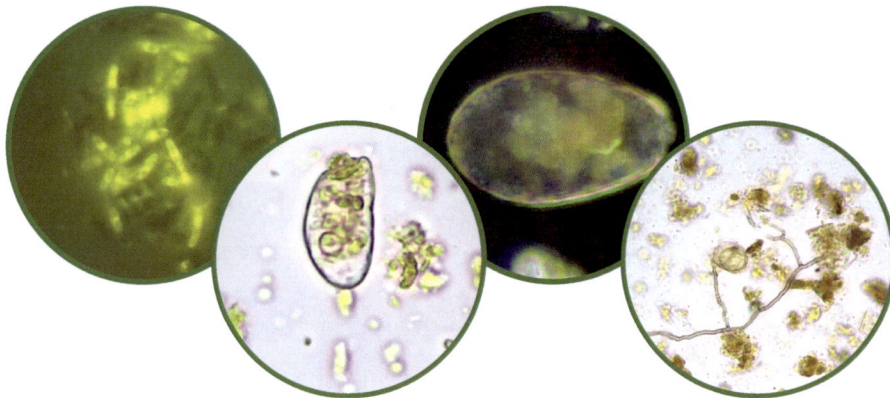

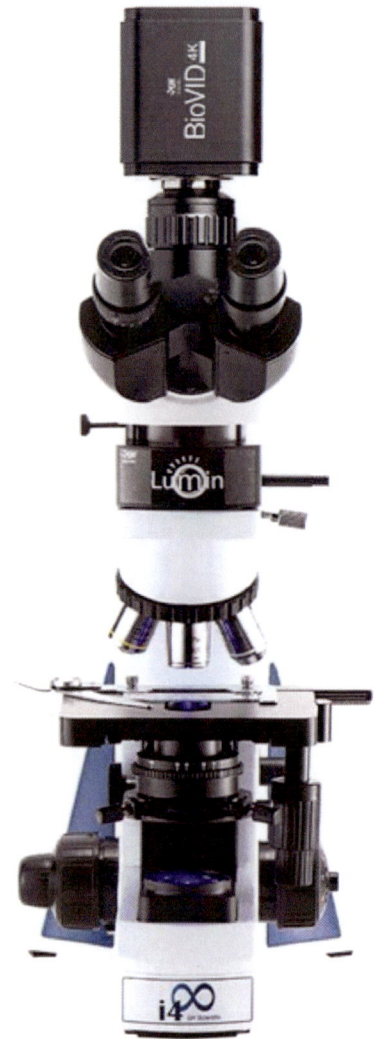

258

Sustainable Growing SOLUTIONS

Restore Your Soil

Shelf-Stable Liquid Inoculants with extremely High Populations and Diversity (20,000 + species)

The MetaGrow™ product line includes 7 different organic inoculants, each using "Directed Biology" for targeted plant and soil beneficial functions.

MetaGrow 5X+	5X Concentrated, all-purpose
MetaGrow C	Chitin metabolism
MetaGrow DCOMP	Crop residue decomposition
MetaGrow F	Fungal dominant
MetaGrow G	Glucan metabolism
MetaGrow ST	Max. diversity, all-purpose
MetaGrow FF Seed Coat	Powdered seed treatment

MICROBE FOODS wake up microbes from stasis, provide food, habitat and UV protection, improve colonization success, and grow and sustain native and applied microbe populations.

MetaGrow MFOOD	Max. diversity microbe food
MetaGrow CFOOD	Chitin-based microbe food

SPECIALTY FERTILIZERS formulated for peak nutrient demands of critical crop growth stages.

MetaGrow SynthesisESSENTIALS	Optimize photosynthesis
MetaGrow BloomESSENTIALS	Critical bloom micronutrients

At SGS we believe that crops achieve their highest potential and farms achieve sustainable profitability by following these principles:

Healthy plants <u>direct</u> healthy soil biology to actively <u>deliver</u> what nutrients it needs, when it needs them and in the proper proportions;

<u>Broad spectrum</u> microbe inoculants and foods will re-establish healthy microbial diversity, population size and beneficial plant functions;

<u>Balanced</u> plant and soil nutrition is key — paying attention to peak nutrient demand by plant growth stage and avoiding nutrient antagonisms that interfere with nutrient uptake;

When plants achieve proper biologically supported nutrition, most pest and disease problems go away.

Plant Nutrition
Plant Directed and Microbe Delivered

Sustainable Growing Solutions LLC
Clarksburg, CA   www.SGS-Ag.com

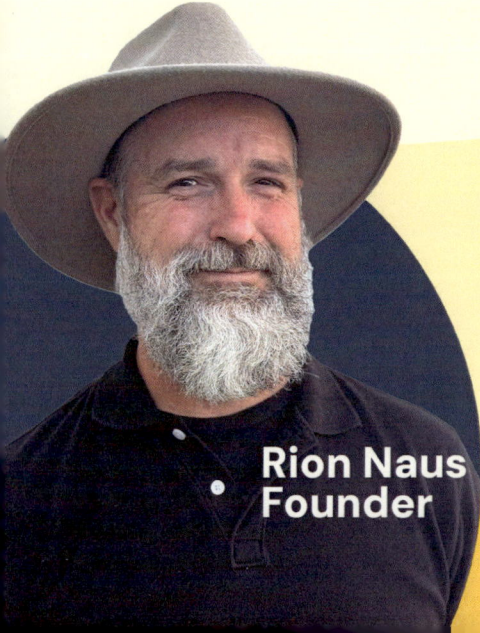

eleV∧te AG

Recapturing A Land of Milk & Honey

Let us assist in promoting a balanced biodiversity within your soil.

Elevate Ag presents a range of beneficial microbiology products tailored for your plants and soil. These microbes remain protected in stasis from extreme conditions such as temperature variations, high osmotic pressures, extreme pH levels, caustic materials, and toxicities, making them adaptable for use in any farming operation.

HYPRgERM Max	Biological Seed Treatment
HyprGrow	All-Purpose Biological Inoculant
HyprCycle+	Post-Harvest Nutrient Recycler
aRISE	Beneficial Biological Inoculant
Elevate'd Fungi	Fungal Dominant Inoculant
SURGE	Highly Diverse Concentrated Inoculant

1333 S. 2500 Rd
Herrington, KS 67449

eleV∧te AG
Recapturing A Land of Milk & Honey

(785) 422-7807
elevateag.com

– REFERENCES –

#

- (2022). Mathbench.org.au. https://mathbench.org.au/wp-content/uploads/2015/11/Microbiology-Methods-for-Counting-Bacteria.pdf
- (2023). Microbiologyjournal.org. https://www.microbiologyjournal.org/wp-content/uploads/2019/05/fg01-3.jpg

A

- Arashida, H., Kugenuma, T., Watanabe, M., & Maeda, I. (2019). Nitrogen fixation in Rhodopseudomonas palustris co-cultured with Bacillus subtilis in the presence of air. Journal of Bioscience and Bioengineering, 127(5), 589–593. https://doi.org/10.1016/j.jbiosc.2018.10.010
- Arbuscular Mycorrhiza - an overview | ScienceDirect Topics. (n.d.). Www.sciencedirect.com. Retrieved April 17, 2023, from https://www.sciencedirect.com/topics/earth-and-planetary-sciences/arbuscular-mycorrhiza
- Arbuscular mycorrhizas. (n.d.). Www.davidmoore.org.uk. http://www.davidmoore.org.uk/assets/mostly_mycology/diane_howarth/am.htm
- Aryal, S. (2018, June 11). Different Size, Shape and Arrangement of Bacterial Cells. Microbiology Info.com. https://microbiologyinfo.com/different-size-shape-and-arrangement-of-bacterial-cells/
- Asiegbu, F. O., & Kovalchuk, A. (2021). Forest Microbiology Volume 1: Tree Microbiome: Phyllosphere, Endosphere and Rhizosphere. Elsevier Science & Technology.
- Aspergillus versicolor. (n.d.). Doctor Fungus. Retrieved April 17, 2023, from https://drfungus.org/knowledge-base/aspergillus-versicolor/
- Aspergillus | Johns Hopkins ABX Guide. (n.d.). Www.hopkinsguides.com. https://www.hopkinsguides.com/hopkins/view/Johns_Hopkins_ABX_Guide/540036/all/Aspergillus

B

- Baek, J., Weerawongwiwat, V., Kim, JH. et al. Paenibacillus arenosi sp. nov., a siderophore-producing bacterium isolated from coastal sediment. Arch Microbiol 204, 113 (2022). https://doi.org/10.1007/s00203-021-02735-3
- Barceló M, van Bodegom PM, Tedersoo L, den Haan N, Veen GF(, et al. (2020) The abundance of arbuscular mycorrhiza in soils is linked to the total length of roots colonized at ecosystem level. PLOS ONE 15(9): e0237256. https://doi.org/10.1371/journal.pone.0237256
- Barnett, H. L., & Hunter, B. B. (2010). Illustrated genera of imperfect fungi. Aps Press.
- Beneficial Microbes In Agro-Ecology. (2020). Elsevier Academic Press.
- biotech simplified. "Acridine Orange Staining - Principle, Method and Result." Acridine Orange Staining - Principle, Method and Result - Biotech Simplified, YouTube, 29 June 2020, https://www.youtube.com/watch?v=CfjFlbr9vfE.
- Boone, D. R., & Castenholz, R. W. (2012). Bergey's Manual of Systematic Bacteriology. Springer Science & Business Media.

- Briggs, G. M. (2021). Rhizobium: nitrogen fixing bacteria. Milnepublishing.geneseo.edu. https://milnepublishing.geneseo.edu/botany/chapter/rhizobium/

C

- Cialdella, M.S. (2022). Fungal Spore Identification and Information Guide. Wonder Makers Environmental.
- Conidial Density and Viability of Beauveria bassiana Isolates from Java and Sumatra and Their Virulence Against Nilaparvata lugens at Different Temperatures - Scientific Figure on ResearchGate. Available from: https://www.researchgate.net/figure/Conidia-of-Beauveria-bassiana-a-viable-conidia-at-24-hour-b-and-48-hours-c_fig3_333870517 [accessed 29 Dec, 2022]

D

- Davison, J., Moora, M., Semchenko, M., Adenan, S. B., Ahmed, T., Akhmetzhanova, A. A., Alatalo, J. M., Al-Quraishy, S., Andriyanova, E., Anslan, S., Bahram, M., Batbaatar, A., Brown, C., Bueno, C. G., Cahill, J., Cantero, J.J., Casper, B. B., Cherosov, M., Chideh, S., & Coelho, A. P. (2021). Temperature and pH define the realised niche space of arbuscular mycorrhizal fungi. New Phytologist, 231(2), 763–776. https://doi.org/10.1111/nph.17240
- Dias Nunes, Mateus & Cardoso, Willian & Luz, José & Kasuya, Maria Catarina. (2014). Lithium chloride affects mycelial growth of white rot fungi: Fungal screening for Li-enrichment. African Journal of Microbiology Research. 8. 2111-2123. 10.5897/ajmr.2014.6619.
- Diplocladiella scalaroides (a dematiaceous anamorphic fungus). (n.d.). Www.bioinfo.org.uk. Retrieved April 17, 2023, from https://www.bioinfo.org.uk/html/Diplocladiella_scalaroides.htm
- Dugan, F. M. (2006). The Identification of Fungi.

E

- Earthworms. (2019, February 15). Penn State Extension. https://extension.psu.edu/earthworms
- ECACC. 7. Counting of Cells Using Trypan Blue and a Haemocytometer, YOUTUBE.COM, 11 Feb. 2016, https://youtu.be/qfT9uqqme8c. Accessed 16 Feb. 2022.
- Eckersley, K., & Dow, C. (2017). Rhodopseudomonas blastica sp.nov.: a Member of the Rhodospirillaceae. Microbiology. https://www.semanticscholar.org/paper/Rhodopseudomonas-blastica-sp.nov.%3A-a-Member-of-the-Eckersley-Dow/4bba71a4a2a33f166d271b66c38298a9c94f1929
- Effect of pH and Temperature on Bacillus subtilis FNCC 0059 Oxalate Decarboxylase Activity. (n.d.). Science Alert. https://scialert.net/fulltext/?doi=pjbs.2017.436.441
- EPI2METM. (n.d.). Epi2me.nanoporetech.com. Retrieved April 17, 2023, from https://epi2me.nanoporetech.com/report-312727

F

- Factors That Affect Streptomyces Growth Biology Essay. (n.d.). Www.ukessays.com. Retrieved April 17, 2023, from https://www.ukessays.com/essays/biology/factors-that-affect-streptomyces-growth-biology-essay.php
- Fang, L. C., Li, Y., Cheng, P., Deng, J., Jiang, L. L., Huang, H., Zheng, J. S., & Wei, H. (2012). Characterization of

Rhodopseudomonas palustris strain 2C as a potential probiotic. APMIS, 120(9), 743–749. https://doi.org/10.1111/j.1600-0463.2012.02902.x

- Fisher, M. R., & Editor. (2017). 2.3 A Cell is the Smallest Unit of Life. Pressbooks.pub; Pressbooks. https://openoregon.pressbooks.pub/envirobiology/chapter/2-3-a-cell-is-the-smallest-unit-of-life/
- Francois Buscot, & Varma, A. (2006). Microorganisms in Soils: Roles in Genesis and Functions. Springer Science & Business Media.
- FungiAdmin. (2022, January 1). Trametes versicolor. All about Growing & Hunting Mushrooms. https://funginomi.com/trametes-versicolor/

G

- Genus: Nitrosarchaeum. (n.d.). Lpsn.dsmz.de. Retrieved April 17, 2023, from https://lpsn.dsmz.de/genus/nitrosarchaeum
- Glawe, Dean. (2006). Synopsis of genera of Erysiphales (powdery mildew fungi) occurring in the Pacific Northwest. 1. https://www.researchgate.net/publication/228656219_Synopsis_of_genera_of_Erysiphales_powdery_mildew_fungi_occurring_in_the_Pacific_Northwest
- Glomus intraradices. (2023). Zut.edu.pl. http://www.zor.zut.edu.pl/Glomeromycota/Glomus%20intraradices.html
- Gluconacetobacter Diazotrophicus - an overview | ScienceDirect Topics. (n.d.). Www.sciencedirect.com. https://www.sciencedirect.com/topics/agricultural-and-biological-sciences/gluconacetobacter-diazotrophicus

H

- Hartmann, A., & Baldani, J. I. (2006). The Genus Azospirillum. The Prokaryotes, 115–140. https://doi.org/10.1007/0-387-30745-1_6
- Heitman, J., Howlett, B. J., Crous, P. W., Stukenbrock, E. H., Timothy Yong James, & Gow, N. A. R. (2018). The fungal kingdom. Asm Press.
- Herbaspirillum seropedicae. (n.d.). Memim.com. Retrieved April 17, 2023, from https://memim.com/herbaspirillum-seropedicae.html
- Histopathology - Aspergillus and Aspergillosis. (2014). Aspergillus and Aspergillosis. https://www.aspergillus.org.uk/diagnosis/histopathology/
- Holt, W. S. (2000). Bergey's Manual Of Determinative Bacteriology. 9th ed. Lippincott Williams & Wilkins.
- *Home*. (n.d.). Invam.ku.edu. Retrieved April 17, 2023, from https://invam.ku.edu

I

- Iglewski, B. H. (2013). Pseudomonas. Nih.gov; University of Texas Medical Branch at Galveston. https://www.ncbi.nlm.nih.gov/books/NBK8326/
- Introduction to Spirochetes - Microbiology - Medbullets Step 1. (n.d.). Step1.Medbullets.com. https://step1.medbullets.com/microbiology/121562/introduction-to-spirochetes

J

- Johann F. Osma, Ulla Moilanen, José L. Toca-Herrera, Susana Rodríguez-Couto, Morphology and laccase production of white-rot fungi grown on wheat bran flakes under semi-solid-state fermentation conditions, FEMS Microbiology Letters, Volume 318, Issue 1, May 2011, Pages 27–34, https://doi.org/10.1111/j.1574-6968.2011.02234.x
- Johann F. Osma, Ulla Moilanen, José L. Toca-Herrera, Susana Rodríguez-Couto, Morphology and laccase production of white-rot fungi grown on wheat bran flakes under semi-solid-state

fermentation conditions, FEMS Microbiology Letters, Volume 318, Issue 1, May 2011, Pages 27–34, https://doi.org/10.1111/j.1574-6968.2011.02234.x
- Judith A. Narvhus, Roger K. Abrahamsen, in Encyclopedia of Dairy Sciences (Third Edition), 2022

K

- Kh, Otgonjargal. (2018). Optimum and tolerance pH range, optimal temperature of the local strain Beauveria bassiana-G07.
- Kim, S. B. (2015). Thermobispora. Bergey's Manual of Systematics of Archaea and Bacteria, 1–3. https://doi.org/10.1002/9781118960608.gbm00213
- Krings, M., Harper, C. J., Cúneo, N. R., Rothwell, G. W., & Taylor, T. N. (2018). Transformative paleobotany : papers to commemorate the life and legacy of Thomas N. Taylor. Academic Press.

L

- Laurence M, Hatzis C, Brash DE. Common contaminants in next-generation sequencing that hinder discovery of low-abundance microbes. PLoS One. 2014;9(5):e97876. Published 2014 May 16. doi:10.1371/journal.pone.0097876
- Laurence M, Hatzis C, Brash DE. Common contaminants in next-generation sequencing that hinder discovery of low-abundance microbes. PLoS One. 2014;9(5):e97876. Published 2014 May 16. doi:10.1371/journal.pone.0097876
- Lee EH, Eom AH. Growth Characteristics of Rhizophagus clarus Strains and Their Effects on the Growth of Host Plants. Mycobiology. 2015 Dec;43(4):444-9. doi: 10.5941/MYCO.2015.43.4.444. Epub 2015 Dec 31. PMID: 26839504; PMCID: PMC4731649.
- li, Baozhen & Liu, Na & Li, Yongquan & Jing, Weixin & Fan, Jinhua & Li, Dan & Zhang, Longyan & Zhang, Xiaofeng & Zhang, Zhaoming & Wang, Lan. (2014). Reduction of Selenite to Red Elemental Selenium by Rhodopseudomonas palustris Strain N. PLoS one. 9. e95955. 10.1371/journal.pone.0095955.
- Li, Q., Chen, X., Jiang, Y., & Jiang, C. (2016). Morphological Identification of Actinobacteria. In D. Dhanasekaran, & Y. Jiang (Eds.), Actinobacteria - Basics and Biotechnological Applications. IntechOpen. https://doi.org/10.5772/61461
- Lloyd-Price, J., Abu-Ali, G., & Huttenhower, C. (2016). The healthy human microbiome. Genome Medicine, 8(1). https://doi.org/10.1186/s13073-016-0307-y
- Lo, K.-J., Lee, S.-K., & Liu, C.-T. (2020). Development of a low-cost culture medium for the rapid production of plant growth-promoting Rhodopseudomonas palustris strain PS3. PLoS ONE, 15(7), e0236739. https://doi.org/10.1371/journal.pone.0236739

M

- Malherbe, Stephanus. (2011). Exploring the relationship between microscope-based soil biology measurements and tomato yield in South Africa using Principal Component Analysis. Conference: Soilborne Plant Diseases Symposium At: Agriculture Research Council Plant Protection Research Institute, Stellenbosch, South Africa.
- Margesin, R., & Franz Schinner. (2005). Manual of soil analysis : monitoring and assessing soil bioremediation. Springer.
- Mata, Moises & Barquero, Miguel. (2022). Evaluación de la fermentación sumergida del hongo entomopatógeno "Beauveria bassiana" como parte de un proceso de escalamiento y producción de bioplaguicidas. Boletín Promecafé, ISSN 1010-1527, Nº. 122, 2010, pags. 8-19.
- Mehrabi, S., Ekanemesang, U. M., Aikhionbare, F. O., Kimbro, K. Sean., & Bender, J. (2001). Identification and characterization of Rhodopseudomonas spp., a purple, non-sulfur bacterium from

microbial mats. Biomolecular Engineering, 18(2), 49–56. https://doi.org/10.1016/s1389-0344(01)00086-7

- Microscope Objectives: What do Plan, Semi-Plan, and Achromat mean? | Pathwooded. (n.d.). Www.pathwooded.com. Retrieved April 17, 2023, from https://www.pathwooded.com/post/microscope-objectives-what-do-plan-semi-plan-and-achromat-mean
- Micscape Microscopy and Microscope Magazine. (n.d.). Www.microscopy-Uk.org.uk. http://www.microscopy-uk.org.uk/mag/indexmag.html?http://www.microscopy-uk.org.uk/mag/artfeb01/rhizo.html
- Morphology of the Archaea. (2020). Berkeley.edu. https://ucmp.berkeley.edu/archaea/archaeamm.html
- Mousavi B, Hedayati MT, Hedayati N, Ilkit M, Syedmousavi S. Aspergillus species in indoor environments and their possible occupational and public health hazards. Curr Med Mycol. 2016;2(1):36-42. doi:10.18869/acadpub.cmm.2.1.36

N

- No Atlas Page Present. (n.d.). Linnet.geog.ubc.ca. Retrieved April 17, 2023, from https://linnet.geog.ubc.ca/Atlas/Atlas.aspx?sciname=Phellinus%20igniarius

O

- Oh S, Rheem S, Sim J, Kim S, Baek Y. Optimizing conditions for the growth of Lactobacillus casei YIT 9018 in tryptone-yeast extract-glucose medium by using response surface methodology. Appl Environ Microbiol. 1995 Nov;61(11):3809-14. doi: 10.1128/aem.61.11.3809-3814.1995. PMID: 8526490; PMCID: PMC167683.
- Ohno, M & Shiratori, H & Park, M & Saitoh, Y & Kumon, Y & Yamashita, N & Hirata, A & Nishida, Hiromi & Ueda, Kenji & Beppu, T. (2000). Symbiobacterium thermophilum gen. nov., sp. nov., a symbiotic thermophile that depends on co-culture with a Bacillus strain for growth. International journal of systematic and evolutionary microbiology. 50 Pt 5. 1829-32. 10.1099/00207713-50-5-1829.
- Ouzounidou, Georgia & Skiada, Vasiliki & Papadopoulou, Kalliope & Stamatis, Nikolaos & Kavvadias, Victor & Eleftheriadis, Eleftherios & Gaitis, Fragiskos. (2015). Effects of soil pH and arbuscular mycorrhiza (AM) inoculation on growth and chemical composition of chia (Salvia hispanica L.) leaves. Brazilian Journal of Botany. 38. 10.1007/s40415-015-0166-6.

P

- Paenibacillus alvei - microbewiki. (n.d.). Microbewiki.kenyon.edu. https://microbewiki.kenyon.edu/index.php/Paenibacillus_alvei
- Pelinescu (maiden name Smarandache), Diana & Sasarman, Elena & Chifiriuc, Mariana & Stoica, Ileana & Tanase, Ana-Maria & Avram, Ionela & Şerbancea, Floarea & Vassu, Tatiana. (2015). Isolation and identification of some Lactobacillus and Enterococcus strains by a polyphasic taxonomical approach. Romanian Biotechnological Letters.
- Perry, R. N., Hunt, D. J., & Subbotin, S. A. (2020). Techniques for Work with Plant and Soil Nematodes. CABI.
- Phellinus igniarius. (2020). Englishfungi.org. http://englishfungi.org/Species/Phellinus%20igniarius

R

- Ramzan. (2023, January 19). Arbuscular Mycorrhiza - Structure, Development & Functions. BIOLOGY TEACH. https://biologyteach.com/arbuscular-mycorrhizas/
- Rhizophagus irregularis articles - Encyclopedia of Life. (n.d.). Eol.org. Retrieved April 17, 2023, from https://eol.org/pages/11932698/articles?lang_group=en

- Rhodopseudomonas - microbewiki. (n.d.). Microbewiki.kenyon.edu. https://microbewiki.kenyon.edu/index.php/Rhodopseudomonas
- Rhodopseudomonas palustris CGA009. (n.d.). Genome.jgi.doe.gov. https://genome.jgi.doe.gov/portal/rhopa/rhopa.home.html
- Rijal, Nisha. "Acridine Orange Staining: Principle, Procedure, Results and Applications." Learn Microbiology Online, 13 Aug. 2015, microbeonline.com/acridine-orange-staining-principle-procedure-results-applications/.

S

- Sapkota, A. (2021, April 24). Lactobacillus acidophilus- An Overview. Microbe Notes. https://microbenotes.com/lactobacillus-acidophilus/
- Sauder LA, Engel K, Lo CC, Chain P, Neufeld JD. "Candidatus Nitrosotenuis aquarius," an Ammonia-Oxidizing Archaeon from a Freshwater Aquarium Biofilter. Appl Environ Microbiol. 2018 Sep 17;84(19):e01430-18. doi: 10.1128/AEM.01430-18. PMID: 29959256; PMCID: PMC6146995.
- Schinner, F., Öhlinger, R., Kandeler, E., & Margesin, R. (1996). Methods in Soil Biology. Springer Berlin Heidelberg.
- Shah, A. S., Wakelin, S. A., Moot, D. J., Blond, C., Noble, A., & Ridgway, H. J. (2022). High throughput pH bioassay demonstrates pH adaptation of Rhizobium strains isolated from the nodules of Trifolium subterraneum and T. repens. Journal of Microbiological Methods, 195, 106455. https://doi.org/10.1016/j.mimet.2022.106455
- SHAHID, Dr. MOHAMMAD. (2014). Optimal Physical Parameters for Growth of Trichoderma species at Varying pH, Temperature and Agitation. Journal Virology & Mycology.. 3. 127.
- Shin, H.-D., & Mulenko, W. (2004). The Record of Erysiphe azaleae (Erysiphales) from Poland and Its Anamorph. Mycobiology, 32(3), 105. https://doi.org/10.4489/myco.2004.32.3.105
- Soesanto, Loekas & Mugiastuti, Endang & Rahayuniati, Ruth. (2011). MORPHOLOGICAL AND PHYSIOLOGICAL FEATURES OF Pseudomonas fluorescens P60.
- Sorangium - an overview | ScienceDirect Topics. (n.d.). Www.sciencedirect.com. Retrieved April 17, 2023, from https://www.sciencedirect.com/topics/biochemistry-genetics-and-molecular-biology/sorangium
- Soudzilovskaia NA, Vaessen S, Barcelo M, He J, Rahimlou S, Abarenkov K, Brundrett MC, Gomes SIF, Merckx V, Tedersoo L. FungalRoot: global online database of plant mycorrhizal associations. New Phytol. 2020 Aug;227(3):955-966. doi: 10.1111/nph.16569. Epub 2020 May 20. PMID: 32239516.
- St-Germain, G., & Richard Charles Summerbell. (2011). Identifying fungi : a clinical laboratory handbook. Star Publishing Company.
- Sung, Moon-Hee & Bae, Jin-Woo & Kim, Joong-Jae & Kim, Kwang & Song, Jae-Jun & Rhee, Sung-Keun & Jeon, Che-Ok & Choi, Yoon-Ho & Hong, Seung-Pyo & Lee, Seung-Goo & Ha, Jae-Suk & Kang, Gwan-Tae. (2003). Symbiobacterium toebii sp. nov., commensal thermophile isolated from Korean compost. Journal of Microbiology and Biotechnology. 13. 1013-1017.

T

- Takov, Danail & Draganova, Slavimira & Toshova, Teodora. (2013). Gregarine and Beauveria bassiana Infections of the Grey Corn Weevil, Tanymecus dilaticollis (Coleoptera: Curculionidae). Acta Phytopathologica et Entomologica Hungarica. 4800. 309-319. 10.1556/APhyt.48.2013.2.12.
- Taxonomy. (n.d.). Taxonomy browser (Nitrosarchaeum sp. AC2). Www.ncbi.nlm.nih.gov. Retrieved April 17, 2023, from https://

www.ncbi.nlm.nih.gov/Taxonomy/Browser/wwwtax.cgi?
mode=Info&id=2259673

- Taxonomy. (n.d.). Taxonomy browser (Thermobispora bispora
  DSM 43833). Www.ncbi.nlm.nih.gov. Retrieved April 17, 2023,
  from https://www.ncbi.nlm.nih.gov/Taxonomy/Browser/
  wwwtax.cgi?mode=Info&id=469371
- Tetrads. (n.d.). Www.environmentalleverage.com. Retrieved April
  17, 2023, from https://www.environmentalleverage.com/
  Tetrads.htm
- Thaumarchaeota - an overview | ScienceDirect Topics. (n.d.).
  Www.sciencedirect.com. Retrieved April 17, 2023, from https://
  www.sciencedirect.com/topics/biochemistry-genetics-and-
  molecular-biology/thaumarchaeota
- Tolar, Bradley & Mosier, Annika & Lund, Marie & Francis,
  Christopher. (2019). Nitrosarchaeum.
  10.1002/9781118960608.gbm01289.
- Trametes versicolor: The Turkey Tail (MushroomExpert.com).
  (n.d.). Www.mushroomexpert.com. https://
  www.mushroomexpert.com/trametes_versicolor.html
- Trichoderma - Beneficial Microbes. (n.d.). Fertilizer New Zealand.
  Retrieved April 17, 2023, from https://www.fertnz.co.nz/
  trichoderma/
- Trichoderma. (2019). Cornell.edu. https://
  biocontrol.entomology.cornell.edu/pathogens/trichoderma.php

V

- Verbruggen, E., Struyf, E., & Vicca, S. (2021). Can arbuscular
  mycorrhizal fungi speed up carbon sequestration by enhanced
  weathering? PLANTS, PEOPLE, PLANET. https://doi.org/
  10.1002/ppp3.10179
- Vibrio Bacteria Overview - Examples, Shape, Structure and
  Infection. (n.d.). MicroscopeMaster. https://
  www.microscopemaster.com/vibrio-bacteria.html

W

- Wang et al. J Pure Appl Microbiol, 12(4), 1679-1687 Dec. 2018
- Watanabe K, Nagao N, Yamamoto S, Toda T, Kurosawa N.
  Thermobacillus composti sp. nov., a moderately thermophilic
  bacterium isolated from a composting reactor. Int J Syst Evol
  Microbiol. 2007 Jul;57(Pt 7):1473-1477. doi: 10.1099/
  ijs.0.64672-0. PMID: 17625178.
- Watanabe, T. (2010). Pictorial Atlas of Soil and Seed Fungi. CRC
  Press.

X

- Xiaoli Wang, Weixin Zhang, Yuanhu Shao, Jie Zhao, Lixia Zhou,
  Xiaoming Zou, Shenglei Fu. Fungi to bacteria ratio: Historical
  misinterpretations and potential implications. Acta Oecologica,
  Volume 95, 2019, Pages 1-11,ISSN 1146-609X, https://doi.org/
  10.1016/j.actao.2018.10.003.

Y

- Yoon, J.-H., & Park, Y.-H. (2006). The Genus Nocardioides. The
  Prokaryotes, 1099–1113. https://doi.org/
  10.1007/0-387-30743-5_44
- Yoon, J.-H., & Park, Y.-H. (2006). The Genus Nocardioides. The
  Prokaryotes, 1099–1113. https://doi.org/
  10.1007/0-387-30743-5_44

Z

- ZHU, LIN & Ji, Xing & Si, Jing & Cui, Bao-Kai. (2018).
  Morphological characters and phylogenetic analysis reveal a new
  species of Phellinus with hooked hymenial setae from Vietnam.
  Phytotaxa. 356. 91. 10.11646/phytotaxa.356.1.8.

- Glossary -

A

ACC Deaminase - a plant stimulating enzyme

Actinobacteria - a group of bacteria that looks like fungi somewhat but are typically thinner and clear - easily distinguishable.

Aerobic - with air

Aggregate - a grouping of particles

AMF - arbuscular mycorrhizal fungi

Amoeba (amoebae, testate amoeba) - a protozoa, a microscopic organism, that is like a blob but often in compost it forms a protective shell called a test. Those are called testate amoebae.

Anaerobic - without air

Anoxic - without oxygen

Ascomycota - a phylum of fungi

Asexual - the ability to reproduce without another organism or mate - single parent

Autoflouresce - to glow without added stain

B

Bacilli - rod-shaped bacteria

Basidiomycota - a phylum of fungi

Binary Fission - the process by which bacteria reproduce by splitting into two distinct organisms from one

Biocontrol - a microbe that prevents or controls the spread or colonization by other microbes of a given substrate, plant, or surface

Biofertilizer - a microbe that fertilizes plants

Bright field - a microscope technique where light shines through the sample towards the viewer (typically light shines from below).

C

Ciliate - a protozoa, a microscopic organism, covered with hairs in a diverse range of species and expressions.

Cocci - round shaped bacteria

Condenser - the section of the microscope below the stage but above the light source: it controls and directs the light. It usually has a iris diaphragm for even closer control of the light.

Conidiaspores - packets of spores

Conidia - spores

Conidiophores - the conidia stalk

Cysts - a dormant protective form some microbes can take to protect themselves when the environment is no longer viable.

D

Dark Field - a microscope method where the condenser has the center blacked out, so only indirect light hits the sample.

E

Endophyte - a microorganism that lives and thrives beneficially inside plants

Enzyme -

Exudate - carbonaceous root secretions and excretions specifically sugars, organic acids, amino acids, enzymes, phenolic compounds, hormones, and secondary metabolites.

Epifluorescence - a type of microscopy where light is shone on a subject and excites specific molecules and they shine. This is a proven technique for visualizing fungal activity, fungi, root inoculation, and phosphorus minerals and compounds.

F

FOV or Field of View -what you see when you look through a single frame of view with your microscope

Filamentous - thread-like but can be more like roots

Flagella (flagellated) - whip-like hairs that extend off of microbes

Focus (coarse & fine) - the knobs we use to move the stage closer and further away from the objectives: so we can bring things into focus.

G

Gibberellins - a group of well-documented and researched plant growth promoting hormones

Gram Negative/Positive - gram staining is a test with two different stains that allows for differentiation of bacteria into two groups based on their cell wall composition.

H

Hemocytometer - a special slide used to count microbes with a very small grid on it.

HGT - horizontal gene transfer

Hyaline - clear, transparent

Hyphae - the root or twig-like branches of fungal (sometimes bacteria)

I

IAA - indole-acetic-acid

IMO - indigenous microorganisms

ISR - induced systemic resistance

K

KNF - Korean Natural Farming

L

LAB - lactic acid bacteria i.e. lactobacillus

M

Methylene Blue - a stain, dye, and miraculous medicine

Metula - another form of extensions off the stalk heads (conidiophores)

Microplastics - fragments of plastic found in soil and water

Morphological - a classification system based on how things appear

Motile - mobile i.e. it moves

Mycelia - the network or mass of hyphae

Mycorrhizae (mycorrhizal) - fungi that lives in symbiosis with plant roots

N

Nematode - a round worm, usually microscopic

O

Objective - the movable lens used to view the sample on the slide. There are a series of objectives that you can choose from.

Oomycete - a false fungi that is harmful and disease causing

P

Pathogenic - disease causing, harmful

PCR - polymerase chain reaction

Peritrichous - evenly distributed

PGPR - plant growth promoting rhizobacteria

Phialides - extensions off the stalk heads phytohormone

Pleomorphism - the ability for bacteria to modulate their morphology in response to their environment

Protozoa - a grouping of microbes that feed on bacteria and fungi primarily but can be parasitic to mammals as well. The ones in the soil we are focused on are amoebae, ciliates, and flagellates.

R

Resolution - the sharpness of an image

Rhizophagy - the process by which roots digest microbes pulled in through their meristem cells and expel survivors through their root hairs which are formed by this process.

Rhizosphere - the area directly around the roots of plants

S

Saprophytic - feeding upon decaying or dead organic matter

Septa - the division between sections of fungal hyphae; they can be closed, semi-closed, or open, allowing or restricting the movement of endosymbiotic bacteria

Siderophore - iron-chelating molecules that are very small and released by microbes

SOM - soil organic matter

Spirilla - a spiral or twisted form of bacteria

Sporulating (sporulation) - when spores germinate and grow hyphae

Stain - a dye used to make microbes or roots more visible

Stage - the part of the microscope where the slide is placed

Stylet - the spear inside nematode mouths

T

Teleomorph - the sexual form of a fungi

Testate Amoeba (see amoeba)

Trinocular - this is a microscope head with 3 outputs instead of just 2 eyepieces. Typically this is an upright section that we can add our cameras to.

U

µm - one micrometer, 1/1000the of a mm.

V

Vesicles - the conidiophore (stalk) heads

Z

Zygomycota - a phylum of fungi

IF A WORD IS NOT HERE - PLEASE REFER TO THE GLOSSARY IN **REGENERATIVE SOIL**

- Index -

Leaf 9, 74, 110, 164, 183, 206, 211,

Lightfield (see Manual Lighting) 7, 11, 174

M

Manual Lighting 11, 23, 28, **168 - 174**

Microarthropod 8, 9, 75, **122, 124**, 136, 141, 148, 176, 195, 196, 211, 212, 220

Microplastic 76, **124, 27 - 129**, 179,

Microscope

... anatomy **27**

... supplies **14 - 16,** 26

... objectives 25, 29, 132, 133, 137, 173

Minerals **124 - 130**

Mycorrhizae (mycorrhizal) 4, 8, 12, 21, 30, 63 - 64, 67, 70, 72 - 73, 76, **78 - 82,** 93, 104, 157, 161, 164, 168, 176, 178, 180, 183, **184**, 195

N

Nematode

... bacterial feeder 111, **113 - 114**, 119

... fungal feeder 80, 110 - 112, **114**, 115, 119, 211

... omnivore 110, 117, **119 -120**, 185, 195, 211

... predator 80, 110, 115, **117 - 119**, 181

... root feeder 80, 111, 112, 114, **115 - 117**, 119

... switcher 112, 117, 119, 188

... eggs 119

O

Objective (see microscope objectives)

Oomycete 12, 67, 75, 76, 211

Organic Matter **124**, 127, 129 - 130, **160 - 161**

Other Tests **176** , **178**

P

Paenibacillus Polymyxa **60**

Pollen 33, **106 - 107**

Protozoa 4, 6, 8, 32, **97**, 101 - 103, 114, 117, 119, 122, 136, 137 , 141, 148, 151, 181, 182, 195, 196, 204, 211 - 213, 220, 227, 240, 241

Pseudomonas 43, 44, 51

Pseudomonas Fluorescens 43, **51**, 212

R

Ratios 3, 10, 16, 31, 32, 69, 70, 73, 132 - 135, 149, 150, 154, 176, 180 - 182

Rhizobia 44, **57**, 181

Rhizobium 43, 44, **57 - 58**

Rhodopseudomonas Palustris 43, **53**, 151, 228, 234 - 236, 240

www.ingramcontent.com/pod-product-compliance
Lightning Source LLC
Chambersburg PA
CBRC091107210326
41600CB00020BA/606